Inequality and Energy

Inequality and Energy
How Extremes of Wealth and Poverty in High Income Countries Affect CO$_2$ Emissions and Access to Energy

Edited by

Ray Galvin

University of Cambridge, Cambridge, United Kingdom
RWTH Aachen University, Aachen, Germany

ELSEVIER

ACADEMIC PRESS
An imprint of Elsevier

Academic Press is an imprint of Elsevier
125 London Wall, London EC2Y 5AS, United Kingdom
525 B Street, Suite 1650, San Diego, CA 92101, United States
50 Hampshire Street, 5th Floor, Cambridge, MA 02139, United States
The Boulevard, Langford Lane, Kidlington, Oxford OX5 1GB, United Kingdom

Notices
Knowledge and best practice in this field are constantly changing. As new research and experience broaden
our understanding, changes in research methods, professional practices, or medical treatment may
become necessary.

Practitioners and researchers must always rely on their own experience and knowledge in evaluating
and using any information, methods, compounds, or experiments described herein. In using such
information or methods they should be mindful of their own safety and the safety of others, including
parties for whom they have a professional responsibility.

To the fullest extent of the law, neither the Publisher nor the authors, contributors, or editors,
assume any liability for any injury and/or damage to persons or property as a matter of products
liability, negligence or otherwise, or from any use or operation of any methods, products, instructions,
or ideas contained in the material herein.

Library of Congress Cataloging-in-Publication Data
A catalog record for this book is available from the Library of Congress

British Library Cataloguing-in-Publication Data
A catalogue record for this book is available from the British Library

ISBN 978-0-12-817674-0

For information on all Academic Press publications visit our website
at https://www.elsevier.com/books-and-journals

Publisher: Brian Romer
Acquisition Editor: Graham Nisbet
Editorial Project Manager: Ali Afzal-Khan
Production Project Manager: Kamesh Ramajogi
Cover Designer: Matthew Limbert

Typeset by SPi Global, India

Working together
to grow libraries in
developing countries

www.elsevier.com • www.bookaid.org

Contents

Contributors xi

About the authors xiii

Acknowledgments xv

Introduction.. xvii

PART 1 Theory and concepts: Bringing economic inequality into energy research

CHAPTER 1 Recent increases in inequality in developed countries 3
Ray Galvin

1 Introduction...3
2 Economic inequality: Some preliminary issues...........................7
  2.1 Wealth inequality compared with income inequality............7
  2.2 Measuring inequality 10
3 Recent changes in wealth and income distribution 15
  3.1 Estimating economic inequality 15
  3.2 What about the bottom 10%?....................................17
  3.3 Tax havens 19
4 How inequality has changed in the developed countries 21
  4.1 Is extreme inequality the norm in human history?.............21
  4.2 Why were we so egalitarian in 1950–80?........................22
  4.3 The shift to neoliberalism..23
  4.4 Being poor in a society of extreme inequality..................25
5 Conclusions..26
  References..27
  Further reading 30

CHAPTER 2 What is money? And why it matters for social science in energy research........................31
Ray Galvin

1 Introduction..31
2 Money as a relationship..33
3 Myths about money 37
  3.1 Myth and obfuscation...37
  3.2 The myth that all money is the same............................38
  3.3 The myth that money is a neutral veil 40

3.4 The myth that money is a commodity 42

3.5 The myth that money is a creation of the state 44

4   Discussion and conclusions: Some implications for social
science in energy research...45

References...48

CHAPTER 3   **Asymmetric structuration theory: A sociology
for an epoch of extreme economic inequality**...........53

Ray Galvin

1   Introduction...53

2   Structuration theory 57

3   Structuration theory, oligarchs and power 59

4   The power of credit and debt 61

5   Competing discourses: Neoliberalism and welfare politics.........64

6   Implications for energy research...66

7   Conclusions...68

References...70

Further reading 74

CHAPTER 4   **Economic inequality, energy justice and the
meaning of life**...75

Ray Galvin

1   Introduction...75

2   The energy justice movement to date 78

3   Justice, morals and Wittgenstein's reflections............................80

4   Justice as fairness—Rawls, Rorty and pragmatism 85

5   Extending justice globally 87

6   Implications for global energy justice...89

7   Energy justice and economic inequality 91

8   Conclusion 92

References...93

**PART 2 Empirical findings: Energy and economic
inequality in practice**

CHAPTER 5   **Energy poverty: Understanding and addressing
systemic inequalities**...99

Lucie Middlemiss

1   Introduction...99

2   Linking energy poverty and other forms of inequality.............101

2.1 Developments in energy poverty research.........................101

2.2 A socially systemic explanation of energy poverty...........102

2.3 Blurring the distinction between energy poverty
and income poverty 104

3   Learning from the lived experience 104

3.1 John's story 105

3.2 Insights from lived experience research 106

4   The politics of energy poverty 108

5   Conclusions 110

References 112

**CHAPTER 6   Housing tenure and thermal quality of
homes—How home ownership affects access
to energy services** **115**

Nicola Terry

1   Introduction 115

2   Housing tenure in the UK and other countries in the EU 117

3   Trends in UK housing tenure 118

4   Who are the private landlords? 120

5   Comparing heating efficiency of privately rented homes
with other sectors 121

6   Comparing thermal comfort of private sector homes
with other sectors 127

7   Take up of energy efficiency measures 131

8   Reasons for poor take-up 133

8.1 The tenants' view 134

8.2 The landlords' view 135

9   Impact of minimum energy efficiency standard 138

10 Summary and conclusions 139

References 140

Further reading 143

**CHAPTER 7   Cold homes and Gini coefficients in EU
countries** **145**

Ray Galvin

1   Introduction 145

2   Measures of fuel poverty 149

3   The variables 151

3.1 Which variables are relevant? 151

3.2 Panel data or year by year regressions 154

4   Descriptive statistics 154

5   Results 158

　　　　　5.1 Tests for model fit 158
　　　　　5.2 Other statistical tests..159
　　　　　5.3 The coefficients 163
　　　6　Discussion: The role of the Gini index 165
　　　7　Conclusions..167
　　　　　References...168
　　　　　Further reading 171

**CHAPTER 8   Why are women always cold? Gendered
　　　　　　　realities of energy injustice** **173**
　　　　　Minna Sunikka-Blank
　　　1　Introduction...173
　　　2　Gendered household practices and the feminization
　　　　　of demand side response 175
　　　3　Fuel poverty, single parents and intergenerational
　　　　　immobility..179
　　　4　Conclusions..184
　　　　　References...185
　　　　　Further reading 188

**CHAPTER 9   Inequality and renewable electricity support
　　　　　　　in the European Union** **189**
　　　　　Lawrence Haar
　　　1　Introduction and background.....................................189
　　　2　Investment and costing issues with renewable
　　　　　electricity 192
　　　3　Renewable energy support 194
　　　　　3.1 Background..194
　　　　　3.2 Growth in RE capacity 195
　　　　　3.3 Expenditure on renewable electricity.........................195
　　　4　Pricing of electricity 197
　　　　　4.1 Cost components..197
　　　　　4.2 Pricing of electricity by household size.....................200
　　　　　4.3 Relating household electricity consumption to
　　　　　　　household income 209
　　　5　Renewable energy and electricity pricing.......................211
　　　6　Conclusions..214
　　　　　Acknowledgment 217
　　　　　References...218
　　　　　Websites consulted 219
　　　　　Further reading 220

**CHAPTER 10 Energy poverty research: A perspective from
the poverty side** **221**
Ray Galvin

1   Introduction 221
2   Poverty as discussed in energy poverty literature 223
    2.1 Energy inefficient buildings 224
    2.2 The targeting approach to energy poverty 225
    2.3 The fear of increasing CO_2 emissions 226
3   Method and approach 227
4   Using the 10% indicator 229
    4.1 The logic of the analysis 229
    4.2 The cost of adequate household energy services
         using the 10% indicator 230
    4.3 Calculating the additional amounts required, under
         the 10% indicator 231
    4.4 How much would this cost high-income
         households? 233
5   Would this redistribution be tolerated? 234
6   Implications for CO_2 emissions 237
    6.1 The CO_2 implications of more progressive taxation 238
    6.2 CO_2 emissions and increased incomes among
         poorer households 240
7   Conclusions 243
    References 245
    Further reading 248

PART 3 Reflections

**CHAPTER 11 Sustainable energy transition and increasing
complexity: Trade-offs, the economics
perspective and policy implications** **251**
Reinhard Madlener

1   Introduction 252
2   The various energy inequality and justice dimensions
    to be considered 254
    2.1 Spatial heterogeneity and trends of (primary) energy
         use: A global perspective 255
    2.2 Spatial heterogeneity and structural economic change:
         A regional perspective 255
    2.3 Temporal income and energy inequality: Let the "*Gini
         out of the bottle*" 257

2.4 (Social) Life-cycle analysis of energy justice.....................259
2.5 Heterogeneity of energy rebound and sufficiency..............260
3 Broadening our understanding of energy poverty:
The economics perspective 261
3.1 Some reflections on terminology and the energy/fuel
poverty and energy justice debate.....................................261
3.2 Measurement issues...263
3.3 Taxation, transfers, and subsidies: How to best
re-balance the level-playing field?....................................267
3.4 Social/economic welfare considerations 272
4 Smart systems: Efficiency, participation and equity
considerations 275
4.1 Smart grids as enabling technologies...............................275
4.2 Smart meters and real-time pricing (RTP).........................275
4.3 Sustainable energy communities......................................275
4.4 Prosumer households and aggregate constructs
(microgrids, virtual power plants, energy
communities/clouds etc.)...276
4.5 The multi-tenant prosumer concept (MTPC).....................277
5 Economic growth, productivity gains, structural change 278
6 Conclusions...279
References...281
Further reading 286

**CHAPTER 12 Can economic inequality be reduced?
Challenges and signs of hope in 2019**...................**287**
Danny Dorling

1 The roaring 20s...287
2 Modern times 291
3 Hard times...292
4 The crash and the rise of the far right 294
5 The human geography of Brexit 295
6 Inequality, pollution and stupidity 300
7 Inequality extremes in Europe 300
8 Hope..306
References...308
Further reading 310

Concluding remarks..311
Author index 317
Subject index 327

Contributors

Danny Dorling
School of Geography and the Environment, University of Oxford, Oxford, United Kingdom

Ray Galvin
University of Cambridge, Cambridge, United Kingdom; RWTH Aachen University, Aachen, Germany

Lawrence Haar
Oxford Brooks Business School, Oxford Brookes University, Oxford, United Kingdom

Reinhard Madlener
Institute for Future Energy Consumer Needs and Behavior (FCN), School of Business and Economics/ E.ON Energy Research Center, RWTH Aachen University, Aachen, Germany

Lucie Middlemiss
Sustainability Research Institute, School of Earth and Environment, University of Leeds, Leeds, United Kingdom

Minna Sunikka-Blank
Department of Architecture, University of Cambridge, Cambridge, United Kingdom

Nicola Terry
Qeng Ho Ltd; Cambridge Architectural Research Ltd, Cambridge, United Kingdom

About the authors

Ray Galvin's background is in engineering, theology, social psychology and environmental policy, with interests in economics, philosophy and social theory. His post-doctoral research focused on energy consumption behavior and policy. More recently he has been exploring issues of economic inequality and how this affects global energy use and households' access to energy services. Dr Galvin is an associate editor of the journal *Energy Research and Social Science* and does research and teaching for universities in Germany and the UK, and project evaluation for the EU Commission.

Lucie Middlemiss is Associate Professor in Sustainability, and Co-director of the Sustainability Research Institute, at the University of Leeds, UK. She is a sociologist, and takes a critical approach to the intersection between environmental and social issues, with topic interests in sustainable consumption and energy poverty. She has published a number of papers on energy poverty, and her textbook on Sustainable Consumption was released by Routledge in 2018.

Nicola Terry is a consultant to UK government and non-government organizations specializing in energy efficiency in buildings, including aspects of fabric and occupant behavior. She is experienced in modeling energy use and analyzing energy data from field studies. Nicola has 1st class degrees in Environmental Studies (Open University) and in Computer Science and Engineering (Cambridge, UK). She also volunteers for Cambridge Carbon Footprint and Transition Cambridge and is the author of "Energy and Carbon emissions: the way we live today" (UIT Cambridge, 2011).

Dr Minna Sunikka-Blank is a Senior Lecturer at the Department of Architecture at the University of Cambridge. She leads the Behavior and Building Performance (BBP) research group and is a founding member of the interdisciplinary research collective Global Energy Nexus in Urban Settlements. Her research focuses on the effectiveness of energy policy instruments, the application of social theory on energy use in buildings, and gender as an overlooked factor in climate change. She is a Director of Studies and Fellow in Architecture at Churchill College.

Dr Lawrence Haar is a Senior Lecturer in Banking and Finance with Oxford Brookes University. Before academia, Dr Haar was a Director and Managing Director for Risk Management and Valuation with major banks, energy and mining companies. He was a Director for Audit Assurance with Deloitte. Dr Haar has expertise in the regulation of Financial Markets, having worked for the UK Authorities as a Senior Risk Specialist. He regularly appears in the financial press as well as academic journals.

Dr Reinhard Madlener is Full Professor in Energy Economics and Management at RWTH Aachen University and Adjunct Professor with the Norwegian University of Science and Technology, Trondheim. At RWTH he is Director of the Institute for Future Energy Consumer Needs and Behavior (FCN), a part of the E.ON Energy Research Center (est. 2006). He is an Editor of *Energy Policy* and Editorial Board member of other well-known journals. His main research interests are in energy economics and policy, behavioral economics, technological diffusion and sustainable development.

Danny Dorling works at the University of Oxford. He was previously a professor at the University of Sheffield, and before then a professor at Leeds. His earlier academic posts were in Newcastle, Bristol, and New Zealand. His most recent book, with Sally Tomlinson, is "Rule Britannia: Brexit and the end of Empire" concerning what the 2016 EU referendum and 2019 "exit" reveals. In 2018 he published "Peak Inequality" on issues of housing, health, employment, education, wealth and poverty in the UK.

Acknowledgments

I wish to thank the Faculty of Business and Economics at RWTH Aachen University, Germany, for a grant that supported some of the initial work for this book, and for the generous use of university facilities at various times during its writing. I am also grateful to the Department of Architecture at Cambridge University for strategic support. As well as our team of co-authors—who have worked together as a mutually supportive team—a number of individuals have contributed to the book through reviewing chapters, discussing possible interpretations of findings, providing data and research materials, and giving encouragement, feedback and strategic support. I especially mention Fabiola Blum, Tugba Atasoy, Brian Wheeldon and Fred Nahme, but other colleagues and friends in Auckland, Cambridge and Aachen have contributed in innumerable ways. Finally, my thanks to the Elsevier team for making the publication process smooth-running and supportive at every turn.

Introduction

The main aim of this book is to show how energy consumption behavior and policy in high-income countries can be better understood if we take full account of the fact that it happens today in a context of enormous economic inequality. This is especially so if we are approaching energy issues via social science or economics.

A book like this would probably not have been needed in the 1950s–1980s. In those times there were historically low levels of inequality in high-income countries such as Australia, Canada, New Zealand, the United Kingdom, the US and most of Western Europe. As I explain in Chapter 1, that period was probably unique in human history. But since the 1980s most of these countries have reverted back to, or at least toward, the extreme levels of economic inequality that characterized the 19th and early 20th centuries.

Three overarching issues have puzzled me as I have thought about this subject, written chapters for this book and worked with its other authors. One is the fact that we human beings are not naturally equipped to grasp the magnitude of the levels of inequality that characterize our societies today. The numbers are just too big to get our heads around. For example, the richest person in the UK has a fortune of about £25 billion. What does a billion mean? How do we conceptualize it in our minds? We could try a thought experiment: suppose I earned the average annual UK income of £30,000, managed to avoid paying any tax on it, and saved every penny of it, year after year. How long would it take me to save up £25 billion? The answer is: just over 830,000 years. 830,000 years ago homo-sapiens was just beginning to emerge in the evolutionary tree. It would take that long.

But this still presents a problem for our brains, because we have great difficulty conceiving of what a number as big as 830,000 is like. We do not have a chamber in our minds 830 times as big as 1000, to conceptualize it.

So let's avoid the mega-rich for a moment and just think about high-earners. As I discuss in Chapter 10, the person with the highest tax-declared income in the UK in 2018 earned £336 million, or about £6.46 million per week. To just make it into the top 10% of earners in the UK you have to be in a household that earns £1000 per week. The top earner earned 6460 times this much.

Even here our brains give up, because how do we conceive of an amount of money that is 6000 times as high as the weekly income of a very good earner? A suggestion: make a power-point slide of a histogram of the bottom 90% of earners and project it on a screen so that it is 1 m wide: 1 m between those who earn £0 per week and those who earn £1000 per week. You would have to extend this histogram 6.46 km to the right, to include the highest tax-declared earner in it.

But even that does not give a full picture, because this is only tax-declared income. The richest person in the UK most likely increased his fortune by about £2.2 billion in 2018, which equates to an income of £42 million per week. You would have to extend the histogram by 42 km to fit him in our histogram, about as far as

Dover to Calais, right across the English Channel. Yet even this is deceptive because long, line-of-sight distances do not actually look as long as they really are. You would have to try swimming the channel to begin to grasp the size of it.

Perhaps this helps us begin to conceptualize the enormity of today's economic inequality. It relates to the second thing that has puzzled me: why do so many people think there is really not that much money around in countries like the UK, Germany and New Zealand and that therefore we can't ask too much of governments in making life easier for the poor? The idea that there's not enough money around is of course complete nonsense, a silly myth. Let's go back to our histogram. The part that is actually on the screen goes all the way up to those comfortably-off households with incomes of £1000 per week. It includes 90% of the UK population. However, the income represented among these households is only about 38% of total UK personal income (before tax). The remaining 62% is earned by those who are, quite literally, off the chart. Further, after tax they get to keep well over half of this. What do they do with it? Bear in mind this is personal income, not business money that gets recycled to create jobs and material wealth. A lot of recent research shows their use of it causes huge excesses of CO_2 emissions—again a theme of Chapter 10.

The point is that there is lots and lots of money in our economies. But our governments have been depriving themselves of larger and larger portions of it over recent decades, by giving bigger and bigger tax breaks to the highest earners, making it harder for governments to fix problems of poverty, and giving the impression that our society is hard up.

The third issue that has puzzled me is why these issues of economic inequality are not a lot more prominent in energy consumption research—or, to put it less modestly, why they hardly ever appear in this research. When they do appear, it is usually fleeting. They are almost always side-lined, as if they exist in another realm of being and are too hard to tackle. This is not the case in other areas of research, such as housing or health studies. Many research papers in these fields go straight to the issue of economic inequality and relate it directly to issues in housing or health. But not in energy consumption studies.

On the surface this seems extraordinary, since energy costs money. If we want to understand why people do what they do with energy, we have to have a pretty thorough background understanding of what money is doing in the economy, what people are doing with money, who has it, who is deprived of it, and why.

This brings me to a personal note. I have been doing research on energy consumption, policy and behavior since about 2008. Most of my early work looked at policy and consumer practice on thermal upgrades of homes, then I moved on to explore energy and CO_2 emission issues in car transport, computing and electrical supply systems. In this work I hardly took any account of issues of economic inequality, and I was never challenged to look at these by the international research community whose works intersected with mine and who reviewed my papers and books. But about 4 years ago my son-in-law, Kim Martinengo, woke me up to the fact that I was missing out the main theme of the story. I hadn't realized, until then, how far economic inequality has gone. So I started researching the issue of how economic inequality impinges upon and co-determines energy consumption and policy.

This took a long time because there was very little in my field of research to build on. But in July 2018 my colleague Minna Sunikka-Bank and I published our paper 'Economic Inequality and Household Energy Consumption in High-income Countries: A Challenge for Social Science Based Energy Research', in the journal *Ecological Economics*. It was just one article but at least it helped put the issue on the agenda. Then, when Elsevier invited me to write a book, I suggested one on the same theme. I was pleasantly surprised when their appointed reviewers enthusiastically approved the outline I had submitted. One of them even offered to contribute a chapter.

Perhaps, then, the tide is turning and a place may be opening up, in energy consumption and policy studies, for more research on the connections between economic inequality, energy consumption and energy policy. This book is an attempt to insert a little exploratory material into that gap.

It is important to note, however, that the book does not simply accept that there is extreme economic inequality and ask how we can better manage energy consumption given this fact of life. It *questions* this fact of life. It questions whether economic inequality is justifiable. It probes into how it could be mitigated.

There are several themes in current energy research that intersect in this book. The main ones are: the climate emergency; energy poverty; energy justice; and the energy transition. I speak here of 'the climate emergency', but when we started this book in mid-2018 one still spoke sedately of 'climate change'. Some might ask, why waste time on issues of economic inequality, when all our attention is needed to avert a catastrophic collapse of the Holocene climate on which our civilization heavily depends? But, far from being a side-issue, parts of this book argue that economic inequality not only contributes to humanity's assault on the climate. The two ills are parallel effects of the greater problem that a minority of a few insanely rich groups and individuals control large swathes of the economy, if not the major part of the planet's economic edifice.

The second of these themes, energy poverty—often identical with or closely related to fuel poverty—has been heavily, widely and increasingly researched in recent years. One would think it is obvious that the 'poverty' aspect of energy poverty is closely related to economic inequality. In some studies this link is recognized. However, almost all the energy poverty research that recognizes this link holds back from investigating it. It refrains from questioning how and why this inequality got there and is reproduced, thereby allowing energy poverty to flourish. This book offers several ways of breaking through this barrier.

The third of these themes, energy justice, could have the potential to provide a powerful ethical framework for investigating how economic inequality, energy poverty and the climate emergency interact. Some energy justice literature begins to do this.

Several of the chapters in this book offer ways of improving the logic and the economics of energy justice literature. Our argument is that it needs a better philosophical base and a more secure economic foundation.

The fourth of these themes, the energy transition, is all about shifting our energy production away from fossil (and in some cases nuclear) fuels to renewables. This is expensive, hence the need, explored in several chapters in this book, to ensure that the poor are not forced to pay for the energy transition. But there is more to it than that. Current energy production is held in place, to a large extent, by very wealthy vested interests. These are among the economically privileged sections of society that are also implicated in the economic rules and structures that lead to economic inequality and cement it in place. Some of the chapters in this book bring this to light and seek to challenge it.

So the book does not simply identify economic inequality and its relationship to energy issues and leave it there. Instead it looks critically at it, asks why it is allowed to continue and why our societies and even research projects seem to tolerate it so benignly. It also relates it to current, major themes in energy studies.

Because there is so little legacy of critically investigating the economic inequality that determines or influences so much of today's energy consumption, the first four chapters of the book offer background material which energy researchers might find useful if they wish to bring poverty and economic inequality more centrally into their empirical research. The next six chapters are empirically-based studies that look at different aspects of energy consumption in relation to different themes in economic inequality. The chapters in the last section and the Conclusions make more general comments, looking toward future action and research.

I now give a brief introduction to the chapters.

In Chapter 1, I set the scene by offering an outline of how societies in high-income countries have become so unequal over the past few decades. I also give some figures and charts as to what the magnitude of this inequality is in various high-income countries—although the impact of this may be limited by our difficulties, described above, of grasping the differences between very large and very small numbers. I then describe the different ways economic inequality is measured in economics and other research. This includes an explanation of the much-touted Gini coefficient, how it relates to measures such as the income share of the top-earning 10% or 1%, the Robin Hood Index, and other measures. One of the key themes in this chapter is the difference between the Keynesian economic model, which prevailed in high-income countries from the mid-1940s until the late 1970s, and the neoliberal economic model which has prevailed ever since. In this chapter I relate these issues only briefly to energy consumption issues, since my aim here is to introduce energy researchers to concepts and historical shifts that may need to be brought more centrally into their own energy research agendas.

In Chapter 2, I investigate the question: what is money? This may sound a rather abstract topic but it is of vital importance in understanding how wealth gets continually syphoned up to those who already have a great deal of it, leaving very little money and a lot of debt among governments and poorer households. I outline the two main competing views as to what money is: a convenient means of exchange of goods and services, which humanity has invented to make markets work more smoothly; and a relationship of obligation and entitlement between a debtor and a creditor. I argue that the second view fits the facts much better, and explains why wealth accumulates to the super-wealthy and eventually turns the rest of us in to serfs, unless governments step in to mitigate this. I also explain in this chapter how the banking system functions, and the extraordinary fact that private banks are permitted to literally make money out of nothing, dissolve it again into nothing when it has done its rounds, and make a lot of profit out of the cycle from the interest rates they charge. This also helps explain why the world is awash with money today, without it leading to excessive inflation or enriching the poor. In case readers think this whole narrative is non-standard or even oddball, I can report that the chapter has been reviewed and enjoyed by several banking experts.

Chapter 3 is intended for sociologists, though it should be useful to anyone who looks at energy consumption through a social science perspective. Its underlying question is: how does extreme economic inequality affect social structures? My assumption is that energy production, supply and consumption are heavily constrained and enabled by the structures of society. By 'structures' I mean the ensemble of institutions, discourses, practices, beliefs, expectations and rules that we all live by. Drawing on Anthony Giddens' structuration theory, I argue that powerful, well-resourced people play a disproportionately large role in shaping these structures, often to their own benefit. Further, they often shape these structures in such a way that they are free to reshape them to suit their needs. Everybody's lives are constrained and enabled by social structures, but there are huge differences in the amount and potency of the resources different people have, to shape and take advantage of them. I then draw on Jeffrey Winters' studies of 'oligarchs', the people with fortunes in the billions. Winters documents how these people successfully use their enormous resources to influence governments to shape laws and regulations so as to protect their fortunes and interests above all else. Energy production, supply and consumption therefore take place largely under their rules, constrained and enabled by the social structures these oligarchs have heavily influenced. I also argue, however, that this is not the whole story, because there are other resources available to us all, to defy unjust structures and work to create new ones. But this requires active, energetic, committed action and the belief that oligarchic rule is not inevitable.

In Chapter 4, the last of the background and conceptual chapters, I discuss energy justice, a topic that has rapidly expanded in social science over the past 7 or 8 years. Energy justice would seem like an ideal framework to promote economic equality for energy consumers, since its central pillar is the notion of fairness. However, I have long felt uncomfortable with energy justice literature, for an important reason: its arguments often depend on an unspoken assumption that there is a kind of metaphysical realm of just laws and rules, existing in and of itself independently of human society, which researchers can identify and explicate and therefore demand others obey. This tends to alienate a lot of good people, including politicians, who may have come to different conclusions about the matters at hand. Instead, what we need to do—in my opinion—is latch onto people's already well-developed moral character and help them build on this and join with us to find solutions together, to the pressing problems of energy injustice. I develop this view through reflection on the work of 20th century philosopher Ludwig Wittgenstein. The so-called 'post-Wittgensteinian' view is that children develop a moral sense through thousands of everyday incidents of learning to get on with other people. Later they often systematize this into a moral code, but it is, essentially a pragmatic heart-based thing, which may vary from person to person and culture to culture, not a set of abstract laws that exist firm and true for all time and peoples. This, I argue, is the context in which we need to persuade our leaders to provide the means to bring greater economic equality and a fairer distribution of energy resources.

Chapter 5 is the first of the empirical chapters. Here, Lucie Middlemiss develops a 'systemic' understanding of energy poverty. It is common to see energy poverty through a fairly static lens, as a kind of fixed state of being, caused by its classic determinants economic poverty, sub-standard homes, and high fuel prices. Through her in-depth, long-term study of the subject and her direct engagement with hundreds of UK households suffering under it, Middlemiss sets energy poverty in a more dynamic context. It is both caused by, and causes, poverty and its deprivations; it can change over time such that previously comfortably well-off households can descend into energy poverty and get stuck there; the very government policies that lead to energy poverty can then exacerbate it and close off routes of escape; people who become energy-poor can develop mental illness or become socially isolated and lose the means of climbing out of their energy poverty. Middlemiss relates the story of a previously successful middle-class couple living their own, fairly spacious home – reading between the lines, this could be any of us. But then the man is made redundant from his job, suffers a relationship breakup, consequently develops mental health issues, and now has below-subsistence income and has to choose between eating well, heating his large home (which he still lives in) and paying for transport to the distant jobcenter where he must frequently report in order to keep his benefits. A theme running through the chapter is austerity: government policies that skimp on basic services and welfare, exacerbated by policies of austerity which impact on the poor while leaving the wealthy unscathed.

Chapter 6 is contributed by Nicola Terry, an engineer with specialist experience and understanding of the thermal and structural features of homes, especially in the UK context. Here she examines the UK housing stock from the point of view of energy efficiency and tenure. Regarding tenure, UK housing can be divided into three groups: social housing (with low or subsidized rents); private rented housing (with rents that are higher and rising); and owner-occupied homes. Terry compares the thermal and other health and comfort-related performance of these three groups, based on features such as: wall insulation; the UK's official thermal rating system called 'standard assessment performance' (SAP); presence of damp and mold; and type of heating system. Some of the findings are not surprising, for example that most private landlords have no professional qualifications or expertise in being landlords; that social housing is significantly better thermally and health-wise than the other two tenures; and that the proportions of both social and owner-occupied housing are falling—because fewer and fewer young people can afford to buy a home and much social housing has been privatized. What is surprising, however, is that owner-occupied homes are not much better than privately rented homes on average, yet owner-occupiers are generally more satisfied with their thermal comfort levels than renters in both the other tenures. Terry presents her empirical findings along with very accessible narratives, graphs and charts, plus straightforward explanations of subjects like SAP and the Decent Homes Standard, which will give even uninitiated readers a valuable introduction to what the UK's housing stock is like.

In Chapter 7, I present a statistical analysis of the relationship between national levels of income inequality in EU countries, and the percentage in each country suffering energy poverty, measured here by the percentage who cannot heat their homes adequately. Controlling for other influences such as GDP per capita, the number of heating degree days and wealth per household, I find a strong correlation between energy poverty and income inequality—represented by the Gini coefficient. I also find that the impact of the Gini on the percentage in energy poverty is at least as strong as the impact of GDP per capita, wealth per household and heating degree days, and much stronger than other likely influences. One of the reasons for this is most likely that in countries with high income inequality there is a disproportionately high number of poor households. Another could be that where income inequality is high, poor households have to compete against rich households for basic goods and services, like energy and home renovation labor. Another interesting finding is that the higher the Gini coefficient (i.e., the higher a country's income inequality), the more sensitive the percentage of households in energy poverty is, to changes in the Gini. Since the Gini can be changed simply by an act of parliament—by tweaking the tax and welfare systems—large reductions in energy poverty could be brought about quite easily by making tax and welfare more progressive—if there is the political will to do this.

In Chapter 8, Minna Sunikka-Blank explores how gender relates to energy consumption, in the context of economic inequality. Sunikka-Blank's empirical work on women and energy mostly takes place in the UK and India, and she finds some remarkable gender and energy parallels between the two cultures. It is perhaps not surprising to hear that women do most of the day-to-day management of energy consumption in the home in both countries, often in terms set by men. A worrying trend, as countries endeavor to reduce their CO_2 emissions, is that women are the ones who bear the domestic burden of the energy transition. For example, time-of-day energy pricing requires load-shifting, meaning women now have to schedule their household chores around the needs of the electricity grid (which is designed by men). A further interesting finding is that becoming a mother loads women with an extra set of energy responsibilities, alongside the huge drop in income women typically suffer at this time. All this is happening in a social context where, Sunikka-Blank finds, women are systemically disadvantaged in reaching the economic goals that people need to reach to be reasonably well insured against energy poverty—both in India and in the UK. To make matters even worse, this is set amidst increasing economic inequality and reluctance, among many governments, to embrace economic and fiscal policies that could mitigate it.

Chapter 9 is offered by Lawrence Haar, a banking and finance specialist who is closely familiar with the energy production industry worldwide. Here he examines pricing structures for electricity in all EU countries, and in particular, how this has been influenced by the rapid increase in renewable electricity generation. The overall message of this chapter is that in most EU countries the poorest households are being made to pay disproportionately for the energy transition. This is not just because feed-in-tariffs for photovoltaic and wind energy are paid for by surcharges on household electricity bills. The nub of the problem is more subtle. With wind and solar, almost all the producers' costs are in the equipment, and these costs are very high. Each extra kWh of electricity production costs virtually nothing—given free by sun or wind. To recoup their costs, producers like to charge consumers high fixed costs, but are less worried about how many kWh a customer consumes. This means, in effect, that the more kWh a household consumes, the less it pays per kWh. This happens in all 28 EU countries except Malta, where progressive tax policy reverses the trend (the less you use, the less you pay per kWh), and Poland and the Baltic countries, where renewables have not yet penetrated far. Haar's detailed work shows that in some EU countries the regressive pricing structure is extreme, and throughout Europe it has been worsening over time. Pricing systems for electricity grids were designed to be fair, in a deregulated context, when electricity was generated by fossil and nuclear plants. But with the transition to renewables happening in that context, poorer households are bearing the brunt.

My second empirical contribution is Chapter 10. Here I ask a question that has not been explored in energy poverty research thus far: what would happen if we addressed energy poverty in the UK by attempting to eliminate economic poverty? My reasoning is that, although this would not automatically fix thermally leaky homes or bring down the cost of heating fuel, it would at least ensure all households have the means to either keep warm at home or upgrade some of the worst thermal features of their houses, or both. I make the 1960s-style suggestion of increasing the marginal tax rate on the highest incomes, and transferring the money to the lowest-income homes (e.g., via job training and creation, more generous benefits, etc.) so that no-one in the UK is below the EU official poverty line. Using solid UK government statistics I calculate how many percentage points the tax increase on the top 10% of earners would have to be. I compare this with past decades, and find we would still be way below the tax rates of the Thatcher years. I then ask what the increase in CO_2 emissions will be if all the formerly poor households now heat their homes comfortably. The surprise is, the increase would be minuscule compared to the reduced emissions from the wealthiest 10% having less money to spend on their CO_2-intensive indulgences. Chapter 10 is a must if you like to think outside the box, dream dreams and see visions of what could happen if we tweaked the economic system just a bit, like they did in Keynesian times.

The final section of the book looks more widely, reflecting on broader issues. Chapter 11 is contributed by Reinhard Madlener, an economist with decades of experience leading research on the economics of energy consumption, supply and production in the context of climate change mitigation. He brings his detailed knowledge of the German *Energiewende* (energy transition) as well as global energy transition issues to bear on some of the recurring themes of this book. In particular, he asks, how can knowledge and expertise in economics be grafted into the concerns and findings of energy justice research? How would the energy transition look if it were guided by just, ethical principles, informed by what economists have learned in the last 4 decades of trying to shift toward sustainable energy? This is an exploratory chapter that looks toward future research where economists and (other) social scientists work more closely together.

In Chapter 12 Danny Dorling reflects on where we are currently, in the shifts between increasing economic inequality and reactions against it. What struck me about this chapter is that there is huge push and pull in both directions. Many people are feeling dispossessed, tired of standing still or going backward as the total amount of wealth in society increases relentlessly but ends up in the hands of a small privileged percent of the population. Some groups of dispossessed are joining with well-organized and well informed movements for radical change toward a more egalitarian society, based on sound and compassionate economic principles. Others, however, are being co-opted by wealthy elites and their proxies, into hard right reactionary movements that blame everything and everyone for their plight, except the economic distortions that are actually causing it. Meanwhile, the oligarchs are lobbying and manipulating as hard as ever to continue syphoning wealth to themselves. Dorling suggests there is always hope for a fairer society, but only if we all work hard for it.

In the Concluding remarks I reflect on the book in the light of future challenges. In particular, I ask what kinds of new emphases there could be in future research.

Ray Galvin

University of Cambridge, Cambridge, United Kingdom;

RWTH-Aachen University, Aachen, Germany

Theory and concepts: Bringing economic inequality into energy research

Recent increases in inequality in developed countries

1

Ray Galvin

University of Cambridge, Cambridge, United Kingdom
RWTH Aachen University, Aachen, Germany

Chapter outline

1 Introduction 3
2 Economic inequality: Some preliminary issues 7
   2.1 Wealth inequality compared with income inequality 7
   2.2 Measuring inequality 10
3 Recent changes in wealth and income distribution 15
   3.1 Estimating economic inequality 15
   3.2 What about the bottom 10%? 17
   3.3 Tax havens 19
4 How inequality has changed in the developed countries 21
   4.1 Is extreme inequality the norm in human history? 21
   4.2 Why were we so egalitarian in 1950–80? 22
   4.3 The shift to neoliberalism 23
   4.4 Being poor in a society of extreme inequality 25
5 Conclusions 26
References 27
Further reading 30

1 Introduction

Over the past 30–40 years the distribution of wealth and income in almost all the developed countries has shifted dramatically. In the 30-year period immediately following the Second World War these countries were unique in recorded history. They enjoyed strong, persistent economic growth leading to historically high real incomes per person, while at the same time they were remarkably egalitarian, with very low rates of poverty and historically small differences in incomes and wealth across the rich-poor spectrum. Over the past 3–4 decades, however, economic growth has

Inequality and Energy. https://doi.org/10.1016/B978-0-12-817674-0.00001-1

slowed in these countries while in most of them economic inequality has persistently increased. Large and increasing proportions of these countries' populations are poor; their middle-income earners are in many cases worse off than previously; the richest 10% are astonishingly wealthy; while the wealth and income of the richest 1% beggars the imagination (Piketty, 2014; Dorling, 2018; Alvaredo et al., 2017).

This has important implications for energy consumption, because energy costs money. In this book we will present evidence of how this plays out at household level, and also at the levels of infrastructure, energy supply, and macro-level decision making. We will also argue that these enormous changes in income and wealth distribution lead to certain individuals having far more say than others as to who gets to enjoy what energy services, how our future energy infrastructure will develop, and how this can affect CO_2 emission levels.

It also has important implications for social science. While engineers study and develop the technical aspects of energy technology and infrastructure, social scientists conduct vitally important research on the people and groups involved in energy provision and consumption. This involves in-depth research on households, suppliers, relevant social networks, and decision makers at all levels, including national and supranational policymakers. Strangely, however, the social science typically used in energy research has not yet caught up with the huge changes in wealth and income distribution in developed countries over the past 3–4 decades. This has led to the peculiar situation that much of current social science research on energy consumption in these countries ignores one of its most important determinants. Research often proceeds as if these huge discrepancies in wealth and income were not there.

In Chapters 3 and 4 I will look at some of the great advances certain niches in sociology have made over the past few decades, in understanding why these discrepancies have occurred and in critiquing the economic theories that appear to have fostered them. Unfortunately, these branches of sociology have not yet been taken up in earnest by the social science approaches commonly used in energy consumption research. Instead, this research often proceeds as if money functions and is distributed much the same today as it was in the 1950s–80s. It might not be a coincidence that the social science frameworks used in this research were formed, or at least have their roots, in the social and cultural milieu of that golden age of egalitarian plenty.

For many sociologists whose views were formed by the world of those times, relative equality and plenty seemed the normal state of the developed world. For example, the prominent sociologist Margaret Archer (2000), argued that Bourdieu's (1958, 1976) insights on poverty, drawn from his empirical work in France and Algeria in the 1950s, were irrelevant to the social science of 2000 because by that year poverty had been all but eliminated in developed countries. What she did not account for (along with others) was that by the end of the 20th century the egalitarian achievements of the post-War years were already being seriously eroded.

I will argue below that the 25–30 years following the end of the Second World War were certainly not the norm for wealthy societies but a very happy aberration. For most of recorded history, societies throughout the world have been characterized by extreme levels of income and wealth inequality (Winters, 2011, 2017; Piketty,

2014; Graeber, 2011). At the top end, a very small elite of outlandishly rich "oligarchs," to use Winters' terminology, have comprised around one-hundredth of 1% of the population. These people have not only had more wealth than most of us can imagine, but have simultaneously either ruled their societies directly as dictators, monarchs or potentates, or used their wealth to decisively influence the direction of government, especially on economic matters (Winters, 2011; Stiglitz, 2013). The 1–10% of people just below this group have been fabulously rich compared to most of the population but their political power has usually been subject to the whim of the oligarchs. Most of the population has lived in varying degrees of poverty, from utterly destitute to just surviving, and in some societies there has been a middle class, usually of just a few percent but sometimes larger, who live fairly comfortably.

In recent centuries in English speaking and European countries this type of economic inequality reached a peak just prior to the First World War. It then reduced more or less steadily until around 1980. Since then economic inequality has increased again and in some developed countries it is approaching the levels seen just prior to the First World War (Piketty, 2014).

To begin to get an idea of how the distribution of income has changed in recent decades, Fig. 1 shows what I call the "10–50 index" for the US and for Germany. This gives the ratio between the share of national income received by the top 10% of earners, and the share received by the bottom 50%. The raw data for these graphs comes from the World Inequality Database (WID, 2018), which is maintained and continually updated by Thomas Piketty's research team at the Paris School of Economics. I will frequently use data from this source in this book, supplemented

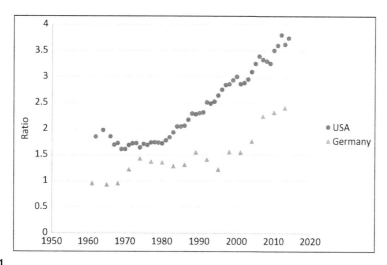

FIG. 1

Ratio between shares of national income of top 10% and bottom 50% of earners, USA and Germany, 1961–2014.

Author's calculations from WID, 2018. World Inequality Database. Available from: https://www.wid.world/.

with data from international bodies such as the OECD and Eurostat. I deliberately chose the US and Germany for the first graph of the book because the US is known to be a highly unequal society, while Germany is often popularly thought of as strongly egalitarian.

Fig. 1 shows that in the US in the 1960s and 1970s, and until about 1982, the highest earning 10% received about 1.7 times as much of total national income as the bottom 50%. The value 1.7 can be misleading because it actually means that the average earnings of those in the top 10% were 8.5 times those of the bottom 50%. This may sound high, but in historic terms it is remarkably low (Winters, 2011). However, after 1982 this ratio increased steadily, to reach 3.7 by 2014. This means the top earning 10% were now receiving, on average, 18.5 times the average income of the bottom 50%. This begins to look more like the income distribution seen in long-range studies of economic history (Graeber, 2011; Winters, 2011, 2014).

Germany shows a surprisingly similar profile, though not so extreme. The 10–50 index was about half that of the US in the 1960s, indicating Germany was a much more egalitarian society than the US during the period when its economy was recovering from wartime devastation—the period of the *Wirtschaftswunder* (economic miracle). Germany's 10–50 index increased to 1.5 in the mid-1970s—about the level of the US—then stayed at that level until 2001, before increasing rapidly to reach 2.5 by 2013. Hence the shift toward high levels of inequality started about 25 years later in Germany than in the US, and has increased since then at about the same rate. Interestingly, the start of this increase in Germany coincides with the Schröder government's introduction of liberalizing economic reforms (though other un-equalizing tendencies were also at work), while in the US the start of the increase coincides with the Regan government's economic reforms. Note that the value 2.5 means that by 2013 the average income of the top earning 10% in Germany was 12.5 times that of the bottom earning 50%. This is not as high as in the US, but is still much higher than in either country in the 1960s–70s.

This brief statistical foray can serve as a broad introduction to the magnitude of the recent changes in economic inequality and how they vary between different countries. The US is widely known to have extremes of inequality, but we might not expect Germany, which is popularly thought of as egalitarian, to be heading in the same general direction.

The main aim of this chapter is to tell the story of the economic changes that have led to today's high levels of economic inequality in developed countries. In Chapter 2 I will outline and build on the niches in sociology that have taken these trends most seriously, focusing on the question "what is money?," in Chapter 3 I will relate these findings to a social science approach to energy studies, and in Chapter 4 I suggest how a relatively new social science approach to energy studies, called "energy justice," could be modified to deal more directly with issues of economic inequality. Chapters 5–11 present a range of empirical findings on how these levels of economic inequality are affecting energy consumption, access to energy services, and the distribution of CO_2 emissions. Chapter 12 engages with the question of how current levels of economic inequality can be reduced.

In Section 2 of this chapter I explain two preliminary issues to smooth the subsequent discussion of economic inequality: the relationship between income and wealth inequality; and the different ways inequality is measured. In Section 3 I outline the changes in wealth and income distribution within developed countries over the past decades. In Section 4 I offer a narrative explanation of how and why these changes came about, including crucially important changes in the banking system and the production of money. I offer conclusions in Section 5.

2 Economic inequality: Some preliminary issues

Before mapping out the changes in economic inequality of the past decades, I draw attention to two important issues: the difference between wealth and income inequality; and different ways of measuring inequality.

2.1 Wealth inequality compared with income inequality

Wealth and income inequality are closely related, but they are two different things. Wealth inequality can increase much faster and be far higher than income inequality. This is largely because low income earners have to spend all or most of their income on living expenses, so their accumulated wealth hardly increases at all and is often zero or negative (i.e., they are in debt), while high earners can save and reinvest a very high percentage of their income. Hence the changes shown in Fig. 1, which deal only with income, tend to underplay the socioeconomic effects of increasing inequality.

Figs. 2 and 3 compare wealth shares with income shares for the US and the UK. Each pair of bars on the graphs compares national private wealth shares with national private income shares. Fig. 2 compares the percentage share of total national private wealth owned by the richest 10%, with the percentage of total national income received by the top-earning 10%. Fig. 3 gives the corresponding comparisons for the richest and top earning 1% (based on data from WID, 2018). In each case, the economically privileged group's portion of national private wealth is far higher than their portion of national income, in some cases twice as high.

In the US, for example, the wealthiest 10%'s share of total national private *wealth* increased from 61.4% in 1986, to 73.0% in 2014, while their shares of total *income* increased from 36.5% in 1986 to 47.0% in 2014. Hence in the US, wealth inequality is far greater than income inequality. The increase in wealth inequality was more extreme for the wealthiest 1% (see Fig. 3), increasing from a low of 22% of national private wealth in 1978 to around 40% in 2014, while their share of total income increased from 10.6% to 20.1%. Note also that these figures mean the richest 1% have more net wealth between them than the next wealthiest 9%: a ratio of 40% to 33%.

The picture is similar for the UK but not so extreme. Using the most up to date figures available at the time of writing, the richest 1% had 20.6% of national wealth

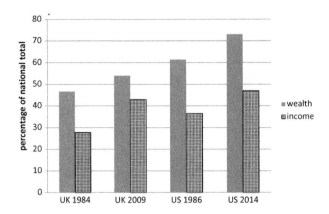

FIG. 2

Percentage share of total national private wealth and income of top 10% in the UK and US in mid-1980s and recently.

Author's calculations from WID, 2018. World Inequality Database. Available from: https://www.wid.world/.

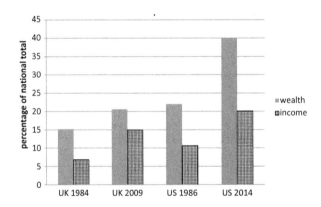

FIG. 3

Percentage share of total national private wealth and income of top 1% in the UK and US in mid-1980s and recently.

Author's calculations from WID, 2018. World Inequality Database. Available from: https://www.wid.world/.

in 2009 while the richest 10% owned 54%. This compares with figures for their shares of national income of 15% and 43%.

The bars in Figs. 2 and 3 also illustrate two other points. Firstly, the shares of both income and wealth of the wealthiest 10% and 1% have increased markedly from the mid-1980s until recently. This is seen by comparing the heights of every second pair of bars. It illustrates the rapid increase in economic inequality in these countries—measured by both income and wealth—within just 2 decades.

Secondly, these graphs appear to show that economic inequality is not as severe in the UK as in the US. However, this is misleading, due to a further important point about the way economic inequality is measured. So far I have represented economic inequality in two different ways: by the 10–50 index, which compares the income share of the top-earning 10% with that of the bottom-earning 50%; and by simply stating the wealth and income shares of the top 10% and 1%. A weakness of both of these measures is that they say nothing about the distribution of wealth or income at the lowest end. None of the above measures enables us to say anything about how wealth and income are distributed among the poorest 10%, 1%, 0.1% etc. of people in the US or the UK.

This weakness is brought home by an interesting dataset from the German Federal Statistics Bureau, Destasis (2016), plotted on a graph in Fig. 4. This shows the percentage of German households "in danger of poverty." This reached a low of 14% in 2006, then increased steadily to 15.7% by 2015. Both types of measures of inequality are needed, plus others not yet mentioned, to get a picture of the full spectrum of economic inequality.

A further reason inequality statistics often underestimate actual hardship is that they do not take into account the effect of interest payments on what I call "toxic debt," which many poorer households have become trapped into. In Chapter 2, building on the work of Di Muzio and Robbins (2016), Soederberg (2014) and Lazzarato (2015) I look at how the banking system needs to draw people into debt in order to balance its books. There and in Chapter 3 I argue that there are, roughly, two kinds of debt. One of these may be called "constructive debt." It includes mortgages, and

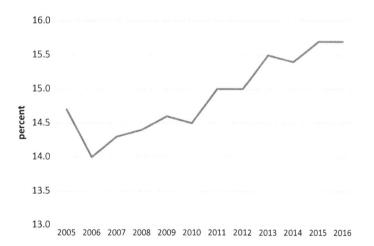

FIG. 4

Percentage of German households "in danger of poverty."

From Destasis, 2016. Statistic_id72188_Armutsgefaehrdungsquote-in-Deutschland-bis-2016. Satistisches Bundesamt Deutschland. Available from: https://de.statista.com/statistik/daten/studie/72188/umfrage/ entwicklung-der-armutsgefaehrdungsquote-in-deutschland/

loans for sensible business ventures and intelligently thought through financial dealing. Most constructive debt is held by wealthy or well-off firms, and usually returns a profit to the borrower even if the interest rate is high.

In contrast, toxic debt is via credit cards, store cards, hire purchase schemes and the like. The interest rates charged on this debt are currently around 18–20% but move higher when central bank interest rates rise. Because low-income households typically have large amounts of toxic debt, their actual disposable income is much less than what shows up in income statistics. Consequently, economic inequality is often much more extreme than the available statistics indicate.

2.2 Measuring inequality

It is almost impossible to gain a clear picture of the extent and meaning of "inequality" from a single figure, such as the 10–50 index, the percentage share of national private wealth of the richest 10%, or the percentage in danger of poverty. This is because of the different ways wealth or income can be distributed both *vertically* (e.g., Piketty, 2014) and *horizontally* (e.g., Kabeer, 2015; Therborn, 2012). Vertical inequality has to do with the different income or wealth shares of the richest, poorest, average, median, etc. groups within a particular country or region. Horizontal inequality has to do with the ways wealth or income are distributed in relation to various social and geographical differences, such as gender, ethnicity, age and location.

Most studies of inequality conducted by academia, think-tanks and policy workgroups are concerned with different ways of measuring vertical inequality. The most common measures are the Gini coefficient and comparisons of percentage shares, but other important measures are the Robin Hood index, the mean-median ratio and the Theil index. I offer a very brief primer on these by way of a "Lorenz" curve, named after the US mathematician Max Otto Lorenz (1886–1959).

Fig. 5 gives a simplified Lorenz curve for income in Australia in 2016. The horizontal axis gives the cumulative percentage of persons, and the vertical axis gives the cumulative percentage of their earnings. The curved line is the actual figures for their cumulative income in 2016 (data from OECD, 2018). The straight diagonal line represents hypothetical perfect equality, where each additional person would add the same amount to the cumulative total of earnings.

With the Lorenz curve we can see at a glance what the distribution of income looks like. The deeper the curve, the more unequal a society is. We can use a Lorenz curve to plot either income or wealth.

One of the most commonly used measures of inequality is the Gini coefficient. This is calculated by dividing the area between the curve and the diagonal line (area A on the graph) by the total area under the diagonal line (A + B on the graph). For a perfectly equal society the Gini coefficient is 0.0; for an infinitely unequal society (where one person has all the wealth) the Gini is 1.0. In EU countries in 2016, Gini coefficients for income after tax and welfare payments ranged from a low of 0.275 for Denmark to middle values of 0.31 for the UK and 0.331 for Italy, to 0.347 for Romania (Eurostat, 2018).

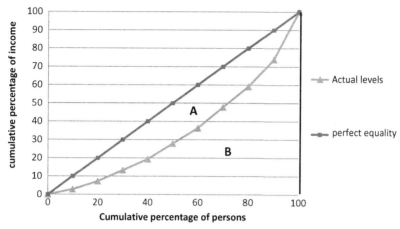

FIG. 5

Lorenz curve for Australia, based on cumulative income shares by decile (hence excludes effect of extremely high income of top 1% and 0.01%).

From OECD, 2018. Key Indicators on the Distribution of Household Disposable Income and Poverty, 2007, 2015 and 2016 or Most Recent Year. Organisation for Economic Cooperation and Development. Available from: http://www.oecd.org/social/income-distribution-database.htm.

It should be noted, however, that Fig. 5 is not a perfect Lorenz curve for income in Australia because I have simplified it into 10 composite steps. The highest step (on the right) can be broken down into the highest earning 1%, then the highest earning 0.1%, etc., producing a very high spike at the upper end—so as to properly capture the extremely high incomes of the group that Winters (2011) calls the oligarchs. The Gini calculated from Fig. 5 is 0.225, but the actual Gini for income in Australia is much higher, at 0.338 (OECD, 2018).

The advantage of the Gini coefficient is that it gives a simple measure of overall inequality. Its main weakness, however, is that it does not take into account various quirks in the shape of the curve. The flatter the curve is at the low end, the higher the percentage of people in poverty. The sharper it is at the high end, the more extreme the highest incomes are. Two countries can have identical Gini coefficients but very different percentages in abject poverty or very different distributions of wealth at the high end.

To avoid this problem Piketty (2014) and his research team (see below) tend to avoid the Gini coefficient and instead use a different method, called "percentage shares." Most of their work plots the percentages of wealth or income owned or earned by the richest 10% or 1%. This is because Piketty's overriding social concern is the potential damage and danger to society of excess wealth in the hands of just a few—a theme repeated frequently in his works and those of his collaborators. This is also a concern in this book, but we are at least as concerned about the shape of the distribution at the low end, as this strongly influences access to energy services.

To test how serious these problems are, I used data from OECD (2018) to plot the Gini coefficients in 2016 for 43 countries, including all OECD countries, the "BRIC" countries and several developing countries, against the percentage of national income earned by the top-earning 10% in each of these countries (Fig. 6), and then against the bottom earning 10% (Fig. 7). As Fig. 6 shows, the percentage earned by the top 10% tracks the Gini very closely (correlation coefficient $R^2 = 0.9636$). Fig. 7, however, shows that Gini coefficient does not track the bottom earnings' percentage of national income as closely. For example, for a Gini coefficient of 0.32, the proportion received by the bottom earning 10% can range from 2% to 3%. This could be the difference between an income of €800 and €1200 per month (in very approximate terms). So we need to be cautious about using the Gini coefficient as a simple measure of poverty.

Another interesting measure of economic inequality, also derived from the Lorenz curve, is the "Robin Hood index." The Robin Hood index is the maximum vertical distance between the straight diagonal line and the curve in Fig. 5. This gives the approximate proportion of the total income of those earning more than the mean, which would need to be transferred to those below the mean, to produce a situation where all have equal income. Among the 43 countries investigated for Gini coefficients in Figs. 6 and 7, the Robin Hood index ranges from a low of 18% for Finland to a high of 45% for South Africa. Among the highest in Western Europe are the UK at 27.5% and Italy at 23.2%.

The Robin Hood index has been successfully used in health studies. For example, Kennedy et al. (1996) found a very high correlation between the Robin Hood index and early mortality rates in US states, and also for different causes of early death.

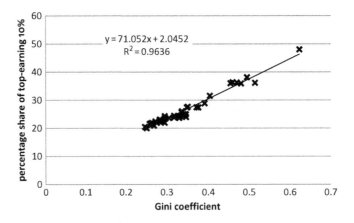

FIG. 6

Percentage share of total national income of top earning 10%, plotted against Gini coefficient, for 43 OECD, BRIC and developing countries in 2016.

From OECD, 2018. Key Indicators on the Distribution of Household Disposable Income and Poverty, 2007, 2015 and 2016 or Most Recent Year. Organisation for Economic Cooperation and Development. Available from: http://www.oecd.org/social/income-distribution-database.htm.

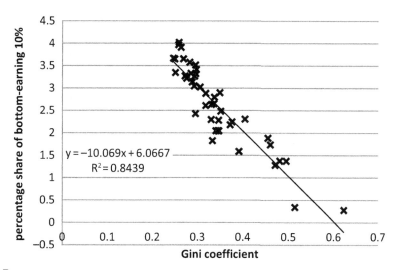

FIG. 7

Percentage share of total national income of bottom earning 10%, plotted against Gini coefficient, for 43 OECD, BRIC and developing countries in 2016.

From OECD, 2018. Key Indicators on the Distribution of Household Disposable Income and Poverty, 2007, 2015 and 2016 or Most Recent Year. Organisation for Economic Cooperation and Development. Available from: http://www.oecd.org/social/income-distribution-database.htm.

Is the Robin Hood index needed in the research for this book? Fig. 8 plots the Robin Hood index against the Gini coefficient for our 43 countries. The two measures track each other almost exactly, with a correlation coefficient of 0.9976. It seems that, given the shapes of income distributions in these countries in 2016, we can estimate the Robin Hood index simply by multiplying the Gini by 73 and subtracting 0.9. The advantage of the Robin Hood index is that it can give a rough indication to governments of the range of tax and welfare changes they would need to make, to produce at least a degree of greater equality.

Another measure of economic inequality is the mean to median ratio: a country's mean income (or wealth) divided by its median income (or wealth). The median inevitably lies below the mean due to the high upper spike of income or wealth distribution among the very highest earning or wealthiest. It is usually assumed, therefore, that the mean to median ratio is a useful measure of the degree of inequality.

OECD (2018) figures show very high mean to median household wealth ratios, i.e., greater than 3.0, for the US, Germany, Austria and the Netherlands; ratios around 2.0 for Luxembourg, Canada, the UK, Spain, Belgium, Italy, France, Portugal, Norway and Finland; and a much lower ratio for Greece, at about 1.3. These are not very informative for the purposes of this book, as a high mean to median ratio can be caused by a very large proportion of very poor people, an extreme upwards curve

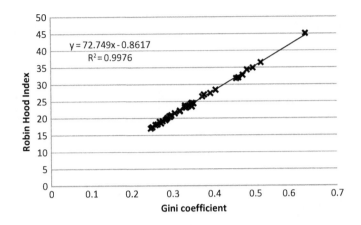

FIG. 8

Robin Hood index against Gini coefficient, for 43 OECD and BRIC countries.

From OECD, 2018. Key Indicators on the Distribution of Household Disposable Income and Poverty, 2007, 2015 and 2016 or Most Recent Year. Organisation for Economic Cooperation and Development. Available from: http://www.oecd.org/social/income-distribution-database.htm.

in the wealth of the richest few percent, or a combination of both. It is not clear simply from the mean to median ratio which of these is implicated.

Another well-known indicator of inequality is the Theil Index, named after its developer Henri Theil (1924–2000). The Theil index is a measure of the entropy, or statistical disorder, of the actual values on the Lorenz curve about the straight diagonal line (see Fig. 5). This can capture variations or wobbles in the Lorenz curve that are not captured by the Gini coefficient. For our purposes, however, it is less useful than other measures because it is less well-known than the Gini coefficient and, like the Gini, does not indicate whether there is excessive inequality at the low (or high) end of the spectrum.

All these measures are concerned with what Kabeer (2015) calls "vertical" inequality: the variations in wealth or income along a spectrum from the poorest to the richest. But Kabeer (2015) and Therborn (2015) also point out the significance of "horizontal" inequality, where income or wealth vary in relation to people's gender, race, ethnicity, health or the district they live in. Therborn distinguishes between four underlying dimensions of inequality: hierarchy, exploitation, distanciation and exclusion. The first three operate vertically, from the top down. The fourth operates horizontally. Hence, for example, a person with high education and work skills, yet who suffers mental illness, may be excluded from earning an average wage because potential employers see her illness as a burden. More generally, women have lower wages and wealth on average than men. Recent research on inequality, on which most of the rest of this chapter is based, does not consider horizontal inequality. However, in Chapter 8, Minna Sunikka-Blank explores this issue in relation to gender, and how it affects access to energy services.

3 Recent changes in wealth and income distribution
3.1 Estimating economic inequality

One of the difficulties in researching economic inequality is that incomes and wealth among the richest are often hidden from public view. In Piketty's (2001) early study on income distribution in France in 1901–98, he revived a methodology pioneered by Vilfredo Pareto (1848–1923) and Simon Kuznets (1891–1985), of using tax data rather than household surveys to estimate top incomes (see discussion in Milanovic, 2014). Using this approach, Atkinson (2003) offered a long-run study of top incomes in the UK, while Piketty and Saez (2003) did the same for the US. Atkinson and Piketty (2007) then made a similar study for continental Europe and all other developed countries, and later studies were offered for developing and emerging economies (Atkinson and Piketty, 2010; Alvaredo et al., 2013).

Wealth inequality is even more difficult to investigate than income inequality, since in most countries people's accumulated wealth does not have to be declared to tax authorities each year, as their income does. Both income and wealth can also be hidden in tax havens, and I will outline below how Zucman (2014, 2015) and his colleagues have made extraordinary progress in beginning to expose these.

By painstakingly mining the available sources of data, Piketty was able, in his *Capital in the 21st Century* (Piketty, 2014), to bring information to light on inequalities in both income and accumulated wealth, going back to the 18th century for the UK and France, and the 19th century for a number of other countries. This revealed persistently high income and wealth inequality in all countries until a turning point after 1914, when inequality began to reduce and continued to do so until the 1970s. In the 1980s the trend reversed and inequality steadily increased toward pre-1914 levels. Findings for the most recent years are continually updated, the most recent being in Alvaredo et al., (2017, 2018).

Fig. 9 displays Piketty's (2014) figures, supplemented with updates from his database at WID (2018), for the income share of the highest-earning 10% in the US, Britain, Germany, France and Sweden, from the 1900s to the 2010s, given at 10-year intervals. In all these countries except the US, the top 10% share peaked in the 1910s then began to fall. In the US the fall began in 1930, just after the Wall Street Crash. All these countries saw a bottoming out of the top 10% share in the decades immediately following the Second World War. This trend reversed sharply in the US and Britain in the 1970s and also in the Netherlands and Ireland (which are not shown on the graph), and more gradually in Germany. The upward trend began 10 years later in Sweden and France, and also in Italy, Australia, New Zealand and Canada, though the upward trend in Denmark has been much less pronounced than in most countries. After another decade a sharp upward trend began in Finland, Norway and Switzerland. In Spain the trends have been more mixed, with several peaks and troughs since the 1970s (Piketty, 2014, p. 323, updated with data from WID, 2018).

Note that these trend lines do not give a complete picture of the increase in income inequality, as they focus only on the top of the income scale. For Germany,

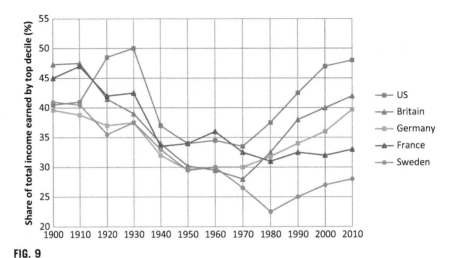

FIG. 9

Top decile income share in US, Britain, France, Germany and Sweden, 1900–2010.
Reproduced from Piketty, T., 2014. Capital in the Twenty-First Century (Translated from the French by Arthur Goldhammer). Belknapp-Harvard University Press, Cambridge, MA, p. 323; WID, 2018. World Inequality Database. Available from: https://www.wid.world/.

for example, the top 10% began to receive a higher share of income as from the 1970s but, as Fig. 1 shows, the ratio between the top 10% and bottom 50% share did not begin to increase significantly until the 1990s. This is because the bottom 50% share plummeted spectacularly from the 1990s (WID, 2018). This illustrates the limitations of using just one measure of inequality. Unfortunately there is very little reliable data on the income share of the bottom 10% in most countries except for recent years. The OECD gives data for all income deciles of 43 countries (see above), but only for 2007, 2015 and 2016 (OECD, 2018). Piketty's online database (WID, 2018) gives bottom 50% shares for a handful of countries from the 1960s and for France from 1900, but not the bottom 10% shares.

Nevertheless, tracking the top 10% share at least shows long-run trends in overall income inequality, even if it does not give clear data on the poorest sections of society. The data available from WID, the OECD, Eurostat, think tanks such as the Joseph Rowntree Foundation and many countries' own statistics bureaus show a clear trend of increasing income inequality in most developed countries over the past 30–40 years. The only exception appears to be Denmark, where the top 1% and 10% income shares fell almost continually and more recently have increased only slightly (WID, 2018).

As noted above, *wealth* inequality is usually much more extreme than *income* inequality, as high income people have to spend less of their incomes on necessities and can therefore accumulate more and increase this further through investment. Robust data on long-run trends in wealth inequality is harder to obtain, but Fig. 10 combines Piketty's (2014) findings with the latest from his database

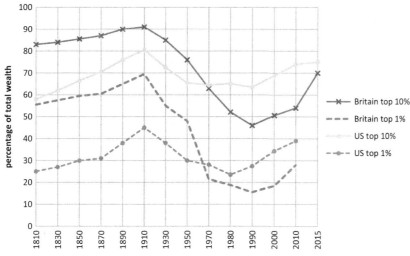

FIG. 10

Shares of total national private wealth of the richest 1% and 10% for the US and the UK from 1810 to 2015.

Reproduced from Piketty, T., 2014. Capital in the Twenty-First Century (Translated from the French by Arthur Goldhammer). Belknapp-Harvard University Press, Cambridge, MA, pp. 344–348; WID, 2018. World Inequality Database. Available from: https://www.wid.world/.

(WID, 2018), on the shares of total national private wealth of the richest 1% and 10% for the US and the UK from 1810 to 2015 (note that the horizontal axis in Fig. 10 changes scale at 1970). The shares of national private wealth of the richest 10% and 1% increased in both countries throughout the 19th century, then began to fall in the 1910s, bottoming out in the US in the 1980s and the UK a decade later. Since then there have been more or less steady increases. The wealthiest 1% in the US owned 40% of US private wealth in 2010 and this measure is again at this level after a small dip after the Great Recession of 2007–2008 (Wolff, 2017). In both the UK and the US the wealthiest 10% own around 75% of their countries' net private wealth. As with income inequality, wealth inequality reduced steadily until the immediate post-World War II years, then increased persistently since the last decades of the 20th century.

3.2 What about the bottom 10%?

Again, however, this does not say much about how poor the poorest are. Australian statistics bring us a little closer to finding this out. The Australian Bureau of Statistics gives the income from 2003 to 2016 at the top of the 10th and 90th percentiles—the richest of the poor compared to the poorest of the rich (ABS (Australian Bureau of

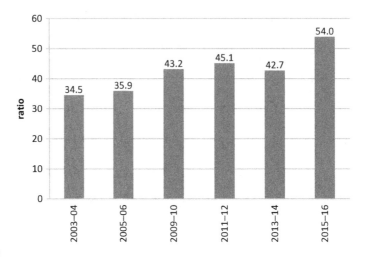

FIG. 11

Australia: ratio between share of national private wealth at top of 90th percentile and at top of 10 percentile, 2003–16.

From ABS (Australian Bureau of Statistics), 2018. Household Income and Wealth, Australia. Available from: http://www.abs.gov.au/ausstats/abs@.nsf/Lookup/by%20Subject/6523.0~2015-16~Main% 20Features~Household%20Income%20and%20Wealth%20Distribution~6.

Statistics), 2018). The income within the bottom few percentiles would be of more interest for insights into poverty, but the data here at least shows how the richest of the poor fare, compared to the poorest of the rich. Fig. 11 displays the ratio between the shares of national private wealth of these two groups.

It is important to note that this ratio has been increasing more or less steadily since 2003, meaning the poorest of the rich have been getting steadily richer in comparison to the richest of the poor. The increase in ratio from 34.5 to 54.0 equates to a 56% relative increase in the wealth share of the poorest of the rich compared to that of the richest of the poor.

Fig. 12 sheds even more light on this. Here I plot the ratio between the weekly *incomes* of those at the top of the 90th and 10th percentiles over the same period. Although this ratio has actually fallen since 2007–08 by about 1.2% per year, the richer group have continued to increase their *wealth* in comparison to the poorer group by about 4% per year (calculated from Fig. 11). In today's financial environment wealth begets wealth at a breathtaking rate even if the discrepancies in wages— the money people get from doing work—stop increasing, a point I discuss further in Chapter 2.

This still does not shed light on the situation of the very poor. A different set of data sources is needed for this, and these are often provided by privately funded foundations such as (in the UK) Oxfam and the Resolution Foundation. We will mention data from these sources at various points in this book.

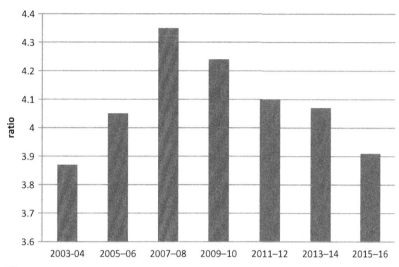

FIG. 12

Australia: ratio between income at the top of the highest earning 90% and the top of the lowest earning 10%, 2003–16.

From ABS (Australian Bureau of Statistics), 2018. Household Income and Wealth, Australia. Available from: http://www.abs.gov.au/ausstats/abs@.nsf/Lookup/by%20Subject/6523.0~2015-16~Main% 20Features~Household%20Income%20and%20Wealth%20Distribution~6.

3.3 **Tax havens**

Before completing this section on recent shifts in economic inequality it is important to note the role of tax havens in hiding the true extent of economic inequality today. Money hidden in tax havens is notoriously hard to bring to light, which is of course the very point of hiding it in tax havens! Gabriel Zucman, a former student of Thomas Piketty at the Paris School of Economics, developed a systematic method of estimating how much of the world's money is hidden in these secretive banks and institutions (Zucman, 2012, 2014, 2015, 2017a,b). Zucman (2012) began by addressing the quandary that, if we add up the nationally recorded debts of all the countries of the world, the total is higher than the sum of all the world's financial assets. Zucman (2015, p. 17) wryly asks, then, whether the world is in debt to a civilization living on Mars! No, he argues, the missing assets are hidden in tax havens on planet earth.

Over the course of several years Zucman obtained data on hidden wealth from Swiss banks. His initial estimate was that around 8% of the world's financial wealth is hidden in tax havens. He was then invited by a consortium of investigative newspapers—the Guardian (UK), Süddeutsche Zeitung (Germany) and the Washington Post and New York Times (US) to assist in analyzing the "Paradise Papers," a huge cache of leaked data from tax havens and secretive investment institutions around the world. This increased his estimate of the world's hidden wealth to 9.8%, and enabled country-by country estimates to be made. These ranged from a low of 1% for South Korea to a high of 76% for the United Arab Emirates (Zucman, 2017a; Alstadsaeter et al., 2018). Estimates for EU countries included Greece (37% of its financial assets are hidden in tax havens), Portugal (21%), the UK (17%); Germany (17%), Italy (12%), Spain (11%), Sweden (7%), Finland (3.5%) and Denmark (3%). Russia had the highest figure for Europe, at 47%. For the US the figure was 8%.

Further, Zucman and his colleagues noted that most of this hidden wealth is owned by the wealthiest 0.1% of each country's population, or an even smaller portion. This means that our estimates for the Gini coefficient would need to be radically revised upwards, as these hidden, extra amounts would produce a sharp spike at the top end of the Lorenz curve. For example, taking hidden wealth into account, Alstadsaeter et al. (2018) estimated the wealthiest 0.1% in the US now own about 12% of total national private wealth.

A further cache of data was leaked to Zucman from an informer at Credit Suisse, giving the names and investment balances of all Credit Suisse's Swedish customers. Because the Swedish government publishes the tax returns of its wealthiest taxpayers, Zucman was able to track which of these investments had been declared according to Swedish tax law. He was able to robustly calculate that the Swedish government was missing about 12% of its tax revenue *just through Swede's assets invested* via *Credit Suisse*. Since this did not take into account assets invested through other channels, Zucman posited that the total loss of tax revenue might be anything up to 30% (Zucman, 2017b). This results from just 7% of Sweden's total private wealth being hidden offshore. In a further analysis of US tax losses, Zucman and colleagues estimated the US tax take would be about 16% higher if the income from the 8% of its wealth hidden in tax havens were declared in tax returns (Alstadsaeter et al., 2018). In general, then, each 1% of private wealth hidden in tax havens could result in a loss of total tax revenue for the relevant country of around 2–4%, though of course more detailed data would be needed to provide robust estimates.

To summarize this section, it is clear that economic inequality has increased enormously in the past 30–40 years in all but a few of the developed countries. This applies to income inequality but even more to wealth inequality. This follows a period of some 25–30 years immediately after the Second World War when inequality was at its lowest levels in recorded history. Further, real economic growth was about twice as high in the post-war years as in the last 30–40 years. Almost all recent growth has been captured by the very rich, leaving the real incomes of the poorest 50% to stagnate or fall back. The wealth at the top end of the spectrum continues to grow apace, even in cases where differences in income have stabilized, thereby increasing the gap between the very rich and most of the rest of society. Meanwhile some 10% of the private wealth of nations is hidden offshore. Most of this belongs to the richest sub-strata of society. Because it is hidden from tax authorities, this could be depriving governments of 20% or more of their rightful tax take.

An important conclusion from this chapter so far is that in high-income countries, energy consumption, production and distribution takes place in an environment of extremely large wealth and income inequalities. Because energy costs money, people do not have equal access to energy services, far from it. Income and wealth distribution are therefore key determinants of energy consumption behavior. Further, we have not yet begun to ask how the privileged position of the wealthiest 1% gives them hugely disproportionate power and sway over energy mix, energy supply and energy policy more generally—themes the recur in later chapters of this book.

4 How inequality has changed in the developed countries
4.1 Is extreme inequality the norm in human history?

This section explores how and why the developed countries have moved so far into inequality, over the past 30–40 years, from their relatively egalitarian state in the decades after the Second World War. We should note, however, that in the long-run of human history the unusual thing is not the recent shift toward inequality but the few decades of stable egalitarianism that preceded it. It is clear from very long-run studies of human economic history such as Winters (2011, 2014, 2017) and Graeber (2011) that the usual state of humanity is extreme economic inequality. As Ingham (2011), Acemoglu and Robinson (2012) and Piketty (2014) argue, certain key factors just happened to converge and many people made extraordinary efforts, to make the immediate post-war years different from the norm.

Winters (2011) found evidence that an outlandishly rich few, whom he calls oligarchs, have exercised overwhelming control over their respective societies in almost every epoch of history. They have often achieved this by playing off the proletariat masses against the very wealthy groups whose fortunes place them just below the oligarchs. Further, until the modern era these oligarchs almost always ruled directly, as monarchs and potentates, with armies and bureaucrats at their disposal to defend their wealth and keep society the way they wanted it. More recently in democratic countries, Winters argues, they rule by controlling politicians through bribes, donations, ownership of the press, or less direct means such as threats of taking their industries offshore.

Nevertheless, argues Winters (2014), the oligarchs do not get an easy ride, because their right to own vast fortunes is constantly, fiercely contested by the less wealthy and vastly more numerous masses, and often also by the very wealthy groups just below the oligarchs on the wealth spectrum. Winters shows, for example, how this contestation played out in the formation of the US constitution. The super-rich in the newly independent states were determined to prevent the masses from wording the constitution in such a way that the wealth of the "few" could be taxed (at all) or otherwise forcibly shared with the "many." They were so successful that it was not until the turn of the 20th century that the US Supreme Court permitted progressive taxation, and the rich were still hardly taxed at all until the government desperately needed funds during the First World War.

Whether these oligarchs are as omni-powerful as Winters claims is open to debate (McCormick, 2012), though Stiglitz (2013) documents many different ways in which recent increases in inequality in the US have led to the super-wealthy having more and more influence on politics and the direction of US public life. Nevertheless, Winters' claim that extreme inequality is the most usual state of society is supported by Graeber's (2011) long-run history of debt, by countless historical studies of particular epochs, and in more recent times by the data being made available by Piketty and his colleagues.

Lest we start to think, however, that there is some kind of cause-and-effect law governing how wealth becomes distributed in societies, Acemoglu and Robinson (2012) bring a further dimension to the discussion. They argue that contingencies—one-off events like the Great Plague, the writing of the Communist Manifesto, the eruption of Krakatoa, the World Wars of the 20th Century, the Great Depression and the invention of the silicon chip—crash into the seemingly regularized patterns of social and political life with unpredictable consequences. There is then new contestation between different interest groups striving to gain advantage from the new situation.

With this framework in mind I make the following observations, firstly as to how the age of egalitarianism came about in developed countries, and secondly how it dissolved and these countries became, once again, far more unequal.

4.2 Why were we so egalitarian in 1950–80?

We saw above how extreme economic inequality had become, by the first decade of the 20th century. But just as this so-called "Belle epoch" was reaching its peak, there came a series of high-impact events. Firstly, the First World War destroyed the accumulated fortunes of many wealthy individuals and families in Europe, in the form of both physical and financial capital. Second, the Great Depression of 1929–35 destroyed or severely reduced vast fortunes among the wealthy elite in the US in particular and in other countries such as Australia, New Zealand and parts of Europe. Thirdly, in the wake of the Great Depression, labor movements became dominant in most high-wealth countries and the New Deal was launched in the US, bringing progressive taxation and universal social welfare. This included a willingness by governments to spend more liberally, partly as a result of the influence of so-called "heterodox" economists such as Joseph Schumpeter (1883–1950) and John Maynard Keynes (1883–1946), and partly because government spending suited the aspirations of labor movements.

Fourthly, the Second World War had two synergistic effects. It repeated the First World War's destruction of private capital wealth but on a much larger scale. It also deepened government expenditure, to finance the war effort. This relied on even more progressive taxation, thus taking more from high-income earners.

Hence the fortunes of the super-rich were assailed from different angles over three decades, and declined precipitously. Because money begets money, it became much harder for a privileged few to continue to multiply their fortunes simply by investing.

Fifthly, leaders of the Allied nations during the Second World War judged that competitive trade wars and opportunist currency devaluations had heavily contributed to the economic disarray that led to the rise of Nazism and Fascism. A conference to forge a new economic world order was held at Bretton Woods, New Hampshire in 1944, resulting in a unique compromise between free market and interventionist economic approaches—heavily influenced by economist John Maynard Keynes. Under the Bretton Woods system, trade would be relatively free of tariffs and barriers so that each country could produce at its most efficient level, but exchange rates would be strictly controlled to prevent cheating by devaluing. Currencies were pegged to the US dollar, and the dollar was pegged to the gold standard at $US34 per ounce. Bretton Woods "sought to establish free markets in everything except money and capital" (Ingham, 2011, p. 85).

Finally, while Bretton Woods was in force, high-wealth countries' economies had little competition from communist and developing countries, and "globalization" had not yet happened as we know it today. Competition was emerging from medium-sized Asian countries like South Korea and Taiwan, but China, India and most of South East Asia were not yet high-tech industrialized, while the European Communist bloc had its own trade pact known as Comecon (Thomas, 2009). The combined effect of these factors was that, for almost 30 years from the late 1940s until the mid-1970s, the high-wealth countries achieved unprecedented, persistent real (i.e., inflation-adjusted) economic growth of around 4% per adult per year in Continental Western Europe and 2.5% per adult per year in the English speaking high-wealth countries (WID, 2018), with historically low levels of economic inequality.

4.3 The shift to neoliberalism

Bretton Woods lasted until the early 1970s, when US President Richard Nixon withdrew the US from the gold standard and exchange rate agreement (Ingham, 2011)—a point I will discuss further in Chapter 2. It was further weakened by the huge oil price increases of the 1970s, the burden of increasing government debt and subsidies to industry, and increasing trade protectionism. About this time "neoliberalism," a reformulated version of classical liberal economics, was being vigorously promoted by the Chicago School, an economic faculty led by Frank Knight (Davies, 2014).

The core of neoliberal thinking is that money is a neutral medium that serves primarily as a means of exchange of goods and services in markets. In order for markets to function efficiently, neoliberals argue, goods and services have to be allowed to find their real values in relation to each other. This also applies to the value of money itself, such as in exchange rates, interest rates, and financial investment instruments composed of different configurations of money such as government bonds, mortgages, etc. In neoliberal thinking, governments should not interfere with the freedom of the markets (Ingham, 2011, p. 196ff), as this can distort prices and introduce dead weight losses into an economy, while "small" is preferred to "big" government, meaning governments should privatize services such as transport, education and health, and social welfare payments should be kept to a minimum. Neoliberal thinkers argued that the economic challenges Bretton Woods countries were increasingly facing in the 1970s were entirely due to the long-term effects of government tampering with the markets.

As I will discuss in greater depth in Chapter 2, for Keynesian and post-Keynesian thinkers, money is anything but a neutral medium designed for market exchange. Instead it is a set of credit-debt obligations, being constantly created and dissolved by human institutions so as to create the kind of trust and obligations that spur people to do work they would otherwise not do. Money can stimulate work as it can be traded by workers for goods and services, in a self-reinforcing cycle that increases the real wealth of society. For example, for post-Keynesians—the main critics of neoliberalism—one route out of economic doldrums is to ensure there is sufficient money available for investment to make use of unused capacity and thereby stimulate economic activity. This implies much more government involvement than neoliberalism would tolerate.

In the 1970s neoliberalism appeared to many to arrive on the scene as a savior, just when developed economies were reeling from the oil price shocks. In a number of countries there was a kind of cultural embrace of neoliberalism along with a reaction against governments' increasing micro-management of their economies (Gould and Robert, 2013). In the US, the UK, Australia and New Zealand there was no sense of working gradually to reform the Bretton Woods framework in specific ways it needed to be reformed. Instead there was a kind of economic revolution, with the new ideology more or less completely replacing the old. Neoliberalism was implemented in the US and UK under President Ronald Reagan and Prime Minister Margaret Thatcher. It soon spread to Australia, New Zealand and Canada, and was progressively implemented in varying degrees in many of the Western European countries (Mundell, 2000).

The advent of neoliberalism is widely seen as the turning point where the unique age of egalitarian economic growth gave way to the current epoch of ever increasing economic inequality. The shifts from high-growth economies with egalitarian plenty to lower growth economies with increasing inequality, as outlined in Section 2, are closely associated with the neoliberal revolution.

Alongside this, other related and contingent factors were leading to the development of a global moneyed elite. Sharp increases in the oil price created huge concentrations of money in the hands of relatively few oligarchs (Novokmet et al., 2017), most of which found its way into the financial markets of high-wealth countries (Graeber, 2011; Ingham, 2011). Several decades of historically low interest rates and, since the credit crisis of 2007–08, a decade of quantitative easing by central banks, created an excess of liquidity which, in a neoliberal institutional environment, finds its way to the already-wealthy (Ingham, 2011, pp. 244–245). Emerging economies, especially China, amassed large positive trade balances which, like oil money, serve to bloat financial markets (Piketty et al., 2017). To this can be added the now enormous pension funds accumulated to the credit of baby-boomers in rich countries. Drug cartels and other organized crime add another 8% to liquid global finance (Von Lampe, 2016).

A trend running in parallel to these events, which I discuss more fully in Chapter 2, is the development of the banking system and in particular, banks' exponentially increasing creation of loan credit and its mirror image, debt (Lazzarato, 2015). The long-term effect of this has been that millions of people now struggle, not just with poverty, but with crippling debt (Di Muzio and Robbins, 2016) and interest payments that often cut their effective income by a huge proportion (Soederberg, 2014). Meanwhile the balance sheets of many governments have become increasingly negative, resulting in moneyed lenders gaining more and more power over governments. Lazzarato (2015) argues that this has created a new, deeply rooted power dynamic in society, whereby people and even governments are turned into "subjects" of the wealthy.

The credit crisis of 2007–08 resulted in financial ruin for large numbers of low-income homeowners while government-sponsored bank rescues shifted more funds upwards in the wealth hierarchy (Ingham, 2011, pp. 227–264). Finally, the increasing sophistication of tax havens (Zucman, 2014) and the wealthy elites' so-called "wealth-defense industry" (Winters, 2017) provide shelter for excess wealth, from governments' tax regimes. The result is an ever-increasing, self-reinforcing tide of money possessed by a very small percentage of the population, alongside stagnating or reducing incomes and wealth, together with increasing debt, among the vast majority (Alvaredo et al., 2017). This is felt both on a global level, and within the now radically transformed developed countries.

These factors merge in the increasingly ubiquitous theme of "financialization" (Kotz, 2010; Fernandez et al., 2016)—the growth of an economy of money as a commodity used for wealth creation, which accelerates the wealth acquisition of those who already have money to spare.

4.4 Being poor in a society of extreme inequality

Where does this leave the poorest residents of these countries? A grim picture has emerged through an increasing number of studies, mostly by charitably funded institutions. A clear and disturbing study of the "precariat" in a UK city today is offered by Savage (2015). A vivid, personal view is shared by McGarvey (2017).

Of course, even without such studies we can see that poverty is becoming an ugly and serious problem. In many developed countries we see the new poverty just by walking down the street. Not only has the number of homeless people and beggars increased markedly, but the characteristics of many of the people now seen to be homeless and begging are different from in previous decades. They include people who have the social and work skills to take a full role in society, but are prevented from doing so by combinations of factors such as: having being made redundant at work; having their unemployment benefit cut off due to welfare department quotas; having their house repossessed due to default on mortgage payments; losing the support of a partner due to the emotional effects of financial stress. These people represent the tip of an iceberg. For every person sleeping on the street there may be a thousand more in inadequate homes barely surviving from day to day in the post-egalitarian era (Savage, 2015).

A further important point is that being among the poorest in a vastly unequal society is different from being just below the average in a modestly poor but relatively egalitarian society. We can see this, for example, in the relationship between house prices and wages. In Auckland, New Zealand, for example, housing was at its most affordable ever in the 1970s. An average priced house then cost about three times an average yearly salary (Eaqub and Eaqub, 2015). This ratio has steadily increased since then, and now stands at about 16 (StatsNZ (Statistics New Zealand), 2018). The Auckland City Council recently calculated that less than half the households in Auckland would be able to afford to buy a modest-priced home worth about 60% of the average house price. Prices for renting have increased by a lesser amount, but are still well above the levels they were in the 1970s (Eaqub and Eaqub, 2015). This type of situation is now strongly evident in many other countries including the US, the UK, Australia, Canada, Germany (Galvin and Sunikka-Blank, 2018) and even Japan (Forrest and Hirayama, 2009). More and more households live in private rental accommodation and therefore have little or no power to make their homes more energy efficient, a theme Nicola Terry touches on in Chapter 6.

In these countries, even a person on an average wage is now poor in relation to the basic need of being housed. But for a person in the 5th or 10th percentile of income, being housed is a huge obstacle, and much more so for people nearer the bottom of the 10th percentile. Although many of the trappings of life are very cheap today compared to earlier years—designer coffee, mobile phones, decent clothes—the effects of economic inequality are vividly seen in the more basic necessity of housing. And housing, of course, is where most of our energy consumption occurs, a theme of vital importance in this book.

5 Conclusions

In this chapter, I have presented key evidence of the persistent increases in economic inequality in developed countries over the past 30–40 years. I began by showing this applies not only to countries like the US, which are widely known to be highly unequal, but also to countries like Germany which are often seen as strongly egalitarian. I then showed how wealth inequality is much higher than income inequality and continues to increase in today's climate even if income differences stabilize or reduce. To provide a clear framework for considering these issues, I discussed different ways of measuring inequality, using the Lorenz curve to show how these are defined. I used this and other displays to show how inequality can have different features, depending especially on the distribution of wealth and income at the lowest and highest ends of the spectrum.

I drew on recent studies to map the changes in wealth and income inequality in Western European and English Speaking countries in recent years, set in their long-run context. I then offered a narrative of how and why inequality began to reduce steadily from extremely high levels in the early 20th century, reached a historic low in the years following the Second World War, then turned upward in the 1980s and continues to increase today. Inequality reduced from its early 20th century peak because of a series of contingencies throughout the first half of the 20th century, which served to massively reduce the fortunes of the very rich while leading governments to actively redistribute wealth and income through progressive tax, generous welfare, and large government spending. The change toward increasing inequality came about when neoliberal economic policies replaced the Keynesian- influenced Bretton Woods economic regime that had held sway during the mid-1940s to mid-1970s. This brought related factors such as an exponential increase of bank credit and societal debt, the financialization of basic assets like housing, the increasing sophistication of the "wealth defense industry," and the persistent growth of tax havens.

I set this in the framework of arguing that the unusual thing in human history is not the rapid, persistent increases in inequality of the past 3–4 decades, but the combination of strong economic growth and relative egalitarianism of the immediate post-World War II decades. We should not be surprised that inequality has returned in force. However, we should also note that it is preventable, or can at least be stringently mitigated.

The broader context of this chapter is the issue of energy consumption. One immediate effect of increased inequality is to put the basic need of affordable housing beyond the reach of increasing numbers of people. Even those who are well-housed, however, may be unable to afford basic energy services. We also need to ask whether the increasing concentration of wealth among the richest 1%, 0.1% or even 0.01% could be influencing energy policy in ways that distort climate-friendly energy policy and exclude more and more households from the basic benefits of energy services, such as in home heating and transport.

Finally, the social science frameworks we use in energy consumption research need to take these massive shifts in wealth and income distribution into account. We no longer live in the egalitarian age in which these frameworks were formed.

References

ABS (Australian Bureau of Statistics), 2018. Household Income and Wealth, Australia. Available at http://www.abs.gov.au/ausstats/abs@.nsf/Lookup/by%20Subject/6523.0~2015-16~Main%20Features~Household%20Income%20and%20Wealth%20Distribution~6.

Acemoglu, D., Robinson, J., 2012. Why Nations Fail: The Origins of Power, Prosperity, and Poverty. Crown Business, New York.

Alstadsaeter, A., Johansen, N., Zucman, G., 2018. Who owns the wealth in tax havens? Macro evidence and implications for global inequality. J. Public Econ. 162, 89–100.

Alvaredo, F., Atkinson, A., Piketty, T., Saez, E., 2013. The top 1 percent in international and historical perspective. J. Econ. Perspect. 27 (3), 3–20.

Alvaredo, F., Chancel, L., Piketty, T., Saez, E., Zucman, G., 2017. Global inequality dynamics: new findings from WID.world. Am. Econ. Rev. Pap. Proc. 107 (5), 404–409. https://doi.org/10.1257/aer.p20171095.

Alvaredo, F., Chancel, L., Piketty, T., Saez, E., Zucman, G., 2018. World Inequality Report 2018. World Inequality Lab. Available at https://wir2018.wid.world/.

Archer, M., 2000. Being Human: The Problem of Agency. Cambridge University Press, Cambridge.

Atkinson, A., 2003. Top Incomes in the United Kingdom Over the Twentieth Century. http://www.nuffield.ox.ac.uk/users/atkinson/TopIncomes%2020033.pdf.

Atkinson, A., Piketty, T. (Eds.), 2007. Top Incomes Over the Twentieth Century: A Contrast Between Continental European and English-Speaking Countries. Oxford University Press, Oxford and New York.

Atkinson, A., Piketty, T. (Eds.), 2010. Top Incomes: A Global Perspective. Oxford University Press, Oxford and New York.

Bourdieu, P., 1958. Sociologie d'Algerie (The Sociology of Algeria). PUF, Paris.

Bourdieu, P., 1976. Outline of a Theory of Practice. Cambridge University Press, Cambridge.

Davies, W., 2014. Neoliberalism: a bibliographic review. Theory Cult. Soc. 31 (7–8), 309–317.

Destasis, 2016. Statistic_id72188_Armutsgefaehrdungsquote-in-Deutschland-bis-2016. Satistisches Bundesamt Deutschland. Available at https://de.statista.com/statistik/daten/studie/72188/umfrage/entwicklung-der-armutsgefaehrdungsquote-in-deutschland/.

Di Muzio, T., Robbins, R., 2016. Debt as Power. Manchester University Press, Manchester.

Dorling, D., 2018. Do We Need Economic Inequality? Polity Press, Cambridge.

Eaqub, S., Eaqub, S., 2015. Generation Rent: Rethinking New Zealand's Priorities. Bridget Williams Press, Wellington.

Eurostat, 2018. Ihr Schlüssel zur europäischen Statistik. European Commission. Online resource available at http://ec.europa.eu/eurostat/de/data/database.

Fernandez, R., Hofman, R., Aalbers, M., 2016. London and New York as a safe deposit box for the transnational wealth elite. Environ. Plan. A. 48 (12), 1–19.

Forrest, R., Hirayama, Y., 2009. The uneven impact of neoliberalism on housing opportunities. Int. J. Urban Reg. Res. 33 (4), 998–1013.

Galvin, R., Sunikka-Blank, M., 2018. Economic inequality and household energy consumption in high-income: countries: a challenge for social science based energy research. Ecol. Econ. 153, 78–88.

Gould, A., Robert, R., 2013. The neoliberal pea and thimble trick: changing rhetoric of neoliberal champions across two periods of economic history and two hypotheses about why the message is less sanguine. Adv. Appl. Sociol. 3 (1), 79–84.

Graeber, D., 2011. Debt: The First 5,000 Years. First Melville House, New York.

Ingham, G., 2011. Capitalism. Polity Press, Cambridge.

Kabeer, N., 2015. Gender, poverty, and inequality: a brief history of feminist contributions in the field of international development. Gend. Dev. 23 (2), 189–205. https://doi.org/10.1080/13552074.2015.1062300.

Kennedy, B., Kawachi, I., Prothrow-Stith, D., 1996. Income distribution and mortality: cross sectional ecological study of the Robin Hood index in the United States. Br. Med. J. 312, 1004–1007.

Kotz, D., 2010. Financialization and neoliberalism. In: Teeple, G., McBride, S. (Eds.), Relations of Global Power: Neoliberal Order and Disorder. University of Toronto Press, Toronto.

Lazzarato, M., 2015. Governing by Debt. Semiotext(e), South Pasadena, CA.

McCormick, J., 2012. Review of oligarchy by Jeffrey Winters. Crit. Dialogue 10 (1), 137–139.

McGarvey, D., 2017. Poverty Safari: Understanding the Anger of Britain's Underclass. Luath Press, Edinburgh.

Milanovic, B., 2014. The return of "patrimonial capitalism": a review of Thomas Piketty's "capital in the twenty-first century". J. Econ. Lit. 52 (2), 519–534.

Mundell, R., 2000. A reconsideration of the twentieth century. Am. Econ. Rev. 90 (3), 327–340.

Novokmet, F., Piketty, T., Zucman, G., 2017. From Soviets to Oligarchs: Inequality and Property in Russia, 1905–2016. National Bureau of Economic Research. Working Paper 23712, http://www.nber.org/papers/w23712.

OECD, 2018. Key Indicators on the Distribution of Household Disposable Income and Poverty, 2007, 2015 and 2016 or Most Recent Year. Organisation for Economic Cooperation and Development. Available at http://www.oecd.org/social/income-distribution-database.htm.

Piketty, T., 2001. Les Hauts Revenus en France au XXe siècle. Inégalités et Redistributions 1901–1998. Bernard Grasset, Paris.

Piketty, T., 2014. Capital in the Twenty-First Century (Translated from the French by Arthur Goldhammer). Belknapp-Harvard University Press, Cambridge, MA.

Piketty, T., Saez, E., 2003. Income inequality in the United States, 1913–1998. Q. J. Econ. 118 (1), 1–41.

Piketty, T., Yang, L., Zucman, G., 2017. Capital Accumulation, Private Property and Rising Inequality in China, 1978–2015. National Bureau of Economic Research. Working Paper 23368 http://www.nber.org/papers/w23368.

Savage, M., 2015. Social Class in the 20th Century. Pelican, London.

Soederberg, S., 2014. Debtfare States and the Poverty Industry: Money, Discipline and the Surplus Population. Routledge, London.

StatsNZ (Statistics New Zealand), 2018. Housing. https://www.stats.govt.nz/topics/housing.

Stiglitz, J., 2013. The Price of Inequality. Penguin, London.

Therborn, G., 2012. The killing fields of inequality. Int. J. Health Serv. 42 (4), 579–589.

Thomas, M., 2009. The Cold War: A beginner's Guide. Oneworld, Oxford.

Von Lampe, K., 2016. Organized Crime: Analyzing Illegal Activities, Criminal Structures, and Extra-Legal Governance. Sage, Thousand Oaks, London.

WID, 2018. World Inequality Database. Available from: https://wid.world/.

Winters, J., 2011. Oligarchy. Cambridge University Press, New York.

Winters, J., 2014. Wealth defense and the limits of liberal democracy. In: Paper for Annual Conference of the American Society of Political and Legal Philosophy, Washington, DC, August 28–31, 2014. Revised 29 April 2015.

Winters, J., 2017. Wealth defense and the complicity of liberal democracy. In: Knight, J., Schwartzberg, M. (Eds.), Wealth, NOMOS LVIII, a Special Issue of the American Society for Political and Legal Philosophy. NYU Press, pp. 158–225. 2017.

Wolff, E., 2017. Household Wealth Trends in the United States, 1962 to 2016: Has Middle Class Wealth Recovered? National Bureau of Economic Research. Working Paper 24085, Available at http://www.nber.org/papers/w24085.

Zucman, G., 2012. The Missing Wealth of Nations: Are Europe and the U.S. Net Debtors or Net Creditors? PSE Working Papers n 2011-07. 2012. halshs-00565224v3.

Zucman, G., 2014. Taxing across borders: tracking personal wealth and corporate profits. J. Econ. Perspect. 28 (4), 121–148.

Zucman, G., 2015. The Hidden Wealth of Nations: The Scourge of Tax Havens. University of Chicago Press, Chicago.

Zucman, G., 2017a. The desperate inequality behind global tax dodging. Guardian. 8 November 2017. https://www.theguardian.com/commentisfree/2017/nov/08/tax-havens-dodging-theft-multinationals-avoiding-tax.

Zucman, G., 2017b. The 2017 Stone Lecture on Wealth Inequality. The Graduate Center, City University of New York. Video recording of lecture available at https://www.youtube.com/watch?v=NigFj5s3M4s.

Further reading

Piketty, T., 2015. Putting distribution back at the center of economics: reflections on capital in the twenty-first century. J. Econ. Perspect. 29 (1), 67–88.

What is money? And why it matters for social science in energy research

2

Ray Galvin

University of Cambridge, Cambridge, United Kingdom
RWTH Aachen University, Aachen, Germany

Chapter outline

1 Introduction 31
2 Money as a relationship 33
3 Myths about money 37
   3.1 Myth and obfuscation 37
   3.2 The myth that all money is the same 38
   3.3 The myth that money is a neutral veil 40
   3.4 The myth that money is a commodity 42
   3.5 The myth that money is a creation of the state 44
4 Discussion and conclusions: Some implications for social science
   in energy research 45
References 48

1 Introduction

In this chapter I explore how a response to the question "What is money?" could affect the way social scientists approach energy research.

Money is almost always an important factor in social science based studies of energy consumption. Energy costs money, which is very unevenly distributed today (Alvaredo et al., 2018). Groups and individuals with a lot of money can influence and control energy infrastructures (Newell and Phillips, 2016) and even governments (Antoniades, 2018). Governments offer monetary incentives to increase energy efficiency (Rosenow and Galvin, 2013) and to produce renewable energy (Toke and Lauber, 2007). Sometimes these pay for themselves but sometimes they do not (Galvin, 2010, 2014). A lack of money is one of the main determinants of energy poverty (Middlemiss, 2017). These are just some of the many ways money comes into discussion of energy in social science research.

Inequality and Energy. https://doi.org/10.1016/B978-0-12-817674-0.00002-3

Nevertheless, social science based energy research has not yet taken a strong interest in the discussion, now expanding among a number of sociologists and economists, on the nature of money and how exploring this makes a difference to social science research. This discussion includes at least three streams of thinking. One is the so-called "new economic sociology," the foundations of which are associated with Granovetter's (1985) work. Granovetter argued that economic actions are "embedded" in, i.e., inseparable from and very much formed and shaped by, the relationships between people that result from how societies are structured. The discussion this led to among sociologists has been well-documented by Swedberg and others (e.g., Swedberg, 1987, 1997; Smelser and Swedberg, 2005; Granovetter and Swedberg, 1992).

A key feature of this "new economic sociology" has been an attempt to resolve the dualism between money and humans that appears to be inherent in classical economics (Ingham, 1996b). On one side of the divide there is money, seen in classical economics as a kind of reified, objectively existing entity against which market values of goods and services can be set (Ingham, 1998). On the other side are human beings, who engage with money in their everyday lives. For orthodox economists, the essential point of engagement is people's rational choice facility. This is the facility human beings use to optimize their utility in the marketplace, while the rest of human nature is "the tosh": the "superfluous rituals, rules of procedure without clear purpose, … needless precautions preserved through habit" (Ingham, 1996b—citation attributed to Oliver Williamson).

Ingham's (1996b) meta-analysis of this discussion within the new economic sociology leads him to recognize that as long as money is seen as an entity different from human beings and existing ontologically separate from them, this will always lead to an irreconcilable dualism. The boundary will be either between money and humans, or between the rational and non-rational parts of the human being, depending on how these are defined by this or that theorist.

This leads us to consider the second stream of thinking on economics among sociologists. Here, Ingham (1996a,b, 1998, 2004, 2011) and others such as Beggs (2015, 2017) and Braun (2016) argue that money is not essentially different or separate from human beings. Instead, it is first and foremost a social phenomenon, namely a credit–debt relationship between two human beings (or groups or organizations of human beings). Money is about one person owing the other something, a fairly universal societal trait that can be seen in many if not all cultures (Graeber, 2011). There is no sense in which money is ontologically alien to humans and needing an explanation of how it interfaces with society. Money *is* social, from start to finish, because it is, essentially, a relationship of obligation and entitlement. This view is shared by many other social theorists (see in Antoniades, 2018) and is also advocated by post-Keynesian economists such as Davidson (1972, 1994, 2002), Harrod (1969), Dow (1993), Chick (1983), Dequech (2014), Sahr (2015, 2017) and Werner (2014, 2016).

A third stream of thinking takes this view further. Building on the notion of money as a credit–debt relationship, a number of social theorists explore how radical power imbalances arise in a society where the economy is very heavily debt-based, as ours is today. The leading theorists among these are Lazzarato (2012, 2013, 2015), Soederberg (2005, 2009, 2014) and Di Muzio and Robbins (2016). While their starting points in social theory differ from each other's, their conclusions regarding debt and power are similar. These are neatly summed up in the title of Antoniades (2018) summary of their work: "Gazing into the abyss of indebted society: the social power of money and debt."

Since the view of money as a credit–debt relation is now widely accepted, I will spend much of this chapter expounding it, and will give less space to alternative theories. Further, I will avoid being sidetracked by claims and counter-claims as to how money arose in ancient societies: how humans came to have money. Anthropologists such as Graeber (2011) have examined the existing empirical evidence from a great range of different kinds of ancient and pre-modern societies and found no concrete examples of a non-money society becoming a money society in any particular setting or by any particular means. The fact is, we find ourselves today to be a society with money, and this is the kind of society and the kind of money we have to reckon with in our social science research, be it on energy or any other subject.

In Section 2 I explain what I understand money to be, setting this in the context of today's neoliberal-dominated financial system. In Section 3 I discuss prevalent myths about money and how these can increase our understanding of the specific types of power relations money is associated with today. In Section 4 I conclude with a discussion of implications for social science based energy research.

2 Money as a relationship

I start with the simple observation that today, money comes into existence by being *created out of nothing* by banks, including both private banks and countries' central banks. Later I will explain how there is a fundamental difference between the monies created by each of these two types of banks, but in both cases money is created out of nothing. Secondly, money is created as a *credit–debt relationship*. Thirdly, it is created *independently of goods and services*.

In its very creation, then, money is a social relationship. Money does not sit alongside society as a neutral, technical medium which interfaces with society's rational side. Rather, everything about money can be explained in terms of relationships between people. As Ingham (2004) expresses it:

> … by a "sociology of money" I intend more than the self-evident assertion that money is produced socially, is accepted by convention, is underpinned by trust, has definite social and cultural consequences and so on. Rather, I argue that money is itself a social relation; that is to say, money is "claim" or "credit" that is constituted by social relations that exist independently of the production and exchange of commodities. Regardless of any form it might take, money is essentially a provisional "promise" to pay whose "moneyness", as an "institutional fact" (see Searle, 1995), is assigned by a description conferred by an abstract money of account. (Ingham, 2004, p. 25).

The following stylized example illustrates how this works in the case of private (commercial) banks. The narrative in this example is based on analyses of the modern banking system in Braun (2016), Sahr (2017), Mundell (2000), Antoniades (2018), Ingham (2011), Werner (2014, 2016), Lazzarato (2015) and others.

Two persons, a banker and a builder, enter a relationship in which the banker produces a sum of money *by crediting this amount to the account of the builder*. Note that the money comes into existence simply by being credited to the builder via his bank account. The banker thereby becomes a creditor of the builder, while the builder becomes a debtor to the bank. At the same instant, however, the builder lends the money back to the bank: since it appears in his bank account, he has effectively lent it to the banker at the same moment she deposited it in his account. Nevertheless, since the banker created the money as her own possession, i.e., to her credit, she writes the amount on the credit side of her ledger (Braun, 2016). The banker thereby has both a credit and a debt that balance each other out. The builder also has both a debt and a credit: he is in debt to the banker for the amount she lent him, while he is in credit to her to the amount in his bank account.

It is important to note, therefore, that although money has been created out of nothing, a kind of anti-money has simultaneously been created, which perfectly nullifies it. Creating money has not made the world richer. It has, however, established a specific type of *power relationship* between the banker and the builder (Lazzarato, 2015; Di Muzio and Robbins, 2016). The builder has an *obligation* to give the money back, plus an agreed amount of interest, by an agreed time, while the banker has an *entitlement* to enforce this payment, if necessary by means of state-sanctioned coercion.

Continuing with the narrative, the builder uses the money to pay (say) a painter. He does so by transferring the money to the painter's bank account. This has four important effects.

Firstly it results in *work being done which otherwise would not have been done*. Even though money has been created out of nothing, it has led to a painter expending time and energy to paint a building.

Secondly, at the moment the builder transfers the money from his bank account to that of the painter, the builder's banker ceases to be in debt to him (though he is still in debt to the banker), as the money has gone out of the builder's account. Instead, the builder's banker is now in debt to the painter's banker. This is because the transfer from the builder's account to the painter's account was actually a *promise to pay* the stated amount, made in the name of the builder's bank. This can be seen, for example, in the fact that the painter's bank is now obliged to pay the painter non-loan money— say cash from reserves—if the painter asks for it.

Thirdly, in the actual world of banking, there are not just two banks and one loan and transfer, but thousands of banks and hundreds of thousands of loans and transfers crisscrossing in all directions on any particular day. The system works because all the banks are networked. They have all agreed to tally up each other's debts to each other at the end of each working day, and make payments to each other to annul the outstanding differences. Usually these differences are very small because all the banks are actively issuing loans, receiving deposits and making payments to customers.

Fourthly, this means that the banks each need to keep only a relatively small amount of reserves of ready cash in comparison to the amount of money they have currently lent out and are awaiting repayment for. They only need enough reserves to guarantee that (a) they can balance up their debts to each other each day, and (b) they can pay out cash (actual notes and coins) if and when their depositors demand it. These reserves are held as deposits in central banks, a point I come to later.

Continuing with the story, the painter then withdraws the money from the painter's bank to pay a car dealer for her new car, *resulting in a sale that would not have taken place without the money being created.* This process may cascade endlessly from firm to firm, buyer to seller, *causing more and more work to be done and goods to be made and traded which would otherwise not have been done or made or traded.*

Partly because there are many banks making many loans to many people, with all this money crisscrossing in various payments for goods and services, the builder eventually finds he has earned enough money (from similar transactions to those above) to repay the banker, plus the interest the banker demanded. When he does so (by electronic transfer from his account to her account), he is thereby nullifying the relationship of debtor to creditor. This power-charged relationship no longer exists.

Because the creditor–debtor relationship no longer exists, the money the bank loaned the builder can no longer exist (since money is a credit–debt relationship). The banker therefore strikes out the amount she had written on the credit side of her ledger. Although the money was created out of nothing and returns to nothingness, its travels through the world have served to increase the amount of goods and services, perhaps by many times the value of the loan. As Dodd (2011) expresses it, "Classically, banks fulfil a vital function in the economy as agents of capitalist entrepreneurship." Up to this point in the narrative, the banking system appears to be sheer genius, provided no one borrows more than their capacity to repay.

But what of the *interest* the builder paid the banker? If all the money in commercial circulation is bank loan money, which gets dissolved when it is repaid, where does the interest come from? The intriguing answer is, it comes from the issuing of *more* debt by private banks. The rate of issuing of loans has to continually increase, to provide the interest for the loans that have already been issued. As Antoniades (2018) comments, in his review of Di Muzio and Robbins (2016):

> *This leads to an unsustainable system where the creation of more debt necessitates faster growth rates to serve this debt, which are achieved through more debt.* (Antoniades, 2018, p. 8)

There is, therefore, a kind of inherent spiral to infinity in the banking system. On the one hand, because money is created out of nothing, the amount of money (and anti-money) being created could, theoretically, increase indefinitely. John Maynard Keynes once commented:

> *It is evident that there is no limit to the amount of bank-money which the banks can safely create provided that they move forward in step. (Keynes, 1930, cited in Ingham, 1998, p. 11; italics in original)*

In one sense this is true. Provided all the banks are well-networked and loans are being transferred between banks at about the same rate, the system remains stable. Two very serious things can go wrong, however. Firstly, borrowers can get over-committed or use loans for investments that fail, and if enough fail all at once, some banks cannot pay their dues to other banks, savers lose confidence in their banks, banks run out of reserves, and the system fails, as in the Great Recession of 2007–2008.

Secondly, even where the banking system is stable and successful, it is forced to continually increase the amount of loan money in circulation, in order to create the money to pay the interest on already existing loans. This is where the insights of Lazzarato (e.g., Lazzarato, 2015), Soederberg (e.g., Soederberg, 2014) and Di Muzio and Robbins (2016) are so important. Such a system demands not only continual economic growth, but also continually increasing indebtedness. On a purely theoretical level it can be shown that, for banks to issue sufficient loan money to cover the interest on their loans, they must issue, in addition to these loans, extra loan money of:

$$U = L \bullet \frac{i - i^{n+1}}{1 - i} \tag{1}$$

where i is the average rate of interest banks charge on loans, L is the value of the loans in any 1 year, and n is the number of years society wants the banking system to survive.

For example, if banks' average interest rate is 5% per year and society wants the banking system to survive indefinitely, the banks must increase the amount they create and lend by 5.26% per year. This amounts to a 70% increase in the total amount of loan money every 10 years. In fact, the world's total indebtedness has increased by 250% since 2003 (SD Bullion, 2018), an annual cumulative increase of 6.3%. This is despite the dip caused by the huge downturn in the banking sector after the Great Recession of 2007–08, not to mention other defaults on loans which happen regularly.

It is important to note that this persistent increase in the amount of loan money, and therefore in total indebtedness, has to continue indefinitely if the banking system is to remain stable. Lazzarato (2014) comments: "Under capitalism, debt rests on and unleashes an infinite process." Banks thereby get caught up in predatory behavior. They have to find new borrowers in order to keep the system flowing. Lazzarato (2014, 2015) argues there is synergy between this and neoliberalism's success in shifting the tax burden away from the rich toward the poor while simultaneously persuading governments to erode social welfare and public services. It creates poorer governments and a poorer class of citizens, who need to borrow money constantly in order to survive—thus keeping the banking system afloat.

This has important implications for social science, including that used in energy studies. The problem is not simply that economic inequality has increased hugely over the past 30–40 years, as Piketty (2014) and others convincingly show and Galvin and Sunikka-Blank (2018) have related to energy issues. It is also that the money system depends on creating more and more indebted people, people who need to borrow to survive.

Further, as governments become poorer due to reduced taxes on the rich, and lose income from state-owned enterprises that have been sold off, governments also become poor and need to borrow.[a] This suits the banking system because of governments' huge appetite for loan moneys. For example, Lazzarato (2015) calculates that effectively, government social welfare programs are funded by government debt to banks, so that banks become the beneficiaries of these programs through the interest they collect. Ironically, the banks then become further potential beneficiaries of government money because they become too big to fail. Social scientists in energy studies need to note that this cluster of factors makes it hard for governments to manage energy issues, such as the mitigation of fuel poverty and the transition toward renewable energy supply, points I return to below.

3 Myths about money

3.1 Myth and obfuscation

Braun (2016, p. 1073) wryly comments: "The history of money consists of a succession of different ways of obfuscating its core feature—namely, money's origins in credit–debt relationships." He argues that certain interest groups, including, at times, central banks, often benefit from fostering various folk myths about what money is. Ingham (1996a, 1998, 2004, 2011) takes a similar perspective, arguing that orthodox economics promotes the theory of money as a neutral medium of exchange because that supports the political program of neoliberalism. He also notes that one of the reasons the myth has survived is that sociologists, who could have readily critiqued it throughout the 20th century, absented themselves from the discussion for decades. There are also mythical views on money that many people or even academic economists tend to default to (Braun, 2016), since many aspects of money are contorted and difficult to see clearly. I deal here with four of the most prominent myths, in each case identifying implications for social science and energy research, which I will return to in Section 4.

---

[a]Governments often do not borrow directly from banks but instead sell bonds—promises to pay—which purchasers often have to bid for in an auction. This also serves to increase the amount of bank-created money in existence, as the money investors pay for the bonds has to come from somewhere. In fact, often it is banks who buy government bonds, putting governments directly in debt to the banking system.

3.2 **The myth that all money is the same**

Cash in one's wallet might seem to be the same thing as a positive balance in one's savings account. But since the establishment of the Bank of England in 1694—the world's first central bank—central banks have worked hard to preserve an important distinction between central bank money, which they call, counterintuitively, "outside money", and commercial bank loan money, called "inside money" (Braun, 2016). Inside money—money created by commercial banks—makes up by far the greater portion of the money circulating in the world, but its net value is zero, because in the moment it is created, a credit and a debt are created simultaneously—or, in more social language, an entitlement and an obligation are created simultaneously. Further, as we saw above, the total amount of credit and debt has to continually increase, to keep the banking system stable.

Central bank money—"outside money"—does not, technically, equate to zero. Like commercial bank money, it is created out of nothing, but not always in relationship to a specific debtor. Often society as a whole is the debtor when central bank money is created. Central bank money is made up partly of cash, which includes all the notes and coins in circulation plus more held in the central bank's vault, called its "cash reserves". The remainder is the reserves (not necessarily cash) which commercial banks deposit in their accounts at the central bank. This is used to reconcile commercial banks' daily imbalances of debts and credits to each other (see above), and buffer them against failures caused by bad loans or a bank run.

The amount of outside money is very small compared to inside money,[b] and central banks only have indirect and somewhat precarious control over how much inside money central banks create[c] (Braun, 2016; Sahr, 2015). Central banks usually demand that commercial banks have, in their reserve deposits, an amount at least as large as a certain percentage of the amount they create as loans. But this does not stop commercial banks increasing both amounts in tandem. In short, it is the private, commercial banks that determine how much money gets created, and central banks are effectively drawn along in train. The tail wags the dog.

An important implication for social science is that commercial banks have a very great amount of power visa vis central banks. Not only do they draw both individuals and governments into relationships of very heavy obligation. They also, in a sense, snub central banks' legitimate role of controlling the money supply. What started out

---

[b]The line between inside and outside money is fuzzy in places. For example, some banks treat some cheque accounts as outside money.

[c]Readers are referred to Braun (2016) for discussion of how banks attempt to use reserve-to-lending ratios to control the money supply; to Sahr (2015) for a nuanced explanation of why, ultimately, commercial banks set the amount of money in circulation; and to Mundell (2000) for a historical account of how, since the 1970s, central banks have become more adept at controlling inflation through setting the inter-bank interest rate.

in 1694 as a government attempt to bring all the diverse types of privately issued loan moneys and scripts into a coordinated system has metamorphosed into the private banking sector gaining a very great amount of power over the government-appointed central banks at the center of the system.

It could also be argued that there is a third type of money, which is not functionally different from commercial bank credit–debt money but operates outside the formal money system. This includes promissory notes, IOUs, gambling house chips, chain store voucher and purchase point systems, and community currencies, which are local, usually temporary folk currencies that are often derived for specific purposes (Seyfang and Longhurst, 2013; Stephanides, 2017; cf. Dodd, 2011). Like "real" money, these are also created out of nothing and come into effect in various types of credit–debt relationships.[d]

What is often not understood is that modern commercial bank money derived, over a period of time, from this type of money. Boyer-Xambeu et al. (1994) describe how, in certain parts of Europe in the 14th century, an importer who had no spare money could obtain a promissory note from a bank—a piece of paper of the form, "I Luigi Medici, promise to pay the bearer of this letter the sum of one thousand Ducats"—which the importer would use to "pay" a foreign exporter for a shipload of goods. The exporter would then give the promissory note to his bank in his own country and demand cash in exchange for it. Because the banks were well networked, the exporter's bank would happily pay cash for the letter, knowing it would be reimbursed by the importer's bank. As in modern banking, there were hundreds of such promissory notes circulating in all directions, and the bankers were well networked and met periodically to reconcile outstanding debts between them. This type of bank credit system spread throughout Europe, and by the 17th century it was ubiquitous in Britain with many different kinds of promissory notes in circulation. The newly formed Bank of England set out to tame the system by buying up these diverse types of note at a discount, and by the 19th century the private bank system was all but a miniature version of today's banking system (Wray, 1990).

But central banks never brought every kind of credit–debt relationship into their orbit. The creation of credit–debt obligations is a basic human activity, and therefore, in a sense, money creation is a "wild" human activity which the state can never tame completely. Social science involved in energy studies could look more seriously at "wild" monies: for example the "crisis community currency movements" investigated by Stephanides (2017), or the "complementary currencies" described by Raworth (2017). In many instances these cause work to be done just as money does, but without the burden of being indebted to the banking system.

---

[d]I omit discussion of crypto-currencies as these raise further issues which space does not permit discussion of. I refer the reader to Bjerg (2016).

3.3 **The myth that money is a neutral veil**

Most economic textbooks proclaim that money is essentially a neutral medium of exchange, a kind of "lubricant" (Ingham, 2006) to enable the efficient exchange of goods and services of differing values (e.g., Jones, 1976; Clower, 1984; and critique in Ingham, 1996a). Economic textbooks usually claim this was as humanity's ancient solution to the problem of barter (see critique of this popular view in Graeber, 2011). A barter economy is very difficult to maintain if different people have different levels of need for many different types of goods and services. Orthodox textbook economists suggest ancient societies therefore invented money as a medium to reflect the values of goods and services, so that these could be easily, fairly and smoothly exchanged.

Money in this view is not part of the "real" economy but, as Schumpeter (1954) observed in his critical account of this view:

> *Money enters the picture only in the modest role of a technical device that has been adopted in order to facilitate transactions . . . so long as it functions normally, it does not affect the economic process, which behaves in the same way as it would in a barter economy: this is essentially what the concept of Neutral Money implies. Thus, money has been called a "garb" or "veil" of the things that really matter ... (Schumpeter, 1954, p. 277).*

Ingham (1998, 2011) notes that one of the reasons this view tends to prevail in textbook economics can be traced to a major split between economics and sociology in the first half of the 20th century. Sociology and economics were not always distinct disciplines which theorized the social and the economic in separate spheres. Early social theorists such as Karl Marx (1818–1883), Max Weber (1864–1920) and Emile Durkheim (1858–1917) made little distinction between economics and sociology (Morrison, 1995). But a rupture in sociologists' theorizing of money came with the so-called "*Methodenstreit*" (dispute over method) between economists and sociologists early in the 20th century. As Ingham explains:

> *As a result, money fell under the jurisdiction of economics, and this fact alone explains sociology's indifference; but it was the particular 'theory' held by the victorious economists that was to have a significant impact on both disciplines' understanding of money. After the* Methodenstreit, *economic thought became dominated by the idea that money was epiphenomenal—that is to say, it was treated as a neutral 'veil' over the underlying 'real' natural economy. (Ingham, 1998, p. 4)*

Hence, for example, although sociologist Talcott Parsons deliberated about money and its effects within society, he assigned the question of what money is, to economists (Parsons, 1937). Giddens (e.g., Giddens, 1979; Giddens, 1984; Giddens, 1990) discussed the role of trust in enabling the money system to operate, and offered lengthy commentary on Marx's theory of money in capitalist society as a source of workers' alienation from the rewards of their labor, but did not explore the ontology of money either in general or in terms of his own enormously influential and insightful structuration theory. Pierre Bourdieu (1979, 1983, 1996) explored how monetary and social capital reinforce people's existing socioeconomic status, but did not turn his tools of analysis to the question of what money is or how money gets there in the first place. These theorists' assumptions as to what money is are those of the mainstream economics of the time. Mizruchi and Sterns' (1994) comprehensive review of the place of money in sociology of the latter half of the 20th century showed that sociology simply took the existence of money for granted.

Three quite decisive critiques can be brought against the view that money is a neutral veil. Firstly, there is no historical evidence that money arose from barter or from any other human activity, as noted above.

Secondly, the barter/neutral veil view presupposes what it purports to argue. As Orléan (1992) points out, it is based on the assumption that there is already a numerically graded market, *logically* prior (even if not historically prior) to the advent of money. Grading the values of different types of goods and services against each other already presupposes there is some neutral scale against which to grade them. You first have to have something that serves as money, before the grading can begin. Hence Aglietta and Orléan (2002) argue that a stable market of mixed goods and services is not possible without something that already serves as money.

But the strongest criticism of this view is that it ignores the empirical evidence of how we see money being created, whether in modern central and private banks, medieval promissory notes, complementary currencies, or informal IOU's. All these involve the creation of credit–debt relationships, obligation and entitlement, which are all basic human activities. Money is a very human phenomenon, not a neutral entity that stands outside humanity and interfaces with people across a human/non-human abyss.

Ingham (2011) and Lazzarato (2015) argue that it suits neoliberalism to foster the neutral veil view of money. If money is a neutral veil that simply reflects the real values of goods and services, then governments should not intervene in markets by taxing high incomes at higher rates than low incomes, by redistributing money to low income households, or by running public services like health, education and transport. They should privatize all services and leave it to the markets to set the most efficient prices for these. This includes energy markets, and it precludes universal benefits designed to lift people out of poverty and therefore out of fuel poverty.

Soederberg (2014) adds that fostering the view that money is a neutral mechanism of exchange between free agents in a fair market draws attention away from the reality that money, based as it is on relationships of obligation and entitlement, "is an all-power disciplinary apparatus" that involves "raw social violence, suppression, exploitation, inequality struggles and class power relations …" (quoted in Antoniades, 2018)—particularly in today's setting of extreme inequality.

3.4 **The myth that money is a commodity**

For much of recorded history in Europe, Asia, the Middle East and North Africa, money has been closely associated with the so-called "precious metals" silver and gold. This has fed the myth that money is essentially a precious, physical commodity which, as in the neutral veil theory, sits alongside the market of useful goods and services. Mundell (2000) offers a succinct account of the gradual demise of the gold standard. Until the early decades of the 20th century, many countries pegged the value of their currencies to the price of gold. From the 1940s until 1971, most developed countries pegged their currencies to the US dollar, which was pegged to gold at $35 per ounce. This meant that anyone in the world who had a dollar could trade it, with the US central bank, for 1/35th of an ounce of gold. After the US left the gold standard in December 1971, gold has been merely a market item like oil, wheat, electricity and apples (but of much less practical use than these!).

The idea that money is a commodity such as gold or silver has obviously been debunked by the fact that money has continued to exist since 1971, though it took central banks several decades to learn how to control inflation without the stability of the gold standard (Mundell, 2000). It is also noteworthy that in epochs when gold and silver coins were extensively used as money, the purchasing power of these coins seldom equaled the market value of the gold or silver in them (Ingham, 2011). Countries, city-states and potentates regularly melted down and reminted their coins to give them face values that reflected the needs of the hour rather than the market value of the metal the coins were made of. This is because, in the end, the value of money is related to the credit–debt relationships of its issuers and users, and not to the quantity of gold or silver owned by the issuer (see Box 1).

Further, being tied to a gold standard can be very inhibiting for economies because the global supply of gold is limited and may grow in fits and starts, irrespective of the needs of the economy.

Nevertheless, having a fixed standard to set the value and quantity of central bank money has a strong advantage. It effectively puts a brake on private bank lending, since every dollar that banks create (out of nothing, by way of a loan) can come back to haunt them in the form of a demand for physical currency, which costs them the price of gold. Going off the gold standard was one of the factors that led the private banking and finance sector to become so large and dominant today. A key insight of Lazzarato (2015) and others is that this freedom, this loss of anchoring to any real commodity value, has led to the debt-driven economies and debt-burdened households and governments we see today.

This points to an interesting implication for social science in energy research. Whether at household level or at the level of national energy supply, one of the factors social scientists need to take account of is the overblown banking system that pervades life today at all levels, and the relationships of power and powerlessness, entitlement and obligation, this causes. Countries are often beholden to international finance for their energy plans (Newell and Phillips, 2016), just as households often

---

Box 1 What kind of being is a dollar?

If money is not gold or the physical notes in our purses and wallets, what kind of being is it? An important clue is provided by the linguistic philosopher J.L. Austin's (1976) concept of the "performative". A performative is a statement that brings into being that which it claims to bring into being. Austin's most fully developed example is a marriage. A marriage comes into being when a couple exchange their vows—they each say something like "I, Chris, take you, Les, to be my lawful wedded husband/wife", in the presence of a properly appointed marriage celebrant at a predetermined time and place, after fulfilling essential preconditions such as not being already married and being over a certain age. Their marriage then exists regardless of what happens to the documents that confirm it, such as the entry in the marriage register. If someone then asks, "What kind of *being* is a marriage; what is its ontology?", Austin would reply that it is a performative, a special class of a socially constructed reality that involves a realigning of people's relationships to each other, in a context where everybody (more or less) has agreed (by default or explicitly) to support this discourse and set of relationships.

Though Austin does not discuss money in this context, I suggest money has the same ontological form. "Inside" money (commercial bank money) comes into being when two parties, a lender and a borrower, enter into a creditor–debtor relationship of a numerically agreed magnitude denominated in an agreed currency, and when various conditions are fulfilled. Austin points out that performatives can go awry in various ways if the societally agreed conditions, actions, words (and in some cases flourishes) associated with them are faulty in any way (Austin, 1976, p. 14ff). In the case of money, the bank manager (or her proxy) and the borrower (or his proxy) have to sign an agreement of a societally agreed type, and the mechanism for issuing the loan has to be performed according to a ritual society has agreed to—in this case by crediting the value of the loan to the borrower's bank account. The money then exists as a credit–debt relationship, and its ontological status is like that of a marriage. It exists as societal discourse and as a realigning of more or less everybody's relationships (the money can be used for transactions between more or less anybody), in a context where society has agreed to support this.

Central bank money is also a performative, though the parties involved and the form of the act of money creation are different. The main party is the head of the central bank, appointed by the government. The act of creation includes this person's signature on a form of an agreed type; the money is produced by entering figures in various ledgers; its coming into being changes many people's relationships, if not everybody's; and it occurs in a context where society has agreed this is the correct procedure.

Other performatives of this type are university degrees and limited liability companies. A person continues to "have" a PhD even if all documentation of it is lost and the university that conferred it is bombed out of existence. A limited liability company has no physical location and continues to exist if all its board members, shareholders and material property and equipment are replaced.

Although a performative has no weight, size or physical location, and although it exists only in societal discourse, relationships and habitual practices, that does not imply it is frail or fragile. The discourse and practices supporting money are some of the strongest on earth, as people who fall into deep debt often find out.

---

have to negotiate personal debt to get access to basic energy services. Solutions have to be sought on the macro-level. Social science needs to challenge governments to develop effective mechanisms to bring the creation of private capital into line with social and national needs.

3.5 **The myth that money is a creation of the state**

The so-called "chartalist" view, closely associated with the early 20th century German economist George Friedrich Knapp (1842–1926), is that money is the creation of the state rather than something that arises spontaneously within society. On the one hand, this view is clearly at odds with the understanding of money I have argued for in this chapter: a credit–debt relationship initiated when a loan is made. On the other hand, it raises important issues about the state's role in the maintenance of a money system.

Although money (of various forms—see Section 3.1) comes into being in a range of situations where relationships of credit and debt, entitlement and obligation are formed, it more often than not needs a competent authority to make it work. Lenders will not lend if there is no enforcement mechanism for debtors who are reluctant to repay. As Ingham (1998) argues, the credit–debt relationship is traditionally backed up by a competent authority who can, if necessary, threaten or use coercion or even violence to enforce it. In a non-state situation the relevant competent authority may not necessarily be a "legitimate" authority: it can be the Mafia or the armed wing of a drug cartel. However, in modern times the state claims a monopoly on violence (Aglietta and Orléan, 2002; Giddens, 1985) and thereby becomes the "legitimate" enforcing agency.

This opens an interesting dimension in the credit–debt relationship in which money resides. Those who have money (i.e., they are in credit) have the state apparatus of coercion on their side against those who are in debt to them. Further, because money tends to beget money (Piketty, 2014), credit tends to accumulate to creditors and debt to debtors. This effectively means the poor are at the butt end of the state's coercive apparatus. Money is highly charged with state-enforced sanctions that can be applied to those who fail to attain enough of it to live on and are thereby perpetually in debt. Beggs (2017) argues that the enforcement function is so important and necessary to the money system that it may be one of the reasons the state came to exist and continues to exist. The ever-developing money system has needed an ever-stronger state to keep repayment of debts orderly and reliable. People with money and wealth spread around in investments, bank accounts and financial instruments do not like to be in a stateless wilderness where others can grab and take.

On the other hand, Beggs notes, the state has been relegated to a rather weak roll in relation to the finance system itself. When the state intervenes to influence the money supply or reduce inflation, it must do so *strategically*, as a careful actor, taking care not to disturb the balances that keep the finance system functioning.

At the same time, Di Muzio and Robbins (2016) remind us that modern states tend to be in huge debt to private banks and the finance system more generally. They talk of the "capturing" of the state by creditors. On the one hand, the state is used by banks to enforce repayment of loans and defend its property rights. On the other hand, the state has little agency in relation to the finance system as it is heavily in debt to the very system it is defending, and its powers to limit this system's behavior are severely curtailed.

Again, the tail is wagging the dog, and again, social scientists involved in energy research need to be aware that this is the milieu in which households, local authorities and governments attempt to develop fair and universal access to adequate energy services.

4 Discussion and conclusions: Some implications for social science in energy research

Social science, as used in energy studies, often has a very specific and localized focus: the household; a local community; a specific group of people such as renewable energy producers; a developing country's electricity supply system. The conceptual frameworks social scientists use for studies in these areas are often very finely honed and even surgical in their application: practice theory; sociotechnical systems theory; actor-network theory; discourse analysis; theories of power. While these can bring very specific concerns into focus, they often thereby bracket out a wider picture.

It will be clear from the above discussion that what money is, and how it operates today, are very wide-ranging, macro-level issues. They are determined and driven by an array of institutions, discourses, behaviors and practices on a very large scale. These include: the private banking system; central banks' attempts to manage the finance industry; the hegemony of neoliberal ideology in government policies, financial arrangements and public discourse; the magnitude of global debt; the increasing indebtedness of governments and households; tax rates and welfare redistribution; international investment practices; and the prevalence of myths and misunderstandings that can impede close, critical engagement with many of these factors. These are set in the context of money as a credit–debt relationship, a type of relationship that occurs regularly and naturally among human beings, but has become formalized, institutionalized, and developed into specific forms via today's dominantly neoliberal financial and governmental structures and practices.

I suggest therefore that the general, overall implication of this chapter for social science in energy studies is that researchers may need to broaden, or even step back from, tried and trusted conceptual frameworks, so as to include in their ambit the effects and influences of these wider, financially related factors on energy supply, consumption and access to energy services.

A closely related point is that frameworks like practice theory, discourse analysis and sociotechnical systems theory are theories of the way the social world more or less always works. They tend to be historically non-specific and designed to be applied in many different and varied situations. This is a problem Michel Foucault faced with his frequent use of the tool of discourse analysis, which he is largely credited with developing. A number of scholars of Foucault have pointed out that it is a mistake to think Foucault relied almost exclusively on this one tool (Braun, 2014; Brenner, 1994; Brigg, 2001; Dumez and Jeunemaitre, 2011; Lazzarato, 2015; Peltonen, 2004). These commentators point out that for Foucault, the *material reality* of the epoch or situation being investigated was the real object of his research.

To bring this into focus, Foucault frequently spoke of the *dispositif*, a French word often somewhat misleadingly translated "apparatus" in English texts. Foucault explained the *dispositif* in these terms:

> *What I'm trying to pick out with this term is, firstly, a thoroughly heterogeneous ensemble consisting of discourses, institutions, architectural forms, regulatory decisions, laws, administrative measures, scientific statements, philosophical, moral and philanthropic propositions—in short, the said as much as the unsaid. Such are the elements of the apparatus. The apparatus itself is the system of relations that can be established between these elements. (Foucault, 1977)*

Lazzarato (2012, 2015) finds this a very useful concept for understanding how money works today. For Lazzarato, the "system of relations" in today's money world is that of domination and subjugation brought about by debt. He focuses on debt, its extent, its power relations, the way households and even governments are controlled by their creditors, the way debt has become central to modern life through the hegemony of neoliberal economics; government debt as a substitute for progressive taxation; debt hand in hand with austerity; the imbalance of debt upon the younger generation; the never-ending spiral of debt due to the effect of interest on loans. He sees subjugation through debt as the guiding principle of the whole, complex apparatus (*dispositif*) of the modern money system. Similarly, Antoniades, drawing on Soederberg (2014), argues that "debt has emerged as the key governing instrument and principle in modern western societies". These authors maintain that governments have less and less power to govern society as creditors gain more and more sway over society.

Social scientists may or may not agree that debt and its subjugating power are the focal point of the modern economic apparatus. But researchers still need some organizing principle, at the very least a heuristic model, to take account of the full impact of today's money, finance and economic system on the great majority of people, including in their role as energy consumers.

Some of the implications for energy consumption studies are as follows:

Firstly, to interpret energy research findings fully, we must keep the wider, macro-situation in mind, no matter how local or atomized our field of energy research is. For example, in investigating thermal comfort practices in the home, we may need to take into account where the householder is positioned on the credit–debt spectrum. This could be strongly affecting how free they feel to turn up the thermostat when the house is cold, or whether they can face going deeper into debt to install a more efficient boiler. If this is the case, then the solution to their thermal discomfort lies beyond the household, in the policy realm that enables the debt-centered regime to continue.

Secondly, we need to look for solutions on the macro-level rather than (only) proposing various sorts of locally targeted interventions. Galvin and Sunikka-Blank (2018) present evidence that macro-level economic inequality is a clear factor in individuals' limited access to energy services; in households' propensity to cause high levels of CO_2 emissions; and in countries' levels of CO_2 emissions. A framework based on Lazzarato's insights would take the issue of inequality further, analyzing the role of debt in subjugating millions of energy consumers while endowing a few with enormous power and entitlement. Whichever way we look at it, key aspects of the solution to energy problems can *only* be found on the macro-level. Subsidizing boilers for specifically selected poor, deeply indebted households is noble, but it would be much better if the debt-focused *dispositif* in which these consumers live were reformed from the top down.

Thirdly, we need to keep calling to mind what money essentially is. For example, although all money is created out of nothing, there is a an unspoken narrative that developing countries dare not create (extra) money out of nothing in order to get jobs done locally, even though wealthy countries do this regularly. Therefore, as Soederberg (2014) shows, developing countries are continually drawn into debt dependence on loan money from wealthy countries.

This applies especially to energy supply projects. Newell and Phillips (2016) give the example of Kenya, which has limited electricity supply based mainly on imported fossil fuels but has huge indigenous potential for renewables. The Kenyan government is committed to increasing these rapidly, but these ambitions "require massive investments that the state is unable or unwilling to provide". Therefore the government looks to "transnational calculations of capital" (Newell and Phillips, 2016, p. 41).

The dominant model of international investment in developing countries is what Newell and Phillips call "disciplinary neoliberalism". This is now prevalent throughout sub-Saharan Africa and involves international finance bodies, including donor NGOs, setting the terms and conditions of project development *and investing their own money in it*. Countries like Kenya can thereby end up with large debts and limited ownership and control of their electricity networks, even though the Kenyan currency, the shilling, has been relatively stable over the past few decades and the local work and resources for such projects has to be paid for in local currency. Although all money is created out of nothing, there is still the myth that western money is better than that created in a developing country (see also Ingham, 2011). Social scientists researching energy projects in developing countries may need to take notice of where the money comes from and why only rich country money is thought good enough for the job.

Fourthly, since money is always created out of nothing, social scientists in energy research could use this understanding to frame research into roles for alternative or community currency projects in the energy domain. Seyfang and Longhurst (2013) review the now copious literature on community currencies in the context of sustainability projects. Many locally bounded, non-official currencies work according to the same logic as official currencies: money is created out of nothing as a credit–debt relationship: "one person's spending equals another person's debt" ((Seyfang and Longhurst, 2013): 883). Community currency regimes can lead to "the tangible achievement of environmental and social sustainability improvements, albeit on a small scale." They also offer a way of breaking free of the subjugation of debt associated with the commercial banking system. Sahr (2015) offers a theoretical study of how and in what institutional structures community currencies can compare with, improve on or be less reliable than official private bank money.

Although schemes such as these show promise, particularly in closely networked groups or communities, most people have to deal daily with the ebb and flow of private bank money. It is therefore all the more important for social science in energy research to be informed by the overall picture of, to use Lazzarato's (2015) term, the "debt model" economy in which energy trading, consumption or deprivation takes place. Many of the distortions of access to energy services are a direct result of the suffocating power of debt and the system which lives off (other people's) debts. Social scientists involved in energy research need to recognize this factor and vigorously challenge it.

In short, building a well-informed sociology of money into our energy research frameworks is a much-needed step to give us a clearer understanding of the objects or our research.

References

Aglietta, M., Orléan, A., 2002. La Monnaie Entre Violence et Confiance [Money Between Violence and Trust]. Odile Jacob, Paris.

Alvaredo, F., Chancel, L., Piketty, T., Saez, E., Zucman, G., 2018. World Inequality Report 2018. World Inequality Lab. Available at https://wir2018.wid.world/.

Antoniades, A., 2018. Gazing into the abyss of indebted society: the social power of money and debt. Polit. Stud. Rev. 16(4). https://doi.org/10.1177/1478929918757135.

Austin, J., 1976. How to Do Things With Words, fourth ed. Oxford University Press, Oxford.

Beggs, M., 2015. Inflation and the Making of Australian Macroeconomic Policy, 1945–85. Palgrave Macmillan, Houndmills.

Beggs, M., 2017. The state as a creature of money. New Polit. Econ. 22 (5), 463–477.

Bjerg, O., 2016. How is bitcoin money? Theory Cult. Soc. 33 (1), 53–72.

Bourdieu, P., 1979. Distinction: A Social Critique of the Judgment of Taste. Trans. Richard Nice, Routledge, London.

Bourdieu, P., 1983. The forms of capital. In: Richardson, J. (Ed.), Handbook of Theory and Research in Education. Greenwood Press, Westport Conn.

Bourdieu, P., 1996. The State Nobility. Polity Press, Cambridge.

Boyer-Xambeu, M., Deleplace, G., Gillard, L., 1994. Private Money and Public Currencies. The Sixteenth Century Challenge. M. E. Sharpe, London.

Braun, P., 2014. A new urban dispositif? Governing life in an age of climate change. Environ. Plann. D Soc. Space 32, 49–64.

Braun, B., 2016. Speaking to the People? Money, Trust, and Central Bank Legitimacy in the Age of Quantitative Easing. Max Planck Institute for the Study of Societies, Cologne.

Brenner, N., 1994. Foucault's new functionalism. Theory Soc. 23, 679–709.

Brigg, M., 2001. Empowering NGOs: the microcredit movement through Foucault's notion of dispositif. Alternatives 26, 233–258.

Chick, V., 1983. Macroeconomics After Keynes. MIT Press, Cambridge, MA.

Clower, R., 1984. A reconsideration of the microfoundations of money. In: Walker, D.A. (Ed.), Money and Markets: Essays by Robert W Clower. Cambridge University Press, Cambridge, pp. 81–89.

Davidson, P., 1972. Money and the real world. Econ. J. 82 (325), 101–115.

Davidson, P., 1994. Post Keynesian Macroeconomic Theory. Elgar, Aldershot, UK.

Davidson, P., 2002. Financial Markets, Money and the Real World. Elgar, Cheltenham, UK.

Dequech, D., 2014. Is money a convention and/or a creature of the state? The convention of acceptability, the state, contracts, and taxes. J. Post Keynesian Econ. 36 (2), 251–273.

Di Muzio, T., Robbins, R., 2016. Debt as Power. Manchester University Press, Manchester.

Dodd, N., 2011. 'Strange money': risk, finance and socialized debt. Br. J. Sociol. 62 (1), 1–20.

Dow, S., 1993. Money and the Economic Process. Elgar, Aldershot, UK. 1993.

Dumez, H., Jeunemaitre, A., 2011. Michel Callon, Michel Foucault and the "dispositif": when economics fails to be performative: a case study. Le Libellio d'AEGIS 6 (4), 27–37.

Foucault, M., 1977. The confession of the flesh (1977) interview. In: Gordon, C. (Ed.), Power/Knowledge Selected Interviews and Other Writings.In: vol. 1980, pp. 194–228.

Galvin, R., 2010. Thermal upgrades of existing homes in Germany: the building code, subsidies, and economic efficiency. Energ. Buildings 42, 834–844.

Galvin, R., 2014. Why German homeowners are reluctant to retrofit. Build. Res. Inf. 42 (4), 398–408. https://doi.org/10.1080/09613218.2014.882738.

Galvin, R., Sunikka-Blank, M., 2018. Economic inequality and household energy consumption in high-income countries: a challenge for social science based energy research. Ecol. Econ. 153, 78–88.

Giddens, A., 1979. Central Problems in Social Theory: Action, Structure and Contradiction in Social Analysis. MacMillan, London.

Giddens, A., 1984. The Constitution of Society. University of Los Angeles Press, Berkeley and Los Angeles.

Giddens, A., 1985. The Nation State and Violence. Polity Press, Cambridge.

Giddens, A., 1990. The Consequences of Modernity Cambridge. Polity Press.

Graeber, D., 2011. Debt: The First 5,000 Years. First Melville House, New York.

Granovetter, M., 1985. Economic action and social structure: the problem of embeddedness. Am. J. Sociol. 91 (3), 481–510.

Granovetter, M., Swedberg, R., 1992. The Sociology of Economic Life. Westview Press, Boulder, CO.

Harrod, R., 1969. Money. Macmillan, London.

Ingham, G., 1996a. Money is a social relation. Rev. Soc. Econ. 54 (4), 507–529. https://doi.org/10.1080/00346769600000031.

Ingham, G., 1996b. Review essay: the 'ne economic sociology'. Work Employ. Soc. 10 (3), 549–564.

Ingham, G., 1998. On the underdevelopment of the 'Sociology of money'. Acta Sociologica 41, 3–18.

Ingham, G., 2004. The Nature of Money. Economic Sociology: European Electronic Newsletter, ISSN 1871–3351. vol. 5. Max Planck Institute for the Study of Societies (MPIfG), Cologne, pp. 18–28 (2).

Ingham, G., 2006. Further reflections on the ontology of money: responsesto Lapavitsas and Dodd. Econ. Soc. 35 (2), 259–278.

Ingham, G., 2011. Capitalism. Polity Press, Cambridge.

Jones, R., 1976. The origin and development of media of exchange. J. Polit. Econ. 84 (4), 757–775.

Keynes, J., 1930. A Treatise on Money. Macmillan, London.

Lazzarato, M., 2012. The Making of the Indebted Man: An Essay on the Neoliberal Condition. MIT Press, Cambridge, MA.

Lazzarato, M., 2013. Il Governo Dell'Uomo Indebitato. Saggio Sulla Condizione Neoliberista. Derive Approdi, Rome.

Lazzarato, M., 2014. Debt, neoliberalism and crisis: interview with Maurizio Lazzarato on the indebted condition. interviewed by Mathieu Charbonneau and Magnus Paulsen Hansen. Sociology 48 (5), 1039–1047.

Lazzarato, M., 2015. Governing by Debt. Semiotext(e), South Pasadena, CA.

Middlemiss, L., 2017. A critical analysis of the new politics of fuel poverty in England. Crit. Soc. Policy 37 (3), 425–443.

Mizruchi, M., Sterns, L., 1994. Money, banking and financial markets. In: Smelser, N., Swedberg, R. (Eds.), The Handbook of Economic Sociology. Princeton University Press, Princeton, pp. 313–341.

Morrison, K., 1995. Marx, Durkheim, Weber: Foundations of Modern Social Thought. Sage, London.

Mundell, R., 2000. A reconsideration of the twentieth century. Am. Econ. Rev. 90 (3), 327–340.

Newell, O., Phillips, J., 2016. Neoliberal energy transitions in the south: Kenyan experiences. Geoforum 74, 39–48.

Orléan, A., 1992. The origin of money. In: Varela, F., Dupuy, J.-P. (Eds.), Understanding Origins. Kluwer, Amsterdam, pp. 133–143.

Parsons, T., 1937. The Structure of Social Action. Free Press, New York.

Peltonen, M., 2004. From discourse to Dispositif: Michel Foucault's two histories. Hist. Reflect. 30 (2), 205–219.

Piketty, T., 2014. Capital in the Twenty-First Century (Translated from the French by Arthur Goldhammer). Belknapp-Harvard University Press, Cambridge, MA.

Raworth, K., 2017. Doughnut Economics: Seven Ways to Think Like a 21st-Century Economist. Business Books, London.

Rosenow, J., Galvin, R., 2013. Evaluating the evaluations: evidence from energy efficiency programmes in Germany and the UK. Energ. Buildings 62, 450–458.

Sahr, A., 2015. Wären wir die besseren Banken?: Zur Debatte um die Repolitisierung des Kreditgeldes. Widerspruch: Beiträge zu sozialistischer Politik, Special issue 'Finanzmacht-Geldpolitik. vol. 66. pp. 103–113.

Sahr, A., 2017. Das Versprechen des Geldes: Eine Praxistheorie des Kredits. Verlag des Hamburger Institutes für Sozialforschung, Hamburg.

Schumpeter, J., 1954. A History of Economic Analysis. Routledge & Kegan Paul, London (1994).

SD Bullion, 2018. Total World Debt Q1 2018. Online resource, https://sdbullion.com/blog/total-world-debt-q1-2018/.

Searle, J., 1995. The Construction of Social Reality. Penguin, Harmondsworth.

Seyfang, G., Longhurst, N., 2013. Desperately seeking niches: grassroots innovations and niche development in the community currency field. Glob. Environ. Chang. 23 (5), 881–891.

Smelser, N., Swedberg, R. (Eds.), 2005. The Handbook of Economic Sociology. Princeton University Press, Princeton, NJ.

Soederberg, S., 2005. The Politics of the New International Financial Architecture: Reimposing Neoliberal Domination in the Global South. Zed Books, London.

Soederberg, S., 2009. Corporate Power and Ownership in Contemporary Capitalism: The Politics of Resistance and Domination. Routledge, London.

Soederberg, S., 2014. Debtfare States and the Poverty Industry: Money, Discipline and the Surplus Population. Routledge, London.

Stephanides, P., 2017. Crisis as Opportunity?: An Ethnographic Case-Study of the Post-Capitalist Possibilities of Crisis Community Currency Movements. Doctoral Thesis, University of East Anglia, UK.

Swedberg, R., 1987. Economic sociology: past and present. Curr. Sociol. 35, 1–221.

Swedberg, R., 1997. New economic sociology: what has been accomplished, what is ahead? Acta Sociologica 40, 161–182.

Toke, D., Lauber, V., 2007. Anglo-Saxon and German approaches to neoliberalism and environmental policy: the case of financing renewable energy. Geoforum 38, 677–687.

Werner, R., 2014. Can banks individually create money out of nothing?—the theories and the empirical evidence. Int. Rev. Financ. Anal. 36, 1–19.

Werner, R., 2016. A lost century in economics: three theories of banking and the conclusive evidence. Int. Rev. Financ. Anal. 46, 361–379.

Wray, R., 1990. Money and Credit in Capitalist Economies. Edward Elgar, Aldershot.

Asymmetric structuration theory: A sociology for an epoch of extreme economic inequality

3

Ray Galvin

University of Cambridge, Cambridge, United Kingdom
RWTH Aachen University, Aachen, Germany

Chapter outline

1 Introduction 53
2 Structuration theory 57
3 Structuration theory, oligarchs and power 59
4 The power of credit and debt 61
5 Competing discourses: Neoliberalism and welfare politics 64
6 Implications for energy research 66
7 Conclusions 68
References 70
Further reading 74

1 Introduction

When social scientists investigate energy issues they bring various sociological frameworks to their approaches, which embody assumptions about what society is and how it functions. Most of the frameworks used today in energy studies have roots in the period before economic inequality began to bite. Examples are practice theory, sociotechnical systems theory and actor-network theory. Of course, these have been further developed in recent years and applied in interesting and novel ways to energy issues, while other frameworks such as "energy cultures" (Stephenson et al., 2010) and "energy justice" (Walker and Day, 2012, and see Chapter 4 of this book) have developed alongside them. But the basic understandings of society that lie behind these approaches come from the profound thinking and detailed empirical field

Inequality and Energy. https://doi.org/10.1016/B978-0-12-817674-0.00003-5

research and socio-historical work of some of the towering academic figures of the latter part of the 20th century.

One cannot discuss the entire field of sociology and energy studies in one chapter. Instead I want to focus on a particular late 20th century theory of society which, it seems to me, lies behind a very great deal of today's energy consumption research. I refer here to Anthony Giddens' (1979, 1984) "structuration theory." Structuration theory was one of the most significant developments in late 20th century sociology. It provided a way of reconciling long-running tensions between views that stressed the solidness or immutability of social structure, and those that emphasized human agency. Giddens posited a "duality of structure" in which society is a recursive phenomenon. On the one hand, the discourse and practices of human agents are constrained and enabled by the routinized, organized, institutionalized structures of society. Meanwhile, however, these structures are continually being produced, reproduced, re-formed and changed, by these human agents through their discourse and practices.

From a purely mechanistic point of view, structuration theory provides a neat meta-account of how societies work: groups of people cohere together with rules, routines and practices which provide a structure for them to act in ways that make sense to each other, but which also both constrain and enable them in acting in such a way as to not only conform to these rules and practices, but also to go against these practices, break the rules or make new rules. This schema takes account of the human propensity to develop order and continuity, but also human creativity and the fact that, as Granovetter (1985) delightfully explains, humans often also like to cheat, lie, embezzle and cause havoc.

One of the most important ways structuration theory has found its way into energy studies is via practice theory. In a nutshell, Schatzki (1996, 1997, 2002) revisited some of the loose ends in Giddens' exposition of how *discourse* (the things people say, write and convey in symbols, including laws and rules) and *practices* (the things people do, often routinely but not always) fit together to make up the duality of structure. Drawing on Wittgenstein's (1953) critique of theories of discourse, Schatzki argued that practices are logically prior to discourse, i.e., that in the end, society is the way it is because of what people do, regardless of what people say and what rules they make (see Galvin and Sunikka-Blank, 2016 for a fuller exposition of Schatzki's argument).

Schatzki's practice theory was adopted for energy consumption studies largely through the pioneering work of Elizabeth Shove and her colleagues (e.g., Shove and Pantzar, 2005; Shove, 2012; Shove et al., 2012). This foothold was broadened and deepened by energy researchers such as Tom Hargreaves (Hargreaves, 2011; Hargreaves et al., 2010), Røpke (2009), Gram-Hanssen (2010) and many others, including some interesting recent applications such as teenagers' mobile phone charging practices (Horta et al., 2016) and how German window making practices clash with energy efficient manual ventilation practices (Galvin, 2013). Some energy researchers have built upon practice theory to form other approaches, such as the "energy cultures" framework (Stephenson et al., 2010), and there are loose connections between practice theory and the "energy justice" approach (Simcock and Mullen, 2016). In a further departure, Schatzki (2010) broadened his conception of practices to place these more explicitly within what he calls "arrangements," which are the locally specific geographical and institutional features within which various practices take place, and Shove and Walker (2014) recently applied this to issues of energy supply. A basic assumption and building block of these approaches is that society is a structurated phenomenon in the terms Giddens proposed.

A key element of practice theory is the notion that practices are not just habits or cultural routines, but are constrained and enabled by the social structures in which people are embedded. We cannot understand, for example, home heating practices, without taking into account things like: the heating technologies available to households and the research and development framework that influences this; the daily and weekly timetabling of employment and the economic factors that determine this; the clothing available in the marketplace and the global trade structures that influence this; the laws, materials and building practices that influence the way houses are built; and of course, the way money works and is distributed, since energy has to be paid for. Research that claims to be based on practice theory but simply investigates peoples' routines and doings apart from the societal structures that constrain and enable them may fail to engage with the factors influencing those routines and doings, and can end up posing solutions that have little chance of working. There is plenty of energy research based on habits and routines (see review in Kurz et al., 2015), but this is not practice theory. For this reason, if we are using a structuration-based framework like practice theory it is important to be acutely aware of the social structural influences on the practices we are seeing, including economic practices, and to be aware of changes in social structure, including economic changes. Practice-based energy research to date tends not to take into account economic-related structural factors, and I aim in this chapter to begin to redress this.

Another important point is that although structuration theory is a compelling and credible notion of the mechanisms of how human societies function, Giddens, in expounding his structuration theory, was not just proposing a general theory of how societies function mechanistically. Together with some of his older contemporaries such as Bourdieu (1976, 1990) and Foucault (1982, 1985), he was also very interested in why some people and groups get the upper hand in particular societies, and why others get trapped at the bottom of the heap. Despite the relatively egalitarian age in which they lived, Giddens and his contemporaries were very much aware that the societies they were investigating were often very unequal and that asymmetries of power *within social structure* played a key role in this.

Giddens argued that the power of different members of a society—different "agents"—varies according to what he called their material and human "resources" and that, as some agents command more resources than others, some people can bring about greater changes in social structure than others, and can also use the existing structures to get more of society's goods and wealth. Giddens said a great deal about resources in this context. Some of his terminology in this regard is not always clear. Sewell (1989) offers a helpful guide to it and attempts to fill some of the gaps in

Giddens' explanation. The important, overriding point is that different people's resources—their ability to change social structure and/or get what they want from it—depend crucially on how much power they have.

In this respect, structuration theory is a *living* theory that changes from epoch to epoch and society to society, because the distribution of power is different in different ages and societies. Hence it is important to review and renew structuration theory for the current epoch in which we live: an epoch of deep and deepening inequality.

I will argue in this chapter that recent insights into power, especially as it derives from money, can increase our understanding of the type of structuration that currently characterizes our society. I will suggest that energy research frameworks with roots in structuration theory need to constantly revisit their understanding of society, to take account of these power relations, how they are changing, and how they are deeply affecting energy access and supply at all levels.

To address this I draw on three complementary sets of insights, after first outlining Giddens' structuration theory in Section 2. The first set of insights comes from the work of Jeffrey Winters (2011, 2014), which I consider in Section 3. Winters shows how small groups of extremely wealthy people in almost all societies in recorded history have controlled or heavily influenced governing structures for their own benefit, and in particular, how they do this with aplomb today. The second, which I discuss in Section 4, concerns the social-economic structural effects of the modern banking system. Here I draw on insights from Geoffrey Ingham (1996a,b, 1998, 2004, 2011) on the nature of money and developments in modern capitalism in relation to the banking system's metamorphosis into a source of subjugation through ever increasing debt (Di Muzio and Robbins, 2016; Lazzarato, 2015; Soederberg, 2014). These streams of thought are concerned with *practices*: how people are *acting* in society—not just engaging in discourse—to wield their resources. However, there is a third set of studies which suggests that *discourse*, in particular the discourse of neoliberalism, is also a very powerful resource—ideological, perhaps even religious—which works to shape the specific types of social structuration we see today. I explore this in Section 5, together with countervailing, deeply entrenched discourses that support universal social welfare.

There are limitations to this analysis, mostly because it focuses on types of power associated with money in today's neoliberal environment. There are many other ways people exercise power in society, either to change it or to get what they want from it. Some examples are through art, science, engineering, institutionalized violence, intellectual creativity, and religion. My focus on money and resistance is intended to describe aspects of today's social structuration that bear on issues of economic inequality. These, I contend, need to be taken up in energy consumption research that investigates energy consumption and supply practices, so that these practices and their determinants are better understood.

This chapter draws on some of the themes in Chapters 1 and 2, which are essential for the argument of this chapter. Rather than repeat long passages here I will mostly just refer to them.

2 Structuration theory

Giddens sought to critically synthesize some of the unresolved issues of sociology as it had developed by the late 20th century. Central among these was the tension between structure and agency. On the one hand, societies appear to be structured such as to severely constrain and limit how people can act. On the other hand, societies are comprised of individuals who appear to have the freedom to act how they will.

Karl Marx had emphasized the apparent solidness and immutability of social structure, referring mostly to how organizations, infrastructures and economic relationships become entrenched and self-perpetuating. For example, in Marx's view the contestation between workers and capital owners was built in to the physical componentry of the social world with its factory production, forms of ownership, class systems, etc. Malm's (2013) study, from a Marxist perspective, of how steam power gradually replaced water power in the British cotton industry, argues that it was the social structural form of the worker-capitalist relation that determined the transition from water to steam, rather than any inherent technological superiority of one of these fuels over the other.

Emil Durkheim took the notion of social structure a step further, speaking of social structures as if they exist somewhat like physical laws, external to human beings and pushing back onto them to drive them inexorably to perform in certain ways. Giddens called this "reification": the notion that "social phenomena become endowed with thing-like properties which they do not in fact have" (Giddens, 1984, p. 180). This has also been called "social determinism" (Lehmann, 1993), as it implies social forces operate of their own accord, much like the regularities of physics. For example, it is sometimes claimed that class struggle is not just a series of actual oppositions and clashes between workers and entrepreneurs, but is also a kind of law of nature written in to the fabric of society, which determines that less affluent socioeconomic classes will always struggle to overthrow wealthier classes. Elements of this are evident throughout Piketty's (2014) account of developments in income and wealth inequality over the past 200 years. Piketty repeatedly claims, without citing evidence, that if inequalities reach a certain extreme, the populace will rebel and take back the wealth to themselves—as if there is a kind of law that determines this. A penetrating and in places entertaining critique of reification is offered by Billig (2013).

The other side of the structure-agency debate concerns agency: individuals' freedom to act how they want to within their societies, and their power, or lack of it, to change these societies. An extreme view, often called "methodological individualism," is that society is merely the sum-total of all the acts of individuals. Methodological individualism is sometimes traced to Max Weber via his student Josef Schumpeter (Zalta, 2015), though Schumpeter maintained it was merely a statement of fact rather than a philosophical position: individual humans, he argued, are the only social entities that can act as agents. If their acting in concert produces structural effects, these are still the products of individuals acting (cf. Harré, 2002).

Giddens (1979, 1984) developed a synthesis of agency-structure dualism with his theory of "structuration," based around what he called the "duality of structure." For Giddens, social structure is the rules, discourses, institutions, practices and relationships that are continually being reproduced by the people who make up a particular society. At the same time, however, people's discourse and practices are constrained and enabled by the characteristics of their society—which is in turn a product of their practices and discourse:

> *The constitution of agents and structures are not two independently given sets of phenomena, a dualism, but represent a duality. According to the notion of the duality of structure, the structural properties of social systems are both medium and outcome of the practices they recursively organize. (Giddens, 1984, p. 25)*

This recursive approach avoids the dualism of structure and agency by arguing they are both, in effect, the product of each other. It avoids positing that social structure is rooted in some reified, abstract realm, even if physical structures like roads and airports have a large influence on it, while simultaneously avoiding the notion that individuals can act in pure freedom, unconstrained by the nature and characteristics of the society they are part of.

Further, Giddens argues that social structure—as he defines it—not only *constrains* individuals in their actions, it also *enables* them. This can be seen in at least three senses. To begin with, almost all human social actions, such a handshake, a wink or a gift of €100,000, make sense to others and carry effective power only because of the social structural context in which they are set. Secondly, in a more materialistic way, depending on where an individual is placed within her society's structures and what those structures are like, these structures may enable her to get what she wants—money, fame, energy services, medical treatment—or they may constrain or even prevent her from getting it. Thirdly, because social structure is itself a product of human action and discourse, existing social structures can serve to empower or prevent certain individuals from acting in ways that *change* these structures.

Giddens further theorizes how it is that different individuals have different degrees of agency to effect changes in social structure or to act how they wish within it. Different agents' power to act, he argues, depends on their different *resources*. Giddens' concept of resources is not always consistently explained, but in a key passage he defines resources as "the media whereby transformative capacity is employed *as power* in the routine course of social interaction" (Giddens, 1979, p. 92, emphasis added). Power is the defining element of what Giddens calls resources. Giddens also argues that these resources, and thereby the power to change society, are unevenly distributed within society—a point that has major implications for studies of economic inequality.

The notion of power has been discussed extensively, often in relation to specific niches of research, such as organizations and institutions (Clegg, 2010); sociotechnical transitions (Geels, 2014); post-Marxist political theory (Laclau and Mouffe, 2014); and developmental studies (Gaventa, 2003). Though these accounts vary considerably, most identify at least three different modes or types of power, which may be summarized as: material power (through money, possessions, physical objects of coercion); discursive power (through winning arguments, policing the terms of debate and discussion); and institutional power (through official position, control of government structures, rule-making). Generally, then, in developing a theory of power such as in this chapter, it is wise to ensure it addresses at least these three different ways power is exercised.

Further, I do not aim to cover the entire field of power here, but to focus on power that relates to economic aspects of social structure that constrain and enable human beings in the context of economic inequality. This will leave out discussion of some features of social structure that are, in fact, crucial for understanding energy consumption practices, such as the diffusion of green technologies (Geels, 2014) and the physical layouts of cities (Ratti et al., 2005). However, these are already taken into account in a great deal of research on energy access and supply that uses frameworks based on social structuration, such as practice theory and sociotechnical systems theory. My aim is to fill in a gap, rather than rewrite an entire story.

3 Structuration theory, oligarchs and power

I first address Giddens' notion of power and resources by building on Winters' (2011, 2014, 2017) studies of oligarchs: the ultra-rich. Winters (2011, p. 12) focuses on *individuals'* power resources, which is entirely consistent with Giddens. Power does not originate in some reified realm of self-acting social structure. It issues from specific human agents acting within their society.

In the context of enormous levels of economic inequality, Winters distinguishes five different types of power resources, which overlap in various ways: (a) power based on political rights; (b) the power of official positions; (c) coercive power; (d) mobilizational power; and (e) material power. I deal only briefly with the first four, as these are well represented in other theories of power, and spend most of the discussion on the fifth.

For Winters, power based on *political rights* is the power various classes of citizens have depending on their society's political system and their standing within it. In a modern western-type democracy most adults have the right to vote and stand for office. Although there are often attempts to erode these rights, for example, by gerrymandering or demanding voters show ID cards, generally this mode of power is fully intact and universal in these countries, i.e., it works against asymmetries of power.

Power based on *official positions* is the power held by elected or appointed officials. In western-type democracies this power is usually held only temporarily and dissolves when a person leaves office (Winters, 2011, p. 15). Hence, apart from offices such as a lifetime appointment as a US Supreme Court judge, this kind of power does not necessarily contribute to asymmetries in social structure.

Coercive power is the power to force others to do what they otherwise would not do. Drawing on Weber, Winters (2011, p. 159) notes that the defining feature of the modern state is its claim to have a monopoly on legitimate coercion and violence. Nevertheless, other actors also successfully use coercion and violence: local gangs; drug barons; physically abusive spouses; mafias; factions within prisons; political revolutionary mass-movements who take up arms. In Chapter 2 I noted that money can only work in a situation where debt obligations can be enforced. Nowadays the state takes on this role. On the one hand, moneyed elites prefer a small, relatively powerless state that will not interfere in the freedom of the markets. On the other hand, they need an extremely powerful, competent state to enforce their property rights, including bringing delinquent debtors to heel (Beggs, 2017).

Mobilizational power is the power afforded by an organized, energized movement of large numbers of people who might individually have little or no political power. This power can be exercised via discourse, challenging the values, moral assumptions and rules that drive the exercise of power by those in official positions, or in a more practice-oriented way by starting new political parties and groupings. It can also be exercised materially, by crowd-funding, or coercively, by blocking streets, staging strikes or, in extremes, storming government buildings or physically protecting or thwarting a military coup.

Material power is power derived from having money. Money is power. In Winters' (2011, p. 18) words: "Material resources in their various forms (the most flexible being cash money) have long been recognized as a source of economic, social and political power." In the US, the poorest 20% of society have, on average, no net wealth at all—their debts cancel out their assets (calculated from Federal Reserve Board, 2018). This means they have no net material power as defined here. The wealth of those above this level increases gradually, then exponentially, as described in Chapter 1. Since only a tiny fraction of the wealth of the hyper-rich is needed for day-to-day living, even at the opulent level of an oligarch, these people have unimaginably large resources left over for their exercise of power in society.

Winters (2011, p. 208ff) catalogues the main ways the very richest of the rich—the oligarchs—exercise this power. Firstly, as Winters outlines in great detail, they employ legions of accountants, lawyers and lobbyists to pressurize, tempt, bribe and cajole politicians to keep income tax rates low for ultra-high income earners. Secondly, they use these same operatives to complicate the tax laws to such a high degree of contortion that the accountants of the super-rich are provided with an array of options for tax breaks. Thirdly, they employ investment experts to hide their money in tax havens, shell companies and complex financial instruments which evade or avoid tax laws (cf. Zucman, 2014, 2015). Fourthly, they finance think-tanks such as the Heritage Foundation, which "supply legitimacy and ideological ammunition to the lobbyists and interest groups ... who work relentlessly, day in and day out, to keep up the tax-cutting pressure on the Hill" (Graetz and Shapiro, 2005, p. 40, cited in Winters, 2011, p. 252). A typical result is that the effective tax rate of the ultra-rich, top earning 400 US oligarchs reduced from 30% in 1994 to 17% in 2007. Hence their effective tax rate—the percentage of their total declared income they have to pay in tax—is now lower than that of the "merely wealthy" top-earning 0.1% who pay 21%, which is lower than that of the "mass affluent" top-earning 1%, who pay 23%. In effect, tax rate progressivity has become reversed. The other English-speaking developed countries and many EU countries have also seen huge reductions in tax rates for top income earners in recent decades (Piketty, 2014).

These power-resourced activities of the ultra-rich are highly relevant to structural increases in economic inequality in recent decades. The reduction of tax rates on high incomes means governments have less to spend on infrastructure, education, medical care, social housing and social welfare, driving low-income households deeper into destitution (Soederberg, 2014). The neoliberal discourse reproduced and reinforced by these oligarch's think tanks increases politicians' inhibitions toward social welfare and other social goods such as environmental protection (Beder, 2001; Jacques et al., 2008).

But these aspects of material power only concern the wealth *defense* activities of the ultra-wealthy. There is a further, vast dimension of practice among the super-wealthy, in which their huge financial resources are used to beget more wealth. In a neoliberal environment this leads to practices such as the selling off of public services and infrastructure (Miraftab, 2004); wealthy elite control of the direction of energy transitions (Galvin, 2018a,b; Newell and Phillips, 2016), and the contracting out of social welfare services (Waquant, 2009). Relating this to structuration theory, we see a relentless assault on the kinds of social structures that were established in past decades to balance out differences in material power between different groups in society. The result is that social structures become more asymmetric in favor of those who already have power in abundance.

With the possible exception of mobilizational power, all these forms of power are exercised by specific individuals, though even mobilizational power only exists where individuals act, and is often dependent on specific, highly motivated leaders. In the case of material power, we see unimaginably highly resourced individuals using their money-power to effect changes in the rules, relationships, dominant discourses and official practices that structure societies. Hence, investigating economic power within a structuration framework leads us directly to focus on individuals, specific people who lead, machinate or manipulate to re-form social structure how they want it.

A further resource of power is continually acting alongside and in concert with the machinations of the very wealthy: the everyday creation of credit and debt by the commercial banking system.

---

4 **The power of credit and debt**

In Chapter 2 I outlined a theory of what money is and how the commercial banking system has a monopoly on its everyday creation out of nothing. I now briefly reiterate and build on some of the key points covered there (cf. Ingham, 2011, p. 33ff), in relation to social structure.

Commercial banks have reserve capital assets deposited with their country's central bank. Using this relatively small amount of "real" or "outside" money as security, these banks create very large amounts of new "inside"[a] money every day by crediting borrowers' accounts with loans at certain interest rates, usually also charging fees. This credit money passes through many hands as payments for goods and services, thus multiplying its effects many times—this is the essential benefit of the capitalist system. At the end of their loan terms the borrowers repay their loans, with interest. The repaid loans are then dissolved back to nothing, but the banks retain the interest and fees. This process is happening continuously, with millions of loans outstanding at any one time, while money that was created as loans (=credit money = debt money) criss-crosses between people in myriads of everyday transactions. Hence outstanding loans make up the bulk of the world's money supply: this is the money people use every day for shopping, paying wages, betting, settling bills, running businesses, etc.

The total amount of loan money—so-called "inside" money supply—is equal to the total amount of debt. Debt and credit are the mirror image of each other. Every dollar I have in my bank account is mirrored by someone else being a dollar in debt.

The key point here concerns the interest that banks charge. Banks use some of the interest and fees they receive, to pay their overheads, some to make up for defaulted loans, and some to pass on as profits to their shareholders. They deposit the rest in their reserve accounts with the central bank, to increase their balance sheets. Central bank rules allow them to create loan money in proportion to their reserve balance: at any one time they can usually have about 10 times as much loan money in circulation as their reserve balance, though other factors often enable them to go beyond this.

Apart from the money banks add to their reserve accounts, the rest of their earnings from interest and fees go back into circulation via their shareholders, employees' wages, etc. Although these moneys are now circulating as non-loan money, they are still mirrored by an equal amount of debt, because they are predicated upon the interest which the borrowers paid. The key point is that to pay her interest, a borrower has to obtain money over and above her loan, through transactions with other borrowers, *thus pushing indebtedness further down the line*. Therefore, over time, the banker's shareholders (the owners of the bank) are accumulating more and more of the total money in circulation to themselves while the pool of borrowers, overall, are getting deeper and deeper into debt (Di Muzio and Robbins, 2016; Lazzarato, 2015).

This creates a dilemma for the banks: they cannot risk beggaring society, or fewer and fewer people will want to take out loans, and more and more of those who take out loans will default on them. The only way they can keep the system going is to continuously increase the total amount out on loan, so that this month's interest payments are financed from next month's new loans, and so on. In Chapter 2 I offered a

---

[a] "Inside" money is money that circulates within the everyday economy. "Outside" money remains within the structure of the central bank, and so is not available for normal everyday transactions.

simple model for this, suggesting that each year the amount of money in circulation (apart from central bank money, which is "outside" the everyday money supply) has to increase by a percentage slightly higher than the average interest rate, in order to keep the system afloat. I also noted that the world's total money supply has increased by an average of 6.3% per year since 2003 (SD Bullion, 2008). The growth rate of money supply is far higher than the growth rate of GDP. According to World Bank statistics, in OECD countries the total amount of debt money in circulation was equal to 65% of GDP 1965, and increased to 135% in 2015 (Sahr, 2015). More and more money is being created over and above that which is used for goods and services.

Since money is created as debt, the world (apart from banks, similar financiers and their shareholders) is more and more deeply in debt in comparison to GDP, i.e., in comparison to the value of all its goods and services. Government debt, household debt, credit card debt, all are continuously increasing (Soederberg, 2014; Di Muzio and Robbins, 2016), because debt has to increase to keep the banking system afloat. The banks thereby become predatory, actively seeking new borrowers. Ross (2014) talks of the "predatory behavior of the finance industry" (cited in Di Muzio and Robbins, 2016).

Although some debt is constructive, such as a household or firm taking out a sensibly proportioned loan to build a house or expand a promising business, much debt is toxic. Excessive government debt leads to crippling austerity programs and a running down of basic public services and social welfare; credit card debt can lead to the effective loss of a large proportion of household monthly income; homeowners can become over-indebted through extending their mortgages to buy cars and other consumer goods. There is now a large and growing research literature on the different ways people get hooked into increasing debt. Achtziger et al. (2015) investigate the links between weak self-control and increased indebtedness in the German population. Aron et al. (2012) compare how UK and Japanese households take on extra mortgage loans to spend on consumption. Walks (2018) shows how car-dependent, low-income Canadian households get deeper into debt through finance packages offered for new cars. Bridges and Disney (2004) investigate how UK households get caught in ever-deepening credit card debt. O'Loughlin and Szmigin (2006) speak of UK and Irish students living in "a credit card environment." They "accuse financial institutions of helping to create this culture of borrowing, credit and debt and of aggressively targeting vulnerable customers like students with offers of credit." Lazzarato (2015) has coined the phrase "homo debitus," denoting a new kind of humanity characterized by being in ever deeper debt.

Soederberg (2014) explores the increasing indebtedness of US households through credit card and student loan debt. She talks of "cannibalistic capitalism," whereby "credit card issuers (US banks) need to constantly ensure a maximum number of workers take on the greatest amount of debt at the highest interest rates and fees possible, to extract ever higher rates of revenue streams." This in itself has become a scientific endeavor. For example, Ha and Krishnan (2012) develop a set of "credit prediction models to recognize the repayment patterns of each segment (of the indebted population) by using a Cox proportional hazard analysis"—a tool banks can use to judge how deeply to get different types of people into debt without too many of them defaulting.

A common theme in the work of Di Muzio and Robbins (2016), Lazzarato (2015) and Soederberg (2014) is "the social power of money and debt" (see review in Antoniades, 2018). Loan finance, says Lazzarato, "functions as a predatory apparatus of capture" (cited in Alia et al., 2012). Citizens are made into subjects of the banking system, through debt. Bankers and other financiers exercise more and more de facto governance over society *and over governments*, through the power of the creditor-debtor relationship. Hence the theme of "governance" through debt (i.e., exercising power over debtors, including governments who are in debt) now appears increasingly in literature on the banking and finance systems.

Relating this to structuration theory, once again we see asymmetric structuration. Money is not a neutral medium of exchange, equally accessible to all. Further, it is not just that inequality is now so great that one group has most of the money. It has gone several steps further. One group has the privilege of creating and dissolving the money all of us use for everyday transactions. This gives them the privileged ability to get rich by doing nothing productive, i.e., by simply collecting interest over and above fair fees for fair work. This makes them the owners of ever-increasing portions of the money supply. To avoid the system imploding they need the money supply to be continually increasing, hence they need more and more ways of drawing people into debt. Hence they become predatory in their lending practices. The outcome is that they gain more and more power over individuals, households, industry and even governments.

We should add, of course, that the banking system needs economic growth so that there is an ever-increasing pool of business ventures and consumer needs and desires to lend money for. There are close historical links between economic growth, the historic growth of the banking system and the historic growth of consumption of fossil fuels (Di Muzio and Robbins, 2016; Huber, 2009; Malm, 2013). Hence ever-increasing debt relates directly to the theme of energy consumption. However, as noted above, the money supply has grown at about twice the rate of global GDP over the past 25 years. The banking system depends on *more than* steady economic growth. It needs an ever more bloated financial sector, with concomitant levels of debt, in order to enjoy stable returns.

5 Competing discourses: Neoliberalism and welfare politics

The banking system described above has been developing for over a century, but it was only with the widespread adoption of neoliberal economic policies from the 1980s onward that commercial banks, with their monopoly on money creation, attained the advantages of a highly deregulated financial environment which effectively allows them to set the pace of monetary growth (Ingham, 2011; Sahr, 2015). Likewise, although oligarchs have been pulling strings on governments for much of recorded history, their ability to do this was much weakened in the Bretton Woods economic environment of the decades immediately following the Second World War (see Chapter 1). However, the deregulation that came with neoliberalism has enabled them to increase their power enormously (Piketty, 2014; Winters, 2014).

Neoliberalism is a set of practices, but these became and remain dominant largely because of the discourse that promotes and supports them (Gould and Robert, 2013). Neoliberal discourse therefore functions as what Giddens would call a "resource": an instrument certain actors can use, to increase their power within social structure.

Neoliberalism claims to be a scientifically based economic theory, but its spectacular failure to live up to its empirical claim that "a rising tide lifts all boats" leads many scholars to label it an "ideology" (see reviews in Gould and Robert, 2013). A number of scholars take this even further. As early as 1999, leading US theologian Harvey Cox argued that neoliberalism had taken on the core characteristics of a religion (Cox, 1999). He pointed out that, like religion, neoliberalism is totalizing, as it claims its approach is the only way to (economic) salvation while the alternatives— such as neo-Keynesianism or zero-growth approaches—are frequently condemned as heresies. Like many religions, it cleverly explains away its repeated failures to produce convincing empirical evidence for its claims, usually saying we just have to apply it more consistently and more purely and for longer. Like religion, it revolves around a god-like being that cannot be questioned: in this case, Cox suggested, "The Market" is God; everything must conform to free market principles.

Later theologians and even economists have taken up this theme. In a recent essay, US-Netherlands theologian Rogers-Vaughn (2013) notes that "many scholars—in the fields of both economics and theology—are asserting that neoliberalism is a faith system and is acting exactly as a new religion." As examples he cites economists Broad and Cavanagh (2000) and George (2000); and theologians Rieger (2009), Sung (2011), and Thistlethwaite (2010).

Like Cox, Rogers-Vaughn notes the global reach of neoliberal ideology, and that it has deeply penetrated many areas of life that were previously associated with the intrinsic value of human relationships and enjoyment. Its tenets of economic efficiency, "best practice" and quantification of outcomes are now embedded in education, business, government services, medical care and even pastoral counselling, Rogers-Vaughn's own specialist field.

Following Carrette and King (2005), Rogers-Vaughn argues that neoliberalism acts as "a new religion" which "is appropriating and marginalizing existing religions." For example, most of the traditional world religions are suspicious of excess wealth and preach the centrality of caring for neighbors and treating them as persons regardless of their utility value, a tenet which leads to support for universal social welfare and equitable redistribution of wealth. Neoliberalism competes directly with these ideals, extolling excess wealth among the few and subjecting welfare to tests of efficiency and utility (cf. Morgan and Gonzales, 2008).

Alston (2018) argues, in his UN-sponsored report on austerity and poverty in the UK, that neoliberal discourse now dominates UK social welfare policy and practice. This has led to severe restrictions on the provision of social welfare and the absurd situation where, in one of the world's wealthiest countries, almost one child in two is poor by official standards. Alston finds that the dominant, neoliberal-inspired discourse blames social welfare beneficiaries for their plight, and espouses targeting only the "most needy" for benefits, rather than providing universal benefits.

It is important to note, however, that public support for social welfare refuses to die. The Conservative UK government is at least beginning to acknowledge the problem, and in his 2018 budget speech the Chancellor claimed the "era of austerity is finally coming to an end." In the 2018 US House of Representative elections, Democrat candidates who openly espoused universal health care made significant gains.

Therefore, despite the apparent hegemony of neoliberal discourse, scholars such as Rehm (2011, 2016) and McDonagh (2015) argue there is deeply entrenched discourse in western democratic societies in support of generous, redistributive social welfare systems. Rehm's (2016) wide-ranging statistical analysis reveals societal support for universal welfare payments running more or less in parallel with the distribution of economic risk in specific societies. McDonagh (2015) relates public support for social welfare to vestiges of monarchy in western countries: she finds that countries which still have a titular monarchy, such as Denmark, Sweden, Spain and the UK, tend to have stronger public discourse in support of social welfare than those which threw off their monarchies violently, such as the US. She suggests this stems from the notion of patrimony, a longstanding, ancient but still existent view that the monarch is the "father of his people" and is responsible for their welfare.

The extent to which these supportive attitudes are actually holding sound welfare policies in place, against the tide of neoliberal discourse, has never been investigated. However, studies such as these offer strong evidence that public support for welfare is indeed there, and has not been expunged by the dominance of neoliberal discourse. There is therefore an important counter-narrative to the dominant discourse of neoliberalism, which survives despite its current weakness.

One the one hand the power of the wealthy—seen for example in the banking system and the oligarchs' sophisticated lobbying apparatuses—is backed up by the overriding, ever-present, totalizing, globally extensive, locally intensive discourse of neoliberalism. On the other hand, countervailing discourses refuse to die, and their manifestations are seen in the failure of neoliberal economics to completely dominate economic and social policy, despite being strongly ascendant.

6 Implications for energy research

What does this mean for social science based energy research?

Firstly, energy research at household level needs to take account of the economic characteristics of the social structures in which households are embedded. The impacts of social structure on different households will vary, depending on where specific households are positioned within this structure. For example, among the lowest income households, toxic debt may be a major factor, especially since many such households now exist from month to month by getting deeper into credit card debt, store card debt or other revolving, high interest debt. But in qualitative interviews used to research energy consumption, the issue of toxic debt may not arise naturally. Debt and energy consumption are often not connected in researchers' minds, as is evident from the paucity of studies in which the two themes come together, despite high levels of household debt. A few probing questions about household debt and repayment levels may bring out issues festering just below the surface.

Another social structural factor is the income of the household in relation to their country's inequality indicators, such as the Gini coefficients of income and wealth. Income and wealth levels are in themselves important factors, but low income and wealth in a context of extreme inequality bring further structural disadvantages for energy consumers because they have to compete harder for certain goods and services. One of these is housing, which tends get appropriated by wealthy investors in unequal societies (Fernandez et al., 2016). Limited access to housing at the bottom of the price range brings huge problems of access to energy services such as adequate home heating, due to poor thermal quality. Inequality also puts labor and material costs for home improvement measures further out of reach of poorer households (BRISKEE, 2018). It can also lead to private landlords underinvesting in maintenance for their rental properties (see Chapter 6) because they cannot recoup their investments by putting the rent up.

Debt among better-off households has different effects. Generally, wealthier households can bear much deeper debt than poor households, and often this debt is for constructive projects like mortgages and business ventures. But in the course of energy research among well-off households it is still worth investigating debt levels, because we simply do not know what effects this is having on energy consumption, whether these debts are the result of predatory lending practices, or how it affects spending patterns (Melzer, 2017).

Secondly, energy researchers need to investigate what effects the lobbying power of the super-wealthy is having on energy supply, access to energy services, and decarbonization of energy. Recent studies in Germany have shown, for example, that local renewable energy producers fear their climate-friendly input to the grid will be phased out or compromised as big business, backed by massive international funding, is tightening its grip on electricity supply while quietly smuggling fossil fuel based generation into the renewables mix (Galvin, 2018a,b). Again in Germany, cities are facing central government opposition to their attempts to exclude heavily polluting diesel cars from city centers, largely because car manufacturers have disproportionate influence on federal government policy (Hey, 2010). Nevertheless, Meckling and Nahm (2018) argue that, while the German government has tended to be captive to the interests of dominant car manufacturers, this has not been so much the case in the US. There, government initiatives have forged a large niche for electric vehicle diffusion despite the lobbies and interests of traditional car manufacturers.

This is an interesting contrast because it indicates the tension between different types of power resources in shaping social structure—big money versus the power of elected officials—and that the outcome of contestations is not determined in advance. One of Giddens' key points was that social structure is not a solid, thing-like structure, nor is it shaped by physics-like, deterministic laws of cause and effect. Real people make and remake social structure, pitting their resources against each other, and the future is by no means decided in advance. Energy consumption researchers need to be aware of this contestation and, although it might seem that one group has it all their way, the future may be as open as the subsequent actions of the people involved. This also applies to the contestation, noted above, between neoliberal discourse and deeply entrenched social support for universal welfare funding.

A third area of interest for energy researchers in relation social structural aspects is the energy habits of the super-rich, since these people command increasingly powerful resources to set agendas for the distribution of energy access and the emission of climate damaging pollution. Recent studies by Jorgenson et al. (2015, 2016) and Chancel and Piketty (2015) indicate a statistically significant relationship between developed countries' levels of economic inequality and their consumption-based CO_2 emissions. Generally, for each 1% increase in economic inequality in these countries (i.e., each increase in the Gini coefficient of one percentage point), consumption-based climate damaging emissions increase by about 0.8%. Among the wealthy swathe of EU countries, economic inequality indices differ by about 7% points. For example, the Gini coefficient for Sweden in 2016 was 26.6%, compared with 33.9% for Portugal, based on Eurostat data (see Chapter 7). Based on the findings of Jorgenson et al. (2015, 2016), this would (at least theoretically) correspond to a difference in consumption-based climate damaging emissions of about 6% between such countries.

One of the reasons for this correlation is the lifestyle of the super-wealthy. Chancel and Piketty (2015) estimate that the consumption practices of the richest in the US cause about 20 times as much climate-damaging pollution, per person, as that of the average US citizen, while for European countries the figure is about 10–15 times. Harrington (2018) interviewed the wealth managers of the world's wealthiest billionaires over a period of 10 years, to find out how the richest 0.1% live. He found extraordinarily opulent lifestyles, ubiquitous tax evasion, and a degree of laziness, promiscuity and drug-taking that is roundly condemned when found among poorer people. Energy researchers should be concerned that today's social structure has been shaped and reshaped by the activities and interests of people such as this. This is vastly different from the more egalitarian-oriented, democratically responsive social structures of the 1950s–70s. Whenever we research any aspect of energy consumption, energy supply, or access to energy services, we need to set our empirical field and findings alongside this very lopsided, billionaire-serving social structure that characterizes today's world.

7 Conclusions

In this chapter I argued that energy access and supply take place within, and are constrained by, social structures. Giddens (1979, 1984) argued that social structure is the product of the discourse and practices of social actors, and that, recursively, people's actions are constrained and enabled by the social structure in which they are embedded.

Further, Giddens argued that different individuals have different resources, or types of power, to act so as to re-shape social structures into forms that suit them, and to get what they want from already existing social structures. Because there are vast gradations of power and resources, social structure is asymmetric: it suits and enables some people far more than others.

Some of the dominant streams of today's energy research attempt to take into account the social structural features that impinge on energy access and supply. They do so by focusing on energy consumers' and suppliers' *practices*, rather than merely their habits, cultural routines, attitudes or social norms. Practices, in this context, are people's doings and sayings that are enabled and constrained by the social structures in which these people are set.

However, this research generally fails to take account of the economic aspects of social structure, how this is influenced by certain privileged actors, and how it deeply affects people's possibilities for access to energy services, and other people's ability to set the terms of energy supply.

In this chapter, therefore, I explored how social structure is being influenced and shaped by actors in the economics field, in particular: the very wealthy; commercial banks; and the purveyors of neoliberal discourse. Drawing on Winters' extensive research on oligarchs—the hyper-rich—I argued that these people have inordinate material power to influence the political process, and do so systematically. They thereby continually reshape financial policy-related social structures so that wealth flows upward to the already wealthy while the incomes and wealth of others stagnates or declines. I then discussed how the modern commercial banking system not only is oriented to draw ever-greater proportions of the world's wealth into the hands of bankers, but also depends, for its continued functioning, on an ever-ballooning money supply and ever-greater levels of debt among households and governments. Thirdly, I suggested that neoliberal ideology, which legitimizes and paves the way for these factors, has come to have some key features of a world religion, serving as an all-pervasive discourse that buttresses these lop-sided social-economic structures. Nevertheless, I suggested the persistence and tenacity of pro-social welfare discourse indicates that neoliberalism does not have it all its own way.

In order to understand household energy practices, we need to probe and explore how householders are constrained and enabled by these economic aspects of social structure—along with other aspects of social structure such as technological and spatial, which are already well covered in energy research. We also need to explore how energy supply developments (or lack of) are constrained and enabled by the same or similar structural-economic factors.

Finally, I also noted other types of power-resources that are more evenly distributed: political rights, such as voting and standing for office; official positions, such as in government and the civil service; and citizens' mobilization movements. Oligarch-dominated social structure can severely constrain the weaker members of society but it does not fully disable them. Human agents are resourceful, creative, and can spring surprises that leave commentators aghast. Perhaps, as energy researchers help to reveal the huge asymmetry in social structure that besets their research subjects, this may serve as a contribution to building new resources of countervailing power.

References

Achtziger, A., Hubert, M., Kenning, P., Raab, G., Reisch, L., 2015. Debt out of control: the links between self-control, compulsive buying, and real debts. J. Econ. Psychol. 49, 141–149.

Alia, A., Narda, L., Boccanfuso, V., 2012. In: Translated and Edited by Lanci, Y., Vandeputte, T. (Eds.), Maurizio Lazzarato: Subverting the Debt Machine. Interview with Maurizio Lazzarato. http://thenewreader.org/Issues/1/SubvertingTheDebtMachine.

Alston, P., 2018. Statement on Visit to the United Kingdom, by Professor Philip Alston, United Nations Special Rapporteur on Extreme Poverty and Human Rights. United Nations, Office of the High Commissioner. Available from: https://www.ohchr.org/EN/News-Events/Pages/DisplayNews.aspx?NewsID=23881&LangID=E (accessed 01.02.2019).

Antoniades, A., 2018. Gazing into the abyss of indebted society: the social power of money and debt. Polit. Stud. Rev. https://doi.org/10.1177/1478929918757135.

Aron, J., Duca, J., Muellbauer, J., Kurata, K., Murphy, A., 2012. Credit, Housing Collateral and Consumption: Evidence from the UK, Japan and the US. University of Oxford, Department of Economics Discussion Paper Series.

Beder, S., 2001. Neoliberal Think Tanks and Free Market Environmentalism. University of Wollongong, Research Online.

Beggs, M., 2017. The state as a creature of money. New Polit. Econ. 22 (5), 463–477.

Billig, M., 2013. Learn to Write Badly: How to Succeed in the Social Sciences. Cambridge University Press, Cambridge.

Bourdieu, P., 1976. Outline of a Theory of Practice. Cambridge University Press, Cambridge.

Bourdieu, P., 1990. In Other Words: Essays towards a Reflexive Sociology. Translated by M. Adamson, Stanford University Press, Stanford, CA.

Bridges, S., Disney, R., 2004. Use of credit and arrears on debt among low-income families in the United Kingdom. Fiscal Stud. 25 (1), 1–25.

BRISKEE, 2018. Behavioural Response to Investment Risks in Energy Efficiency: EU-Funded Research Project in Horizon 2020. https://www.briskee-cheetah.eu/briskee/.

Broad, R., Cavanagh, J., 2000. The death of the Washington consensus? In: Bello, W., Bullard, N., Malhotra, K. (Eds.), Global Finance: New Thinking on Regulating Specula-tive Capital Markets. Zed Books, New York, pp. 83–95.

Carrette, J., King, R., 2005. Selling Spirituality: The Silent Takeover of Religion. Routledge, New York.

Chancel, L., Piketty, T., 2015. Carbon and Inequality: From Kyoto to Paris Trends in the Global Inequality of Carbon Emissions (1998–2013) & Prospects for an Equitable Adaptation Fund. Paris School of Economics. http://piketty.pse.ens.fr/files/ChancelPiketty2015.pdf.

Clegg, S., 2010. The state, power, and agency: missing in action in institutional theory? J. Manag. Inq. 19 (1), 4–13.

Cox, H., 1999. The market as god: living in the new dispensation. The Atlantic, March 1999, 1–12.

Di Muzio, T., Robbins, R., 2016. Debt as Power. Manchester University Press, Manchester.

Federal Reserve Board, 2018. Inequality in 3-D: Income, Consumption, and Wealth. Finance and Economics Discussion Series, Divisions of Research & Statistics and Monetary Affairs, Federal Reserve Board, Washington, DC.

Fernandez, R., Hofman, R., Aalbers, M., 2016. London and New York as a safe deposit box for the transnational wealth elite. Environ. Plan. A. 48, 2443–2461.

Foucault, M., 1982. The subject and power. Crit. Inq. 8 (4), 777–795.

Foucault, M., 1985. The Archaeology of Knowledge. Tavistock, London.

Galvin, R., 2013. Impediments to energy-efficient ventilation of German dwellings: a case study in Aachen. Energ. Buildings 56, 32–40.

Galvin, R., 2018a. 'Them and us': regional-national power-plays in the German energy transformation: a case study in lower Franconia. Energy Policy 113, 269–277.

Galvin, R., 2018b. Trouble at the end of the line: local activism and social acceptance in low carbon electricity transmission in lower Franconia, Germany. Energy Res. Soc. Sci. 38, 114–126.

Galvin, R., Sunikka-Blank, M., 2016. Schatzkian practice theory and energy consumption research: time for some philosophical spring cleaning? Energy Res. Soc. Sci. 22, 63–68.

Gaventa, J., 2003. Power After Lukes: An Overview of Theories of Power Since Lukes and Their Application to Development. Draft Paper, http://www.powercube.net/wp-content/uploads/2009/11/power_after_lukes.pdf.

Geels, F., 2014. Regime resistance against low-carbon transitions: introducing politics and power into the multi-level perspective. Theory Cult. Soc. 31 (5), 21–40.

George, S., 2000. A short history of neoliberalism: twenty years of elite economics and emerging opportunities for structural change. In: Bello, W., Bullard, N., Malhotra, K. (Eds.), Global Finance: New Thinking on Regulating Speculative Capital Markets. Zed Books, New York, pp. 27–35.

Giddens, A., 1979. Central Problems in Social Theory: Action, Structure and Contradiction in Social Analysis. MacMillan, London.

Giddens, A., 1984. The Constitution of Society. University of Los Angeles Press, Berkeley and Los Angeles.

Gould, A., Robert, M., 2013. The neoliberal pea and thimble trick: changing rhetoric of neoliberal champions across two periods of economic history and two hypotheses about why the message is less sanguine. Adv. Appl. Sociol. 3 (1), 79–84.

Graetz, M., Shapiro, I., 2005. Death by a Thousand Cuts: The Fight Over Inherited Wealth. Princeton University Press, Princeton.

Gram-Hanssen, K., 2010. Residential heat comfort practices: understanding users. Build. Res. Inf. 38 (2), 175–186.

Granovetter, M., 1985. Economic action and social structure: the problem of embeddedness. Am. J. Sociol. 91 (3), 481–510.

Ha, S., Krishnan, R., 2012. Predicting repayment of the credit card debt. Comput. Oper. Res. 39, 765–773.

Hargreaves, T., 2011. Practice-ing behavior change: applying social practice theory to pro-environmental behaviour change. J. Consum. Cult. 11 (1), 79–99.

Hargreaves, T., Nye, M., Burgess, J., 2010. Making energy visible: a qualitative field study of how householders interact with feedback from smart energy monitors. Energy Policy 38, 6111–6119.

Harré, R., 2002. Social reality and the myth of social structure. Eur. J. Soc. Theory 5 (1), 111–123.

Harrington, B., 2018. The bad behavior of the richest: what i learned from wealth managers. The Guardian. 19 October 2018, https://www.theguardian.com/us-news/2018/oct/19/billionaires-wealth-richest-income-inequality.

Hey, C., 2010. The German paradox: climate leader and green car laggard. In: Oberthür, S., Pallermaerts, M. (Eds.), The New Climate Politics of the European Union. Brussels University Press, Brussels, pp. 211–230.

Horta, A., Fonseca, S., Truninger, M., Nobre, N., Correia, A., 2016. Mobile phones, batteries and power consumption: an analysis of social practices in Portugal. Energy Res. Soc. Sci. 13, 15–23.

Huber, M., 2009. Energizing historical materialism: fossil fuels, space and the capitalist mode of production. Geoforum 40, 105–115.

Ingham, G., 1996a. Money is a social relation. Rev. Soc. Econ. 54 (4), 507–529. https://doi.org/10.1080/00346769600000031.

Ingham, G., 1996b. Review essay: the 'ne economic sociology'. Work Employ. Soc. 10 (3), 549–564.

Ingham, G., 1998. On the underdevelopment of the 'sociology of money'. Acta Sociol. 41, 3–18.

Ingham, G., 2004. The Nature of Money. Economic Sociology: European Electronic Newsletter, ISSN 1871-3351, vol. 5 (2). Max Planck Institute for the Study of Societies (MPIfG), Cologne, pp. 18–28.

Ingham, D., 2011. Capitalism. Revised Edition with Postscript, Polity Press, Cambridge.

Jacques, P., Dunlap, R., Freeman, M., 2008. The organisation of denial: conservative think tanks and environmental skepticism. Environ. Polit. 17 (3), 349–385.

Jorgenson, A., Schor, J., Huang, X., Fitzgerald, J., 2015. Income inequality and residential carbon emissions in the United States: a preliminary analysis. Hum. Ecol. Rev. 22 (1), 93–105.

Jorgenson, A., Schor, J., Knight, K., Huang, X., 2016. Domestic inequality and carbon emissions in comparative perspective. Sociol. Forum 31 (S1), 770–786.

Kurz, T., Gardner, B., Verplanken, B., Abraham, C., 2015. Habitual behaviors or patterns of practice? Explaining and changing repetitive climate-relevant actions. WIREs Clim. Change 6, 113–128.

Laclau, E., Mouffe, C., 2014. Hegemony and Socialist Strategy: Towards a Radical Democratic Politics, second ed. Verso, London.

Lazzarato, M., 2015. Governing by Debt. Semiotext(e), South Pasadena, CA.

Lehmann, J., 1993. Deconstructing Durkheim: A Post-Structuralist Critique. Routledge, London.

Malm, A., 2013. The origins of fossil capital: from water to steam in the British cotton industry. Hist. Mater. 21 (1), 15–68.

McDonagh, E., 2015. Ripples from the first wave: the monarchical origins of the welfare state. Perspect. Polit. 13 (4), 992–1016.

Meckling, J., Nahm, J., 2018. When do states disrupt industries? Electric cars and the politics of innovation. Rev. Int. Polit. Econ. 25 (4), 505–529.

Melzer, B., 2017. Mortgage debt overhang: reduced investment by homeowners at risk of default. J. Financ. 72 (2), 575–612.

Miraftab, M., 2004. Public-private partnerships: the Trojan horse of neoliberal development? J. Plan. Educ. Res. 24, 89–101.

Morgan, S., Gonzales, L., 2008. The neoliberal American dream as daydream: counter-hegemonic perspectives on welfare restructuring in the United States. Crit. Anthropol. 28 (2), 219–236.

Newell, O., Phillips, J., 2016. Neoliberal energy transitions in the south: Kenyan experiences. Geoforum 74, 39–48.

O'Loughlin, D., Szmigin, I., 2006. "I'll always be in debt": Irish and UK student behaviour in a credit led environment. J. Consum. Mark. 23 (6), 335–343.

Piketty, T., 2014. Capital in the Twenty-First Century (Translated from the French by Arthur Goldhammer). Belknapp-Harvard University Press, Cambridge, MA.

Ratti, C., Baker, N., Steemers, K., 2005. Energy consumption and urban texture. Energ. Buildings 37 (7), 762–776.

Rehm, P., 2011. Social policy by popular demand. World Polit. 63 (2), 271–299.

Rehm, P., 2016. Risk Inequality and Welfare States: Social Policy Preferences, Development, and Dynamics. Cambridge University Press, Cambridge.

Rieger, J., 2009. No Rising Tide: Theology, Economics, and the Future. Fortress Press, Minneapolis, MN.

Rogers-Vaughn, B., 2013. Pastoral counseling in the neoliberal age: hello best practices, goodbye theology. In: Sacred Spaces: The E-Journal of the American Association of Pastoral Counselors. vol. 5. American Association of Pastoral Counselors, pp. 5–45.

Røpke, I., 2009. Theories of practice—new inspiration for ecological economic studies on consumption. Ecol. Econ. 68, 2490–2497.

Ross, A., 2014. Nine Arguments for Debt Refusal. Strike Debt. February 7, http://strikedebt.org/nine-arguments/.

Sahr, A., 2015. Wären wir die besseren Banken?: Zur Debatte um die Repolitisierung des Kreditgeldes. Widerspruch: Beiträge zu sozialistischer Politik 66, 103–113. Special issue Finanzmacht-Geldpolitik.

Schatzki, T., 1996. Social Practices: A Wittgensteinian Approach to Human Activity and the Social. Cambridge University Press, New York, NY.

Schatzki, T., 1997. Practices and actions: a Wittgensteinian critique of Bourdieu and Giddens. Philos. Soc. Sci. 27 (3), 293–308.

Schatzki, T., 2002. The Site of the Social: A Philosophical Account of the Constitution of Social Life and Change. Pennsylvania State University Press, University Park, PA.

Schatzki, T., 2010. Materiality and social life. Nat. Cult. 5 (2), 123–149.

Bullion, S.D., 2008. Where does money come from? Available from: https://sdbullion.com/blog/where-does-money-come-from/.

Sewell, W., 1989. Toward a Theory of Structure: Duality, Agency, and Transformation. Center for Research on Social Organization. Working Paper Series 29. (Accessed 15 May, 2019).

Shove, E., 2012. Putting practice into policy: reconfiguring questions of consumption and climate change. Contemp. Soc. Sci. 9 (4), 415–429.

Shove, E., Pantzar, M., 2005. Consumers, producers and practices understanding the invention and reinvention of Nordic walking. J. Consum. Cult. 5 (1), 43–64.

Shove, E., Walker, G., 2014. What is energy for? Social practice and energy demand. Theory Cult. Soc. 31 (5), 41–58.

Shove, E., Pantzar, M., Watson, M., 2012. The Dynamics of Social Practice: Everyday Life and How It Changes. Sage, London.

Simcock, N., Mullen, C., 2016. Energy demand for everyday mobility and domestic life: exploring the justice implications. Energy Res. Soc. Sci. 18, 1–6.

Soederberg, S., 2014. Debtfare States and the Poverty Industry: Money, Discipline and the Surplus Population. Routledge, London.

Stephenson, J., Barton, B., Carrington, G., Gnoth, D., Lawson, R., Thorsnes, P., 2010. Energy cultures: a framework for understanding energy behaviours. Energy Policy 38, 6120–6129.

Sung, J., 2011. The Subject, Capitalism, and Religion: Horizons of Hope in Complex Societies. Palgrave Macmillan, New York, NY.

Thistlethwaite, S., 2010. Dreaming of Eden: American Religion and Politics in a Wired World. Palgrave Macmillan, New York.

Walker, G., Day, R., 2012. Fuel poverty as injustice: integrating distribution, recognition and procedure in the struggle for affordable warmth. Energy Policy 49, 69–75.

Walks, A., 2018. Driving the poor into debt? Automobile loans, transport disadvantage, and automobile dependence. Transp. Policy 65, 137–149.

Waquant, L., 2009. Punishing the Poor: The Neoliberal Government of Social Insecurity. Duke University Press, Durham and London.

Winters, J., 2011. Oligarchy. Cambridge University Press, New York.

Winters, J., 2014. Wealth defense and the limits of liberal democracy. In: Paper for Annual Conference of the American Society of Political and Legal Philosophy, Washington, DC, August 28–31, 2014. Revised 29 April 2015.

Winters, J., 2017. Wealth defense and the complicity of liberal democracy. In: Knight, J., Schwartzberg, M. (Eds.), Wealth: NOMOS LVIII. In: A Special Issue of the American Society for Political and Legal Philosophy, NYU Press, pp. 158–225.

Wittgenstein, L., 1953. Philosophical Investigations (G.E.M. Anscombe, Trans.). Blackwell, Oxford.

Zalta, E. (Ed.), 2015. Methodological Individualism. Stanford Encyclopedia of Philosophy. Center for the Study of Language and Information (CSLI), Stanford University.

Zucman, G., 2014. Taxing across borders: tracking personal wealth and corporate profits. J. Econ. Perspect. 28 (4), 121–148.

Zucman, G., 2015. The Hidden Wealth of Nations: The Scourge of Tax Havens. University of Chicago Press, Chicago.

Further reading

Bridges, S., Disney, R., Henley, A., 2004. Housing wealth and the accumulation of financial debt: evidence from UK households. In: Bertola, G., Disney, R., Grant, C. (Eds.), The Economics of Consumer Credit. MIT Press.

Forrest, R., Hirayama, Y., 2009. The uneven impact of neoliberalism on housing opportunities. Int. J. Urban Reg. Res. 33 (4), 998–1013.

Giddens, A., 1990. The Consequences of Modernity. Polity Press, Cambridge.

Shove, E., 2017. What is wrong with energy efficiency? Build. Res. Inf. https://doi.org/10.1080/09613218.2017.1361746.

Economic inequality, energy justice and the meaning of life ☆

4

Ray Galvin

University of Cambridge, Cambridge, United Kingdom
RWTH Aachen University, Aachen, Germany

Chapter outline

1. Introduction 75
2. The energy justice movement to date 78
3. Justice, morals and Wittgenstein's reflections 80
4. Justice as fairness—Rawls, Rorty and pragmatism 85
5. Extending justice globally 87
6. Implications for global energy justice 89
7. Energy justice and economic inequality 91
8. Conclusion 92
References 93

1 Introduction

An interesting development in social science based energy research over the past decade has been the growing body of literature on "energy justice." Broadly speaking, this is an attempt to provide an ethical framework for discussing issues of fairness between people, on both local and global scales, in relation to energy supply, production and consumption.

The energy justice literature represents a consensus among an increasing number of energy researchers, mostly from social science backgrounds, that there needs to be a joined-up ethical framework for bringing the findings of energy research into the policy realm. For example, in one of the most recent summaries of the energy justice movement to date, McCauley et al. (2019), p. 619 maintain that the global transition toward sustainable, renewable energy systems "must take into consideration

---

☆ This chapter is based partly on my article "What does it mean to make a moral claim? A Wittgensteinian approach to energy justice," published in Energy Research & Social Science.

Inequality and Energy. https://doi.org/10.1016/B978-0-12-817674-0.00004-7

questions of energy justice to ensure that policies, plans and programs guarantee fair and equitable access to resources and technologies."

These researchers see themselves as not just presenting policymakers with facts and figures about society and energy that emerge from their research, but as combining these facts and figures with a systematic framework of reflections on their ethical implications.

The topic of energy justice is directly relevant to issues of economic inequality. A thread running through the energy justice literature is that justice, in relation to energy supply and consumption, has three main dimensions: "*distributional* justice, justice as *recognition* and *procedural* justice" (Jenkins et al., 2017, italics added). This thread is present in the earliest energy justice publications, such as Eames (2011), the most recent, such as McCauley et al. (2019) and almost every energy justice study in between (see reviews in Jenkins et al., 2016; Jenkins et al., 2017, p. 631ff and Bartiaux et al., 2018, p. 1221ff; and the works cited in this chapter).

"*Procedural justice*" is concerned with how decisions about energy supply, planning, transition, production and access are made. It approves of procedures that enable all affected parties to participate in decision making.

"*Recognition justice*" is concerned to afford "cultural and political respect and recognition of vulnerable and marginalized social groups" in relation to energy issues (Bartiaux et al., 2018) and avoid bypassing certain groups who are relatively powerless or appear to stand in the way of what others define as progress.

These aspects of energy justice thinking are relevant to issues of economic inequality. For example, low income groups are least likely to have the means to participate in public consultations about the future of energy systems, and certain immigrant groups are more likely to be poor and powerless due to economic inequality. The specific energy needs of some groups, such as single mothers, who are often deeply preoccupied with child care and are over-represented among the fuel poor, can easily get overlooked in energy planning—as Minna Sunikka-Blank discusses in Chapter 8.

"*Distributional justice*" is about the distribution of costs and benefits of energy supply and consumption—who pays, who can afford to pay, what benefits they get and what disadvantages they suffer. Distributional justice has the most direct relevance to issues of economic inequality. It has a *spatial* dimension, in that people in certain geographical regions or districts may get far greater or lesser benefits from energy than others (Bouzarovski and Simcock, 2017). It has a *temporal* dimension, in that energy for future societies may be much scarcer or more expensive than energy today (Jones et al., 2015, p. 162ff; Sovacool and Dworkin, 2014, p. 288ff). And it has a general *social* dimension in that, in any given society, certain classes or groups of people have greater access to much-needed energy services than other classes or groups, mostly due to economic inequalities (Bartiaux et al., 2018; Walker et al., 2016).

Considering this "distributional" side of the energy justice concept, it would seem that the notion of energy justice could provide a strong intellectual framing for a critique of economic inequality. A clear example is seen with the issue of fuel poverty (or energy poverty). The energy justice studies that investigate fuel poverty argue it is caused by one or more of three factors: households having low income; dwellings being energy inefficient; and fuel prices being high (e.g., Bouzarovski and Simcock, 2017; Sovacool, 2015; Walker and Day, 2012). It is a simple step of logic to argue that, therefore, there would be a much higher level of energy justice in a society if far fewer of its households had low incomes—much like, for example, in Danish society today. Since low incomes are strongly associated with economic inequality (see Chapter 1), it would follow that mitigating economic inequality would tend to reduce energy poverty and serve to increase the level of energy justice.

The same conclusion could be arrived at from a different angle. Large-scale surveys in Europe indicate that poorer households are less likely to be able to increase the energy efficiency of their dwellings and appliances, than households with higher incomes (BRISKEE, 2018). Modelling of survey results suggests up to 10% of potential energy savings could be lost, by 2030, if households continue to be too poor to implement basic energy efficiency measures. It would seem, then, that a reduction in economic inequality (resulting in far fewer low-income households) would serve the ends of energy justice in two ways: by enabling more households to afford the energy to warm their homes in winter; and by reducing CO2 emissions through energy efficiency measures, thereby supporting international efforts to mitigate climate change.

The energy justice movement is relatively new in academia and its concerns are very broad and global, so it is not surprising that it has not yet worked through the specific implications of energy justice for mitigating economic inequality. However, if the energy justice concept gains more adherents and, more to the point of this chapter, if it increases the robustness and credibility of its claims, this could be a very fruitful area for ongoing research.

I make this last point advisedly because it seems, from my reading of energy justice literature, that it has yet to deal with a key issue of credibility in its basic claim, namely, what it means to hold a moral or ethical[a] position, or to invoke the authority of a moral or ethical rule or principle, such as justice. When we ask policymakers or business leaders to pursue a particular course of action regarding some aspect of energy supply, production or consumption *because justice demands it*, what precisely are we saying?

The notion of "justice" is fraught with conundrums. To claim something is "just" or "unjust" is to make a very strong claim that goes way beyond saying one finds it distasteful, upsetting, infuriating or unpalatable. It is more like claiming there is some kind of rightness or wrongness about it, which any "reasonable person" (to use moral philosopher John Rawls' phrase) would recognize. To demand politicians and business leaders stop a course of action because it is "unjust" is to ask them, implicitly, to accept one's moral framing as universally fair and reasonable or even, perhaps, as some kind of universal truth.

---

[a] I use the words "moral" and "ethical" interchangeably in this paper.

I do not think the energy justice literature has yet presented a strong enough case to address policymakers in this way. Indeed, what is probably the most overtly philosophical energy justice paper do date says its aim is "not to have readers ponder what justice is" (Jones et al., 2015, p. 147). Yet at the same time its authors want to "persuade energy analysts, business leaders, politicians, utility managers, and energy consumers to consider or reconsider the ethical dimensions of our current energy system." It then outlines its view of what these ethical dimensions are. This could seem a little unfair on these leaders. They are being asked to take action on issues based on ethical claims, but not being told how these ethical claims are grounded, what makes them credible, what makes them different from merely the authors' preferences and wants. Unless we "ponder what justice is," and whether it even makes sense to speak of moral values like justice, and how we could support a claim that one person's understanding of "justice" is better than someone else's, it is difficult to see how we could convince politicians and other leaders to accept our conclusions.

In this chapter, I want to make a modest attempt to move this discussion forward. I want ask what we are doing when we claim something is just or unjust, morally right or wrong, ethically good or bad. I will attempt to show that we have to be very clear about what we are doing when we make such claims and that our "moral" case might not be as strong as we assume. I will argue it is unwise (as well as intellectually questionable) to claim or imply there is some objective, universal standard of what is "just" and "unjust" and that our studies of energy justice are applying this objective standard to energy issues. Nevertheless, I will argue that, if we work toward a clearer understanding of what we are doing when making moral claims (such as that this or that action is just or unjust), we may be more successful in winning the support of politicians, other leaders and the public.

In Section 2 I offer a brief overview of the energy justice movement and the main energy-related themes it has so far investigated. This section is not essential to my argument; it is merely to help readers who are not familiar with energy justice literature to gain a general overview of it. The real argument begins in Section 3, where I draw on Wittgenstein's later work to scrutinize ideas of "justice" and "ethics". I then apply this to Rawls' philosophy of justice in Section 4. In Section 5 I draw on Rorty and Walzer to suggest how this Wittgensteinian approach can be extended globally. In Section 6 I discuss the implications of these ideas for the energy justice approach, and relate this to economic inequality issues in Section 7. In Section 8 I offer conclusions.

2 The energy justice movement to date

Prior to the emergence of the energy justice movement in academia there were already two broadly based, globally oriented "justice" movements in energy research literature: "climate justice" (Pettit, 2004; and see review in Schlosberg and Collins, 2014) and "environmental justice" (Lester et al., 2001; Schlosberg, 2009). Academic energy justice literature acknowledges its debt to these movements, but has also noted their limitations for dealing with energy issues (Jenkins, 2018; McCauley and Heffron, 2018).

The phrase "energy justice" first appeared among NGOs and citizens' groups early in the 21st century, providing inspiration for a number of scholars to bring the concept into the academic orbit. Papers exploring the notion were presented at a seminar entitled "Energy justice in a changing climate" at the InCluESEV (Interdisciplinary Cluster on Energy Systems, Equity and Vulnerability) conference in London in November 2011 (Eames, 2011; Saunders, 2011; Skea, 2011). In one of these, Eames (2011) explored how issues of justice arise in the transition toward renewable and sustainable energy systems. He asked questions such as: How is the social distribution of costs and benefits of different energy-related technological options evaluated? Whose vision—knowledge interests and values—will drive the energy transition? Who will participate? How will access to information work? He concluded: "The conceptual frameworks and policy oriented tools of sustainability transitions research need to incorporate more explicit consideration of (distributive, procedural & epistemic) justice," and added, "But then so do many more established fields of energy research."

This could be seen as a challenge to energy researchers to bring considerations of justice more explicitly into the center of their research. Issues of equity and fairness had been regularly arising in the energy sphere, many of which are more specific to energy than to environment or climate more generally. Yet up until a decade ago, it seemed, energy research had hardly begun to consider these systematically, and certainly not under a common ethical rubric such as "justice".

The earliest published academic paper using the phrase "energy justice" appears to have been Goldthau and Sovacool (2012), which included a subsection discussing disparities of access to and costs of energy, and externalities in energy supply. It argued these need to be addressed under the heading of energy justice. One of the first academic papers to offer a comprehensive account of the theme of energy justice, which appeared in 2013, challenged "researchers to address *justice*-based concerns within *energy* systems, from production to consumption" (McCauley et al., 2013, emphasis in original). In the same year Hall (2013) offered a study linking the "emerging concept of energy justice" with already existing literature dealing with ethical consumption issues. Meanwhile Walker and others had begun relating issues of fuel poverty to the notion of justice (Walker and Day, 2012).

Since 2013 studies on energy justice have proliferated. This has included special issues on the topic in the journals Energy Policy (editorial by Jenkins et al., 2017) and Applied Energy (editorial by McCauley et al., 2019), a special issue in the journal Energy Research and Social Science on transport and energy justice (editorial by Simcock and Mullen, 2016), and articles in that journal monitoring and assessing the progress of the energy justice movement (Jenkins, 2018; Jenkins et al., 2016).

A large number of energy justice studies have the form of global overviews of ethical implications of the interconnectedness of energy systems and their effects across the planet (e.g., LaBelle, 2017; Goldthau and Sovacool, 2012; Sovacool and Dworkin, 2014; Sovacool et al., 2016). Some explore energy justice principles for specific countries or territories such as the US (Hernández, 2015), Chile (Alvial-Palavicino and Ureta, 2017) and the UK (Chatterton et al., 2016; Walker et al., 2016) and the Arctic region (Sidortsov and Sovacool, 2015).

Studies on energy justice have also emerged in relation to specific themes in energy research. One of the most common is energy poverty (also called fuel poverty). Day et al. (2016) explored energy justice in relation to energy poverty in general, while Berry et al. (2016) did so in relation to the transport sector. Energy justice themes have been employed to investigate energy poverty in homes (Bartiaux et al., 2018; Gillard et al., 2017; Sovacool, 2015; Walker and Day, 2012), including the spatial-geographic dimensions of energy poverty (Bouzarovski and Simcock, 2017) and its racial dimensions (Reames, 2016).

Energy justice motifs have also been used to explore justice issues that arise in global and national transitions to sustainable energy production and supply (Healy and Barry, 2017; Miller et al., 2015; Sareen and Haarstad, 2018) and in specific large scale energy projects (Pesch et al., 2017). Closely related to this is the transition to renewable energy. Justice issues in this sphere are explored at household level in relation to photovoltaic policies, by Poruschi and Ambrey (2019), and in relation to the phase-out of nuclear power, by Rehner and McCauley (2016).

Energy justice has also been used as a conceptual framework for critiquing inadequate policy relating to cycling as an alternative to cars (Smith, 2016) and policies for car sharing (Cheyne and Imran, 2016). There are also studies calling for a more interdisciplinary focus in energy justice studies (Heffron and McCauley, 2017; McCauley et al., 2019).

Hence the theme of energy justice has been applied to a very wide range of issues in energy production, supply and consumption, and this is continually expanding. But what of the more fundamental issue of what these authors mean when they say something is just or unjust? What is energy "justice", and how can such a notion be supported robustly?

3 Justice, morals and Wittgenstein's reflections

The notion of justice has a deep and complex history in moral philosophy. Several energy justice studies discuss or touch on aspects of energy justice that relate to this tradition (Jenkins et al., 2016; Jones et al., 2015; Sovacool et al., 2017; Sovacool and Dworkin, 2015), but this side of the energy justice literature is not yet well-developed. Most of these studies are eclectic rather than engaging with academic moral philosophy, i.e., they draw upon, and pick and choose from, a wide range of different ideas about justice, mostly from within western thought but to some extent also from different cultures and traditions (e.g., McCauley et al., 2019; Sovacool and Dworkin, 2014). It is probably fair to say that the theory of justice developed by US moral philosopher John Rawls (1921–2002) in his 1971 essay "Justice as fairness" (Rawls, 1971) is the favorite and most frequently referred to source of philosophical support for the energy justice frameworks being put forward.

Despite their acknowledgement of non-western and indigenous ethical systems, in the end most of these frameworks fall back on Rawlsian reasoning to decide what counts as justice. I will discuss Rawls' approach in more detail below.

Because energy justice literature does not overtly explore what it means to make a moral claim, we are left to assume or imagine what is in the authors' minds when they do this. There are two opposite poles of understanding as to what a moral claim is. At one end of the spectrum, a moral claim is seen as a statement about rules or laws or imperatives of moral right and wrong, which exist in and of themselves in some reified metaphysical realm and can be discovered and explored by rational thinking beings and then applied to specific empirical situations. This position is represented by post-enlightenment thinkers such as Emmanuel Kant (1724–1804), Arthur Schopenhauer (1788–1860), and more recently by transcendental realists such as Roy Baskhar (1944–2014) and the early writings of John Rawls. For brevity I call this the "rational metaphysical" approach.

At the other end of the spectrum is the "pragmatist" view, that moral claims are really nothing more than statements about human affect (feelings, wants, preferences), perhaps most overtly represented by early US pragmatists Dewey (1920, 1934) and James (1907) and in recent times by neopragmatists such as Rorty (1997) and Price (2011) (see discussion in Pihlström (2005, 2012)). This view is often also called "deconstructionist" (Mouffe, 2000; Pihlström, 2012), though this can be confusing due to other usages of that word.

It seems clear that energy justice authors are not generally putting forward a pragmatist view, as they claim far more for their findings on justice than that these are merely their own personal preferences or feelings. On the other hand, although a lot of the language they use resonates with a rational metaphysical approach and they draw heavily on Rawls, they do not overtly claim moral laws exist in and of themselves waiting to be discovered by rational thinkers.

No doubt many energy justice advocates would be comfortable with something between these two poles. Such an approach—or more accurately, a family of approaches—has been developed by scholars informed by Ludwig Wittgenstein's (1953, 1965, 1969) linguistic philosophy, such as Mouffe (2000), Pihlström (2012), Lovibond (1983) and others. I will attempt to expound this approach, but will also draw on features of pragmatist literature to fill this out in ways which, I suggest, can make it more applicable to a global ethic such as energy justice. I introduce these concepts by way of a practical example.

In Cambridge UK, the city where I live, which has a booming economy, there are thousands of households living in energy poverty, some of whom I know personally. Suppose I say to our Member of Parliament, "It is morally wrong, reprehensible, that these households have to live in energy poverty, in our city, especially since it is such a wealthy city," the following statement is *not* what I mean:

(a) "There is a moral law which decrees it is wrong for some households to have to live in energy poverty in a wealthy city, and I am reminding you of this law (or at least attempting to make you aware of it)."

Nor am I *merely* saying:

(b) "I strongly dislike the fact that some households have to live in energy poverty in our city, a wealthy city."

Instead, what I am saying is:

(c) "Moral commitments and beliefs are among the central things that give meaning to our lives as human beings. One of my such commitments and beliefs is that households should not have to live in energy poverty in a community where there is abundant wealth, and it troubles me especially that this is occurring in my own community and among people I personally know. I know you to be a morally mature person so I would think that, if you know the facts of the matter, you will be motivated to act to correct it."

Statement (c) includes three important features. Firstly, it grounds my moral belief and commitment in the *meaning of my life as a human being*, rather than in an abstract, universal moral law existing in a reified metaphysical realm. Secondly, it indicates this moral commitment pertains most strongly in respect of people *in the community I see myself as belonging to*. Thirdly, it makes an appeal to the person I am trying to influence, based on my recognition and affirmation of his[b] own moral maturity.

I will cover all three of these points in the course of the ensuing discussion.

The first point arises out of Wittgensteinian scholars' discussion on Wittgenstein's reflections on moral statements. Wittgenstein can be fiendishly difficult to understand, but one of his most delightfully readable lectures, published posthumously but given in the 1930s, lays the groundwork for this idea (Wittgenstein, 1965; for a detailed commentary on this see Redpath, 1972). In a series of simple but devastating logical steps, Wittgenstein argues that moral statements cannot refer to real, independently existing entities, but instead are expressions of an aspect of human life which, he says, "I personally cannot help respecting deeply and I would not for my life ridicule it."

Wittgenstein later developed his understanding of what human beings are doing when they make philosophical statements about things that are purported to exist in reified, abstract realms, and a number of scholars have systematically explored what this means for moral statements (see, e.g., the essays by Diamond, Garver and Stroud in Sluga and Stern, 1996). The core of Wittgenstein's argument is that, although there are no such things as moral laws existing in and of themselves with power to make imperative claims on us, people's discourse about, and commitment to, moral values are nevertheless meaningful. As Wittgensteinian scholar Savina Lovibond[c] explains:

"… it is only in so far as (the individual) can conceptualize his life in the terms laid down by some real system of moral intuitions that the individual will succeed in

---

[b] The Member of Parliament for Cambridge at the time of writing is male.

[c] For a critical discussion of aspects of Lovibond's interpretation of Wittgenstein (see Diamond, 1996). I interpret Lovibond's use of the term "real system" loosely here rather than as system of absolute moral demands laid upon a person from some reified realm.

finding a meaning in life as a whole." (Lovibond, 1983, p. 223, and see Pihlström's, 2005 comments on this).

In other words, moral commitments and beliefs go toward giving meaning to one's life and are thereby a central part of being human. A woman or man without a moral compass is a disintegrated being, a being who cannot participate effectively in the practices of her or his community. This has countless everyday forms, many of which are so mundane as to hardly be noticed, such as how we take turns in everyday conversation, how we buy and sell things, how we listen when someone tells us their fears or hopes, how we know when to physically touch another person. It also has very strong social-critical dimensions, such as how we learn our government is acting worthily or not. A striking example comes from the ancient Hebrew prophet Jeremiah. When Jeremiah was told by the king to keep quiet about the extremes of wealth and poverty the king's policies were producing, he replied that he simply had to speak out, because:

> *There is in my heart, as it were, a burning fire, shut up in my bones*
> *And I am weary with holding it in*
> *And I cannot. (Jeremiah 20:9)*

Jeremiah saw that he would disintegrate as a human being if he did not stay true to his moral commitments. Having moral commitments, and having meaning in one's life, go hand in hand.

Moral statements, then, are neither proclamations of moral truths existing in a reified universal realm, nor simply statements of personal feeling. Scholars informed by Wittgenstein's insights argue there is a very important middle ground here, between these two extremes. To begin with, apart from in his very early writings, Wittgenstein was at pains to emphasize that moral statements are meaningful in their own terms. Unlike the so-called logical positivists of the 1920s, Wittgenstein argued that a statement does not have to be backed up by a physical or metaphysical reality, in order to be meaningful. Instead, he argued, different kinds of statements belong to different *language games*. The statement, "It is wrong to make Cantabrians live in energy poverty" is meaningful, even though it is a different *"language game"* from utterances like "Gravity causes light to bend"; or "Hide!"; or "I name thee the SS Titanic!" or "Grandma's ghost is in the house again". All are meaningful, but belong within different language games.

Further, a language game is a *social* practice. Communities of people engage in it together, where they produce and share meanings, not just by talking, but by their everyday actions. Moral commitments and beliefs emerge out of, and are reproduced and modified in, the actions and discussions of communities of people doing everyday things.

In her comments on Wittgenstein, Mouffe (2000) points out that notions like "justice" emerge out of the many different overlapping, intertwining, and sometimes contradictory language games (including the practices associated with them) within each particular community. Different types of language game have what Wittgenstein called "family resemblances": there is a family of practice-based language games that involve, depend on, produce, reproduce and modify a community's moral commitments and beliefs. These practices are often characterized by involving explicit or implicit use of moral-type notions, such as obligation, duty, rights, responsibilities, fairness, good and bad, evil, pernicious, and so on (Searle, 1964; Pihlström, 2005). However, as Diamond (1996) notes, morality is often implied in human speech even without explicit use of such words.

Mouffe (2000) points out that an individual's moral commitments, including her notion of justice, may develop characteristics unique to her, but overall it is formed within and reflects the mix of moral or justice-related ideas and practices in her community. It is also likely to contain many inconsistencies, because it is not based on a straightforward logical process but on the "hurly-burly" (to use another of Wittgenstein's phrases) of real life in a real social community.

Further, a person's moral framework may be substantially modified as she moves from one community or life situation to another and as she thinks critically about it, but it is grounded in her *learning to be a human being*, in a *meaningful way*, through participating in societal practices.

Returning to Lovibond's point about the meaning of one's life, we would say individuals' lives are meaningful at least partly because of the moral commitments and beliefs they absorb and contribute to within their community (or communities, since many people today belong within a number of different communities at the same time). In a further step we might say that a community (such as the City of Cambridge) derives its meaningfulness, in part, from its moral commitments—such as (hopefully!) that no Cambridge household should have to live in energy poverty.

It should also be noted that someone in a different epoch or culture might have quite different moral priorities, some of which may seem strange to a British person. For example, a Maori person might say, "It is morally wrong, reprehensible, that these *manuhiri* (foreigners or people from a different district) are coming to live in our village (or work in our firm or institution) without being given a *powhiri* (a formal, ceremonial welcome in which the *manuhiri* are grafted into the local community with full rights, responsibilities and feelings of belonging)." The moral issue is very different but the human process is the same. People express a sense of moral outrage; they may use language like 'This is morally repugnant,' or 'This is unjust,' and in doing so they are being true to their humanness, to key aspects of the meaning of their life.

Much of the energy justice literature is clearly aware of the different moral frameworks that can develop in different cultures, and some studies mention particular non-western moral frameworks and how these could throw light on various 'energy justice' issues (e.g., Sovacool and Dworkin, 2014; Jones et al., 2015). Nevertheless, there is a tendency in the energy justice literature to fall back on the moral philosophy of John Rawls. This not only subjects non-western views on justice to the criteria of a western approach. It also reproduces Rawls' residual attachment to a rational metaphysical approach to justice. I will therefore discuss Rawls' approach at some length.

4 **Justice as fairness—Rawls, Rorty and pragmatism**

Rawls attempted to bridge the gap between a rational, universalist grounding for the ethic of justice, and a practice-based, community-derived understanding of justice. Rawls overtly agreed with critics of the rational metaphysical approach who argue that mainstream western moral and political philosophers had wrongly grounded their understanding of ethics on the enlightenment notion of universal rational truth—an edifice of moral truth standing on its own merits waiting to be discovered. When religion ceased to be universally accepted as a basis on which to build moral values, western philosophers of the enlightenment replaced this with the idea that universal truth could be known through clear-headed reasoning. They argued there are universal truths about what is right and good for human beings, which apply to all human beings everywhere, regardless of culture or historical setting, and that these can form the basis of an ethic of justice. This is perhaps most overtly seen in Immanuel Kant's "categorical imperative" which is frequently cited in energy justice literature and which, in a nutshell, says it is ethical to perform only those actions which can be universalized, i.e., which all human beings everywhere could perform without compromising others' wellbeing. Further, the Kantian approach envisages a social contract, in which, first, the meaning of "justice" is established by rational reflection (presumably by post-enlightenment philosophers); then people are required to conform to that definition if they want to be members of society.

Rawls' approach turned this the other way round. In his words:

> Rather, the guiding idea is that the principles of justice for the basic structure of society are the object of the original agreement. (Rawls, 1971, p. 664, emphasis added)

In other words, in logical sequence, the potential members of society *first* decide among themselves what justice is, *then* enter into a social contract to be members of that society *on that basis.*

This appears to remove the deliberation and decision-making on what justice is, from philosophers and their fixation on rational universal principles, and hand it back to the people who will have to live in the society so devised. On the surface this seems to be fully context-situated and historically contingent.

A difficulty, however, comes with the *process* by which these potential members of society perform their deliberations. For Rawls this must be strictly controlled. Firstly, as people discuss their future preferred type of society they do not know where they would be placed socially, economically, or in terms of their abilities, assets, psychological characteristics or circumstances, in the society they would be joining. Their deliberations must take place behind a "veil of ignorance" of future personal situations. Rawls concludes that the kind of society such a group devises would be such as to enable every individual to have her or his basic needs met and to have routes to happiness and fulfilment. In such a society, he claimed, inequalities would only be tolerated as long as they did not reduce the wellbeing of the poorest and weakest members of society. In this sense it would be a just society, where "justice is fairness" (another phrase found in most energy justice studies), even though it could be an unequal society in terms of assets and opportunities.

Secondly, and most importantly for Rawls, the deliberations would be structured such that each person would have fair and equal participation in the discussion. This is much the same situation e.g., Habermas (1993) envisages with his "communicative reason," which involves a process of carefully managed discussion between people on an equal footing to deliberate on moral and political issues.

Both Rawls and Habermas strive to acknowledge that moral philosophy today takes place in a post-metaphysical age, where we cannot assume that rational deliberations will identify already-existing truth as to what is good and right for human beings. Instead they claim their approaches are contextually, historically situated. Nevertheless, Rawls (1971) maintains that, no matter where and in what historical epoch these deliberations take place, their outcome would be a society very much like a western liberal democracy.

The difficulty with this approach, as Rorty (1997), Mouffe (2000), Walzer (1994) and others point out, is that the discussions people would engage in *in order to devise their just society*, with all the caveats and rules Rawls applies to these discussions, are very much an ideal of western liberal democratic society in the first place. As Rawls (1971, p. 654) explains, the protagonists "would accept an initial position of equality as defining the fundamental terms of their association." In other words, in order to begin their discussions on how they should associate together as a society, they would first have to associate together as children of the (western) enlightenment, steeped in notions of equal rights, turn-taking and listening without prejudgement. The process Rawls is advocating would have to take place in a western liberal democratic society, or one very much like it, if it were to *produce* a society very much like a western liberal democracy.

In his later writings Rawls (1993a,b) appears to acknowledge the difficulty of starting with western-type rules of discussion, but argues his notion of justice can nevertheless be extended to non-liberal societies provided they are "reasonable peoples." He defines "reasonable" in a number of passages, equating it with impartially protecting "the fundamental interests of all members of society" (Rawls, 1993a, p. 61): the right to life, freedom from slavery and forced labor, protection of private property, treating all juridical cases equally (Rawls, 1993a, p. 62); and "a measure of liberty of conscience and freedom of thought" (Rawls, 1993a, p. 62).

Commenting on these claims, Rorty (1997) notes that:

> *Rawls's notion of what is reasonable, in short, confines membership of the society of peoples to societies whose institutions encompass most of the hard-won achievements of the West in the two centuries since the Enlightenment. (Rorty, 1997, p. 143)*

In short, western liberal democratic assumptions about justice are built in to Rawls' approach in both his early and later work, and he can only apply it globally by implicitly claiming these assumptions are universally ethically correct. We may well applaud and agree with western notions of justice, and may well want these to apply to all peoples everywhere, but that is different from claiming they are morally right and that therefore all peoples are morally obliged to adhere to them.

There is, also, a far deeper problem with Rawls' approach. The idea that a group of people with *no initial moral commitments* who discuss and decide upon the kind of society they want, would then automatically be *morally* committed to it, shows a mistaken understanding of human moral commitment. A post-Wittgensteinian would say this is simply not the way humans come to have moral commitments. They do so by everyday participation in *practices*, which include acting and conversing, which begin in childhood, are very haphazard and are directly related to concrete situations people find themselves in. If there were such a thing as a group of individuals with completely blank moral slates participating in a Rawlsian-type discussion, there is no reason at all to think they would end up with any kind of *moral* commitment to the kind of society they had just devised.

We return, then, to the post-Wittgensteinian notion that moral commitments and beliefs are basic to the meaning of human life. A person has to *have* a moral compass in order to be able to participate meaningfully in deliberations about how people should live justly. She develops this moral compass *only* by growing up in a community (or communities) where moral commitments are held and practiced and talked about. Only moral beings can participate meaningfully in moral deliberations.

5 Extending justice globally

Returning to the Cambridge energy poverty example, we can now consider the second issue in this. The reader will recall that in that example I said that Cambridge, where there are thousands of households in fuel poverty, is my home town and I know some of these households personally.

The moral concern expressed here is biased toward local and personally known households in energy poverty, rather than such households in general or in faraway places. It is very interesting that we tend to get far more morally outraged about injustices close to home among our own people, than about those in distant places among people of different cultures. This is a serious problem with energy and climate change issues because, as energy justice literature often points out, burning fossil fuels in one country has direct effects on the climate globally (Jenkins et al., 2018, p. 632; Jones et al., 2015; Walker et al., 2016). It also bears directly on issues of economic inequality. Studies show that higher levels of economic inequality lead to higher greenhouse gas emissions (see Chapter 10 and also Galvin and Sunikka-Blank, 2018), and these emissions are likely to affect vulnerable peoples thousands of kilometers away.

Rorty (1997) and Walzer (1994) make interesting suggestions as to how this issue can be addressed within a pragmatist approach to justice. Rorty argues that moral impulses are based on *affect* (feelings, passions, desires, wants) rather than rationality. People may use their rational faculties to develop their moral views and relate them to specific situations and dilemmas, but there has to be a moral grounding there in the first place (similar to the post-Wittgensteinian notion that a person's moral commitments and beliefs have to be intrinsic to the meaning of their life). Rorty uses the example of loyalty. He argues that we develop the moral view that children should not be left hungry, by personally knowing the children in our family and friendship circle. Because we know and love them, we act loyally toward them, getting them food even at personal cost or injury, and facing down those who might try to prevent us. Rorty notes that if we are offered the choice to let our own children starve so that ten times as many children of a different culture in a faraway land may be saved from starving, we almost certainly opt for our own children—and it is highly unlikely that anyone would criticize our morals for this. Rorty suggests that affect-based notions like loyalty are a better guide to how our moral commitments actually work, than rational-based notions like justice.

A post-Wittgensteinian approach would say this does not quite fit the bill as it misses an important dimension of moral impulse. Wittgensteinian scholars argue that moral commitments emerge in and through particular types of language games that involve words or ideas about obligation, not just feeling, and which are associated with practices in which these obligations are played out (Mouffe, 2000; Pihlström, 2012). Hence, building on Rorty's example, our *moral* commitment to children's welfare is not just an extension of the love and loyalty we *feel* for the children in our family and extended family. Rather, it also arises from *practices* in which we act out our *obligations* to them and talk about these actions in terms of obligations. Kant was right that morality is not merely an extension of affect. But nor is it merely a matter of the head. The notion of obligation is deeply interwoven in the mix of feelings and practices toward those we love and are loyal to (Pihlström, 2005, 2012).

Nevertheless, Rorty is right to point out that morality begins at home, among those we are closest to. Walzer (1994) takes this a step further with his notion of "thick" and "thin" morality. Like Wittgensteinian scholars, he argues that we become moral beings as members of a close community, including family, school, church, mosque, synagogue, neighbors, sports clubs, etc., so that our moral impulses and aspirations are closely tied to those we know and love. He calls this "thick" morality, as it relates to the people we are thickly connected to.

Walzer notes, however, that "thick" morality is not sufficient for an adult life. Since we have to relate to much wider and different communities as we grow up, we learn how to extend our moral impulses "thinly," in various directions to different sorts of people and situations that are often analogous to our initial community but may be very different and distant from it. Further, in today's world, with globalism, climate change, global energy scarcity and loss of biodiversity, we are challenged to take this much further, extending our moral concerns way beyond our original setting.

Sovacool and Dworkin (2014) offer an interesting example that could be used to illustrate how the notion of "thick" and "thin" morality can extend energy justice to future generations. It comes from sociologist Elise M. Boulding and Iriquois leader Daisaku Ikeda, and is called "the two hundred year present":

> *It begins one hundred years ago today, on the day of the birth of those* among us *who are centenarians, celebrating their one hundredth birthday today. The other*

boundary of this present moment is the one hundredth birthday of the babies being born today … We are linked with both boundaries of this moment by the people among us whose lives began or will end one of those boundaries, three and a half generations away in time. (quoted in Sovacool and Dworkin, 2014, p. 308, emphasis added)

I italicized the words "among us" to draw attention to the fact that this moral commitment starts off "thick," engendered by being in daily close contact with people whose lives extend across this stretch of time. Because we have moral commitments to these people whose lives together span 200 years, and they have moral commitments to us, it is a manageable step to then envision and be touched by our moral commitments to future generations.

Rorty and Walzer both note that moving from thick to thin is a *learning* process. It involves personal development rather than just accepting a grand intellectual globally universal ethical system. This is because a person's morality is intimately bound up with the meaning of her life, rather than just an intellectual position on this or that issue. Again to build on Rorty's example, our sense of moral obligation toward our children (including those of our extended family) is an essential element in the meaning of our life. Part of what it means to be a human being is to have such obligations: they are an intrinsic part of us.

6 Implications for global energy justice

The above discussion has implications for the meaning of a moral claim, but also for winning adherents, including policymakers and business leaders, to the cause of energy justice. I do not think we win adherents to a moral cause by presenting them with a great global moral framework which we claim is universally true for all humanity and demanding they adopt it as their own. To begin with, we would be presenting them with a "chimera," to use Wittgenstein's description of alleged moral universals (Wittgenstein, 1965), a thing that may sound grand and noble but is illusory. There *is* no moral law of justice permeating the cosmos, or even the portion of it occupied by planet earth.[d] Putting forward an illusion will not draw thoughtful people to our cause.

Because we cannot claim universal validity for our moral framework, we need to communicate it in such a way as to *nurture and increase people's moral maturity,*

---

[d] I do not think that even Christian theology supports such an ethical edifice. Rather, as I understand it, Christians believe God has engaged with them personally by loving them just as they are and giving them the pleasure of His/Her presence in their everyday lives. This motivates (or is supposed to motivate) them to love others in a similar way, and this deepens and intensifies (or is supposed to deepen and intensify) their moral commitment to the good of others (their neighbors), starting with those nearest to them and moving outward to all humanity and all living things (see, e.g., Rogers-Vaughn, 2013). In the Wittgensteinian sense, the religious encounter is part of, and enhances, the practices in which moral commitment is grafted into the meaning of people's lives (Pihlström, 2012). In terms of Walzer's notion of thick and thin, the religious encounter thickens up the thick and accelerates the development of the thin.

rather than intellectually press them into our way of thinking by implying our moral framework is the globally correct one. We cannot demand, for example, that a politician in London whose moral compass is almost entirely "thick"—focused on local hardships and needs—suddenly becomes an advocate for the rights of, say, rural people in Peru who are being displaced by the building of a hydroelectric dam. Our intellectual argument may be spotless on its own terms, but it may not connect with the meaning of life for those we are trying to convince, and may simply come across as irritating.

This is the significance of the third aspect of the Cambridge example: where the speaker appeals directly to the moral character of the Member of Parliament. The speaker seeks to nurture and build on the moral maturity the Member of Parliament already has, to win him to the cause.

A similar issue pertains to the local versus global issues of climate change. It has long been argued that western governments are slow to reduce emissions because, in part, they see their own countries as least vulnerable to the effects of climate change while counting the social costs of rapid decarbonization as too high for their own communities—a case of lots of "thick" and very little "thin." Recently, however, the effects of climate change *on western countries* are being increasingly felt and publicized. Recent articles in the German weekly *Der Spiegel* and the British daily *The Guardian* explain quite matter-of-factly how western countries, among others, will be catastrophically ravaged by sea level rise (Der Spiegel, 2018; McKie, 2018). Facts such as these can appeal directly to the already developed, "thick" moral sense of western policymakers and peoples. It might then give them an empathetic sense of the injustice of such events happening in faraway places among different cultures.

A further dimension to winning adherents to energy justice causes is that with justice-type issues, the empirical facts often speak for themselves. For example, United Nations Special Rapporteur Professor Philip Alston recently investigated the effects of a decade of austerity in Britain, finding that austerity policies have radically deepened poverty and enormously increased the number of households suffering its deprivations. In his 28-page report (Alston, 2018) he painted a picture of suffering, exacerbated by the running down of health and public services, presided over by a government which appears as if it couldn't care less. One might say, injustice shouted through every paragraph.

However, the word "justice" only appears once in the text of the report, and once in the title of a book he lists in the references, and the word "injustice" was conspicuously absent. The facts speak for themselves to readers who know the towns and perhaps even the neighborhoods Alston mentions, who stumble over homeless people every day on their way to work, and *who are real human beings who have a well-developed moral sense*. Framing the findings around intellectual arguments about justice would be highly unlikely to win converts.

Many press articles have the same effect. Writing for the Guardian, Davis (2018) describes the development that life expectancy is now falling for women in the poorest areas of England, due to austerity and deteriorating health services. Again, the words justice and injustice do not appear in the article, but the bare facts are likely to tug at the moral consciences of any British or person with a reasonably mature moral sense. Again in the Guardian, Butler (2018) writes that the charity "Shelter" estimates there are now at least 320,000 homeless people in Britain, again without using words like justice or injustice. Looking further at press articles and special reports of think tanks in areas to do with what we might call "justice" or "injustices," I find very few actually use these words, and if so, very sparingly. I presume the writers trust the facts to speak for themselves and the readers to have sufficiently mature moral characters to see the moral import of these facts.

With local energy issues this approach may well be more effective than intellectualizing about justice. This is more difficult in relation to the global impact of energy issues, but winning adherents still depends on appealing to people's moral sense, rather than presenting them with ethical arguments about what they should believe and do. To use Walzer's terms, a challenge for the energy justice movement is to develop ways of thickening up people's already-developed moral sense and leading them to extend this in a global direction.

7 Energy justice and economic inequality

As I noted in Section 1, energy poverty is one of a number of concerns explored in energy justice literature, which affirms that one of its three direct causes is poverty. Another of its causes, energy inefficient buildings and appliances, is also exacerbated by poverty. Poverty becomes deeper and more widespread as economic inequality increases (Galvin and Sunikka-Blank, 2018; Piketty, 2015; Stiglitz, 2013), so it is fair to say that economic inequality leads to energy poverty. How can the understanding of justice developed in this chapter support a more direct involvement of the energy justice movement in mitigating economic inequality?

To begin with, there is a natural alliance between people who are morally awake to the wrongness of the extremes of economic inequality we are witnessing today, and the moral wrongness of energy poverty. This is not to say they believe in an abstract realm of moral rights and wrongs that declares both energy poverty and economic inequality to be unjust. Rather, these people's personal moral development has sensitized them to the wrongness of these phenomena, and they can see the connections between them, or at least their family resemblance, to use Wittgenstein's phrase. Being morally incensed about fuel poverty and economic inequality is part of what it means for these people to be human. For these people, to disregard these phenomena would be tantamount to denying an essential part of their humanity.

This makes for powerful motivation to take up opportunities to act against these deprivations. It also suggests possible links to other people, including political and business leaders, who might not yet feel morally incensed about these phenomena but do feel incensed about similar or related deprivations that touch their own spheres of interest and influence. I doubt very much if we will recruit these people to our cause by showing them that energy poverty and economic inequality are injustices according to a great global scheme of rational moral reasoning. Rather, they might be recruited by appealing to their moral sensitivities, their humanity, the moral side of the meaning of their lives. They are morally mature people who will not be able to rest easy when they see the simple facts of economic inequality and its connection to deprivations such as energy poverty.

I am, therefore, taking the radical step of suggesting that a classic western, enlightenment style approach, where we rationally expound the moral connections between abstract notions of universal justice and their implications for energy issues, may not be an effective way to recruit decision makers to our cause. Instead, *our approach must put the brute facts of economic inequality and its energy-related consequences before people who have reasonably well developed moral humanness.*

This may not be an easy approach for western academics and activists who have deep faith in the power of rational argument to convince sceptics. There is of course a place for rational argument. If a human being already has a strong moral impulse for some cause, rational thinking can help fill out its direct implications—for example, if you are already morally incensed by poverty in a high-GDP country, it helps to rationally explore and critique the economic policies that have led to this strange juxtaposition. This may lead you to then to become morally incensed about, say, regressive tax rates. But presenting a great global rational edifice about justice and its implications is highly unlikely to touch a person who has not yet developed basic moral values of compassion for neighbors in need.

8 Conclusion

My basic question in this chapter was to ask how the energy justice movement could relate to issues of economic inequality in relation to energy. This involved a roundabout route, as I argued that the philosophical basis of energy justice literature needs to be more convincingly developed, as this hinders it from effectively recruiting policymakers and other leaders to its cause. I suggested that, although energy justice literature seeks to include a range of different views of justice drawn from different branches of western philosophy and also from non-western cultures, in the end it falls back on the approach of John Rawls, informed by western enlightenment assumptions, against which it tends to judge the validity of other approaches. Apart from the internal inconsistencies in Rawls' approach, I argued that this locks it into treating moral actions and beliefs as rational responses to reified, metaphysical-like laws of rights and wrongs which are assumed to be simply there, waiting for humans to discover them and be subject to them.

Basing my approach on a post-Wittgensteinian critique, I argued that moral commitments and beliefs are a characteristic of human beings and an essential element in their having meaning in their lives. Moral beliefs are not a rational acceptance of reified moral laws. They are formed in communities through their practices and discourse; individuals develop moral impulses, commitments, beliefs and behavior as part of growing up, learning how to live in relation to others, and becoming functional human beings whose lives have meaning.

Drawing on Rorty and Walzer, I further argued that morally well-developed persons can extend their moral impulses outwards, even globally. But this is a process of human moral development, not simply an assent to a rational global moral edifice.

I argued that in communicating the findings of energy justice to policymakers, business leaders and the public, we need to aim to connect with these people's already developed moral character. The same moral character that is sensitive to human suffering due to poverty and extremes of economic inequality may also be sensitive to the human suffering being caused by runaway energy consumption and greenhouse gas emissions. The connections need to be made within the heartfelt moral compasses of persons, not just in intellectual edifices.

The challenge for the energy justice movement, as for those concerned about economic inequality, is to develop an understanding of morality and moral commitment that grounds justice in the personal moral development of human beings, and to develop strategies to recruit supporters on this basis and help them grow more mature, joined-up moral outlooks.

References

Alston, P., 2018. Statement on Visit to the United Kingdom, by Professor Philip Alston, United Nations Special Rapporteur on Extreme Poverty and Human Rights. Available for download at https://www.ohchr.org/EN/NewsEvents/Pages/DisplayNews.aspx?NewsID=23881&LangID=E.

Alvial-Palavicino, C., Ureta, S., 2017. Economizing justice: turning equity claims into lower energy tariffs in Chile. Energy Policy 105, 642–647.

Bartiaux, F., Vandeschrick, C., Moezzi, M., Frogneux, N., 2018. Energy justice, unequal access to affordable warmth, and capability deprivation: a quantitative analysis for Belgium. Appl. Energy 225, 1219–1233.

Berry, A., Jouffe, Y., Coulombel, N., Guivarch, C., 2016. Investigating fuel poverty in the transport sector: toward a composite indicator of vulnerability. Energy Res. Soc. Sci. 18, 7–20.

Bouzarovski, S., Simcock, N., 2017. Spatializing energy justice. Energy Policy 107, 640–648.

BRISKEE, 2018. Two H2020 Research Projects That Provide Empirical Evidence to Support Energy Efficiency Policies. Findings available at https://www.briskee-cheetah.eu/.

Butler, P., 2018. At least 320,000 homeless people in Britain, says shelter. The Guardian. 22 November 2018.

Chatterton, T., Anable, J., Barnes, J., Yeboah, G., 2016. Mapping household direct energy consumption in the United Kingdom to provide a new perspective on energy justice. Energy Res. Soc. Sci. 18, 71–87.

Cheyne, C., Imran, M., 2016. Shared transport: reducing energy demand and enhancing transport options for residents of small towns. Energy Res. Soc. Sci. 18, 139–150.

Davis, N., 2018. Life expectancy falling for women in poorest areas of England. The Guardian. 23 November 2018.

Day, R., Walker, G., Simcock, H., 2016. Conceptualising energy use and energy poverty using a capabilities framework. Energy Policy 93, 255–264.

Der Spiegel, 2018. London, Paris und Polen sind untergegangen. November 30, 2018. Available from: https://www.spiegel.de/plus/klimawandel-szenarien-london-paris-und-polen-sind-untergegangen-a-00000000-0002-0001-0000-000161087440.

Dewey, J., 1920. Reconstruction in Philosophy, second ed. Beacon Press, Boston. reprint 1948.

Dewey, J., 1934. A Common Faith. Yale University Press, New Haven, CT and London. reprint 1991.

Diamond, C., 1996. Wittgenstein, mathematics and ethics: resisting the attractions of realism. In: Sluga, H., Stern, D. (Eds.), The Cambridge Companion to Wittgenstein. Cambridge University Press, Cambridge.

Eames, M., 2011. In: Energy, Innovation, Equity and Justice.Energy Justice in a Changing Climate: Defining an Agenda, InCluESEV Conference, 10 November 2011, London.

Galvin, R., Sunikka-Blank, M., 2018. Economic inequality and household energy consumption in high-income: countries: a challenge for social science based energy research. Ecol. Econ. 153, 78–88.

Gillard, R., Snell, C., Bevan, M., 2017. Advancing an energy justice perspective of fuel poverty: household vulnerability and domestic retrofit policy in the United Kingdom. Energy Res. Soc. Sci. 29, 53–61.

Goldthau, A., Sovacool, B., 2012. The uniqueness of the energy security, justice, and governance problem. Energy Policy 41, 232–240.

Habermas, J., 1993. Justification and Application: Remarks on Discourse Ethics. MIT Press, Cambridge, MA.

Hall, S., 2013. Energy justice and ethical consumption: comparison, synthesis and lesson drawing. Local Environ. 18 (4), 422–437.

Healy, N., Barry, J., 2017. Politicizing energy justice and energy system transitions: fossil fuel divestment and a "just transition". Energy Policy 108, 451–459.

Heffron, R., McCauley, D., 2017. The concept of energy justice across the disciplines. Energy Policy 106, 658–667.

Hernández, D., 2015. Sacrifice along the energy continuum: a call for energy justice. Environ. Justice 8 (4), 151–156.

James, W., 1907. In: Burkhardt, F.H., Bowers, F., Skrupskelis, I.K. (Eds.), Pragmatism: A New Name for Some Old Ways of Thinking. Harvard University Press, Cambridge, MA and London. reprint 1975.

Jenkins, K., 2018. Setting energy justice apart from the crowd: lessons from environmental and climate justice. Energy Res. Soc. Sci. 39, 117–121.

Jenkins, K., McCauley, D., Heffron, R., Stephan, H., Rehner, R., 2016. Energy justice: a conceptual review. Energy Res. Soc. Sci. 11, 174–182.

Jenkins, K., McCauley, D., Forman, A., 2017. Editorial: energy justice: a policy approach. Energy Policy 105, 631–634.

Jenkins, K., Sovacool, B., McCauley, D., 2018. Humanizing sociotechnical transitions through energy justice: an ethical framework for global transformative change. Energy Policy 117, 66–74.

Jones, B., Sovacool, B., Sidortsov, R., 2015. Making the ethical and philosophical case for 'energy justice'. Environ. Ethics 37 (2), 145–168.

LaBelle, M.C., 2017. In pursuit of energy justice. Energy Policy 107, 615–620.

Lester, J., Allen, D., Hill, K., 2001. Environmental Injustice in the United States: Myths and Realities. Westview Press, Boulder, CO.

Lovibond, S., 1983. Realism and Imagination in Ethics. University of Minnesota Press, Minneapolis.

McCauley, D., Heffron, R., 2018. Just transition: integrating climate, energy and environmental justice. Energy Policy 119, 1–7.

McCauley, D., Heffron, R., Stephan, H., Jenkins, K., 2013. Advancing energy justice: the triumvirate of tenets. Int. Energy Law Rev. 32 (3), 107–110.

McCauley, D., Ramasar, V., Heffron, R., Sovacool, B., Mebratu, D., Mundaca, L., 2019. Energy justice in the transition to low carbon energy systems: exploring key themes in interdisciplinary research. Appl. Energy 233–234, 916–921.

McKie, R., 2018. Portrait of a planet on the verge of climate catastrophe. The Guardian. 2 December 2018.

Miller, C., Richter, J., O'Leary, J., 2015. Socio-energy systems design: a policy framework for energy transitions. Energy Res. Soc. Sci. 6, 29–40.

Mouffe, C., 2000. Wittgenstein, political theory and democracy. In: Mouffe, C. (Ed.), The Democratic Paradox. Verso, London, New York, pp. 241–253.

Pesch, U., Correljé, A., Cuppen, E., Taebi, B., 2017. Energy justice and controversies: formal and informal assessment in energy projects. Energy Policy 109, 825–834.

Pettit, J., 2004. Climate justice: a new social movement for atmospheric rights. IDS Bull. 35 (3), 102–106.

Pihlström, S., 2005. Pragmatic Moral Realism: A Transcendental Defense. Rodopi, Amsterdam and New York.

Pihlström, S., 2012. A new look at Wittgenstein and pragmatism. Eur. J. Pragmat. Am. Philos. 4 (2), 1–20.

Piketty, T., 2015. Putting distribution back at the center of economics: reflections on capital in the twenty-first century. J. Econ. Perspect. 29 (1), 67–88.

Poruschi, L., Ambrey, C., 2019. Energy justice, the built environment, and solar photovoltaic (PV) energy transitions in urban Australia: a dynamic panel data analysis. Energy Res. Soc. Sci. 48, 22–32.

Price, H., 2011. Naturalism Without Mirrors. Oxford University Press, Oxford and New York.

Rawls, J., 1971. A Theory of Justice. Harvard University Press, Cambridge, MA.

Rawls, J., 1993a. Political Liberalism. Columbia University Press, New York.

Rawls, 1993b. The law of peoples. In: Shute, S., Hurley, S. (Eds.), On Human Rights: The Oxford Amnesty Lectures. Basic Books, New York, p. 44.

Reames, T., 2016. Targeting energy justice: exploring spatial, racial/ethnic and socioeconomic disparities in urban residential heating energy efficiency. Energy Policy 97, 549–558.

Redpath, T., 1972. Wittgenstein and ethics. In: Ambrose, A., Lazerowitz, M. (Eds.), Ludwig Wittgenstein: Philosophy and Language. George Allen and Unwin, Plymouth, pp. 95–119.

Rehner, R., McCauley, D., 2016. Security, justice and the energy crossroads: assessing the implications of the nuclear phase-out in Germany. Energy Policy 88, 289–298.

Rogers-Vaughn, B., 2013. Pastoral counseling in the neoliberal age: hello best practices, goodbye theology. Sacred Spaces: The E-Journal of the American Association of Pastoral Counselors 5, 5–45.

Rorty, R., 1997. Justice as a larger loyalty. Ethical Perspect. 4 (2), 139–151.

Sareen, S., Haarstad, H., 2018. Bridging socio-technical and justice aspects of sustainable energy transitions. Appl. Energy 228, 624–632.

Saunders, J., 2011. Energy justice—the policy challenges. In: Energy Justice in a Changing Climate: Defining an Agenda, InCluESEV Conference, 10 November 2011, London.

Schlosberg, D., 2009. Defining Environmental Justice: Theories, Movements, and Nature. Oxford University Press, Cary, NC.

Schlosberg, D., Collins, L., 2014. From environmental to climate justice: climate change and the discourse of environmental justice. Wiley Interdiscip. Rev. Clim. Change. https://doi.org/10.1002/wcc.275.

Searle, J., 1964. How to derive "ought" from "is". Philos. Rev. 73 (1), 43–58.

Sidortsov, R., Sovacool, B., 2015. Left out in the cold: energy justice and arctic energy research. J. Environ. Stud. Sci. 5, 302–307.

Simcock, N., Mullen, C., 2016. Energy demand for everyday mobility and domestic life: exploring the justice implications. Energy Res. Soc. Sci. 18, 1–6.

Skea, J., 2011. Energy justice and the transformation to a low carbon society. In: Energy Justice in a Changing Climate: Defining an Agenda, InCluESEV Conference, 10 November 2011, London.

Sluga, H., Stern, D. (Eds.), 1996. The Cambridge Companion to Wittgenstein. Cambridge University Press, Cambridge.

Smith, M., 2016. Cycling on the verge: the discursive marginalisation of cycling in contemporary New Zealand transport policy. Energy Res. Soc. Sci. 18, 151–161.

Sovacool, B., 2015. Fuel poverty, affordability, and energy justice in England: policy insights from the warm front program. Energy 93, 361–371.

Sovacool, B., Dworkin, M., 2014. Global Energy Justice: Problems, Principles, and Practices. Cambridge University Press, Cambridge.

Sovacool, B., Dworkin, M., 2015. Energy justice: conceptual insights and practical applications. Appl. Energy 142, 435–444.

Sovacool, B., Heffron, R., McCauley, D., Goldthau, A., 2016. Energy decisions reframed as justice and ethical concerns. Nat. Energy 1. 16024.

Sovacool, B., Burke, M., Baker, L., Kotikalapudi, C., Wlokas, H., 2017. New frontiers and conceptual frameworks for energy justice. Energy Policy 105, 677–691.

Stiglitz, J., 2013. The Price of Inequality. Penguin, London.

Walker, G., Day, R., 2012. Fuel poverty as injustice: integrating distribution, recognition and procedure in the struggle for affordable warmth. Energy Policy 49, 69–75.

Walker, G., Simcock, N., Day, R., 2016. Necessary energy uses and a minimum standard of living in the United Kingdom: energy justice or escalating expectations? Energy Res. Soc. Sci. 18, 129–138.

Walzer, M., 1994. Thick and Thin: Moral Argument at Home and Abroad. Notre Dame University Press, Notre Dame.

Wittgenstein, L., 1953. Philosophical Investigations (G. Anscombe, Trans.). Basil Blackwell, Oxford.

Wittgenstein, L., 1965. A lecture on ethics. Philos. Rev. 74 (1), 3–12. Also available for download at http://www.naturalthinker.net/trl/texts/Wittgenstein,Ludwig/ethics.pdf.

Wittgenstein, L., 1969. In: Anscombe, G.E.M., von Wright, G.H. (Eds.), On Certainty. trans. G.E.M. Anscombe & Denis Paul.Basil Blackwell, Oxford.

Empirical findings: Energy and economic inequality in practice

Energy poverty: Understanding and addressing systemic inequalities [⁎]

5

<hr>

Lucie Middlemiss

Sustainability Research Institute, School of Earth and Environment, University of Leeds, Leeds, United Kingdom

Chapter outline

1 Introduction 99
2 Linking energy poverty and other forms of inequality 101
   2.1 Developments in energy poverty research 101
   2.2 A socially systemic explanation of energy poverty 102
   2.3 Blurring the distinction between energy poverty and income poverty 104
3 Learning from the lived experience 104
   3.1 John's story 105
   3.2 Insights from lived experience research 106
4 The politics of energy poverty 108
5 Conclusions 110
References 112

1 Introduction

Despite the UK being ranked the sixth richest country in the world, energy poverty, also known as fuel poverty, is a pernicious problem and indeed is widespread throughout other developed nations (Thomson and Snell, 2013). Since Brenda Boardman's foundational work on the topic (1991), calls to address energy poverty have become an established feature of the political agenda in the UK, albeit framed differently over time. Despite the fact that most conceptions of energy poverty associate this experience with the more general condition of income poverty,

<hr>

[⁎] I am grateful to Tom Hargreaves and Harriet Thomson for their helpful comments on an earlier draft of this chapter.

Inequality and Energy. https://doi.org/10.1016/B978-0-12-817674-0.00005-9

energy poverty is rarely talked about in terms of inequality in the UK. Indeed, I have argued elsewhere that in the UK, recent changes to policy marked an attempt to delineate energy poverty from income poverty: to draw boundaries around this particular form of resource scarcity in order to avoid talking about the wider conditions (inequality, austerity, energy market liberalization) in which it occurs (Middlemiss, 2017).

My own engagement with this topic, in the first instance through qualitative work on the lived experience of energy poverty (Middlemiss et al., 2019; Middlemiss and Gillard, 2015), and more recently in critical analyses of national policies on energy poverty (Kerr et al., 2019; Middlemiss, 2017; Middlemiss et al., 2018), suggests that a socially systemic conceptualization of this topic, which starts from the intersecting inequalities faced by energy poor households, is both more reflective of the lived experience and more likely to result in effective solutions. This is a position supported by a number of other authors, including those working with qualitative data on the lived experience (Baker et al., 2018; Bouzarovski and Petrova, 2015; Day et al., 2016; Gillard et al., 2017; Großmann and Kahlheber, 2017; Meyer et al., 2018; Middlemiss and Gillard, 2015; Petrova, 2018; Robinson et al., 2018). These authors see energy poverty as multi-faceted, complex, and intersecting with other forms of poverty. In other words, in our conception of energy poverty, households face inequalities across multiple domains of their lives, which can have compounding effects on each other, and as such form a system of intersecting inequalities. As Großmann and Kahlheber put it: "the inability to pay for energy costs results from a complex state of deprivation, in which a variety of disprivileging characteristics of a household intersect and further interact with external conditions" (2017, p. 29).

In this chapter I argue that the framing of this agenda in the UK and elsewhere has to date been driven by technocratic and economic understandings of the problem of energy poverty, and that it is high time for a more socially systemic conception of this problem, rooted in the experiences of the energy poor. I begin by making the case for linking energy poverty and a wider range of inequalities, arguing for this socially systemic conception which sees the problem of energy poverty as having three key facets: (1) energy poverty has to be understood in the context of social, physical and technological conditions, (2) people's experiences are affected by intersecting challenges according to these conditions, and (3) the nature and the progression of these challenges is variable over time. I then present two strands of work that have been influential in this field: qualitative work on the lived experience of energy poverty, and analytical work on the politics of energy poverty. The story I tell of John's experience of energy poverty exemplifies how the socially systemic understanding has emerged from qualitative data on everyday life. While in the UK the policy response has been rather narrow and technocratic, there are signs of a multi-faceted approach in some nations. I conclude by offering some thoughts on the opportunities to better understand and address inequalities in access to energy.

2 Linking energy poverty and other forms of inequality
2.1 Developments in energy poverty research

In a global North context, energy poverty[a] has roots in the British term "fuel poverty," established by Brenda Boardman (1991). This early work was motivated by a recognition that cold homes were having detrimental impacts on the lives of their inhabitants (especially health impacts), and that these impacts were caused by a combination of drivers which are slightly different to those causing poverty more generally. The three widely recognized drivers of fuel poverty (low incomes, high bills, energy inefficiency) also suggest different policy solutions to those associated with poverty. Energy inefficiency is the most distinctive driver here, as both low incomes and high bills are also recognized as drivers of poverty.

In recent years, there has been a flurry of research activity in this field, moving from discussions of how to measure this problem, to a range of questions around the experience of energy poverty, the nature of vulnerability in this context and ways of identifying and finding the energy poor (Bouzarovski, 2014; Bouzarovski and Petrova, 2015; Day et al., 2016; Großmann and Kahlheber, 2017; Robinson et al., 2018; Thomson et al., 2017; Thomson and Snell, 2013). Some of these new questions have emerged from qualitative (social science) researchers engaging with this topic, drawing on understandings of the lived experience in such a way as to question and reframe some of the starting points in this field (Bouzarovski and Petrova, 2015; Day et al., 2016; Gillard et al., 2017; Großmann and Kahlheber, 2017; Meyer et al., 2018; Middlemiss and Gillard, 2015; Middlemiss et al., 2018; Petrova, 2018). In addition, the political agenda on fuel poverty in the UK has spread to other European countries and beyond, relabeled "energy poverty" in the process, including more recently an engagement of the European Union with this topic. The European Energy Poverty Observatory (2018), launched last year, marks an attempt both to solidify the agenda within the EU, and to think about how countries can learn from each other.

The energy poverty agenda grew out of a relatively narrow conceptualization of this problem: an understanding that cold homes produced distinct health impacts, and that these impacts were not being addressed by contemporary policy measures. The blooming interest in the field by researchers from a range of disciplines, and by qualitative researchers, has complicated this story. Energy poverty is no longer merely a problem of cold homes (Simcock et al., 2016), physical health impacts (Liddell and Guiney, 2015), or indeed energy inefficiency (Middlemiss, 2017; Middlemiss and Gillard, 2015). Indeed, energy poverty is linked with both physical and mental health problems (de Chavez, 2017; Liddell and Guiney, 2015; Liddell and Morris, 2010), impacted by people's tenure status (Ambrose et al., 2016), their social relations (Middlemiss et al., 2019; Schwanen et al., 2015; Willand and Horne, 2018), and their

---

[a]I follow Simcock et al. here to understand energy poverty as the inability of households to access adequate energy services, including home heating, electrical appliance use and mobility (Simcock et al., 2016).

geographical location (Jouffe and Massot, 2013; Mullen and Marsden, 2018; Robinson et al., 2018). These additional dimensions can both exacerbate and be an outcome of energy poverty: for instance, people experiencing some forms of ill-health will need more access to energy services (warmth, warm water), and if they cannot access these, their health is likely to deteriorate further. As we broaden our conception of the problem and potential solutions to it, we also find new populations that are affected differently (see for instance de Chavez, 2017; Snell et al., 2015; Snell et al., 2013 on disability and energy poverty).

2.2 A socially systemic explanation of energy poverty

These new findings and insights have led to a wider conceptualization of energy poverty which I am calling *socially systemic*. It is *social* because it has grown out of understandings of the lived experience and politics of energy poverty, and as such defines the challenges faced by the energy poor in social, rather than economic or technical terms. So, for instance, the lack of access to energy efficient housing is a function of a failure to invest in housing infrastructure, and the resulting inability to access this resource by the energy and income poor. It is *systemic* because it recognizes a range of drivers, outcomes and potential solutions to energy poverty, which are interconnected, and which form a web of causation that incorporates feedback loops, unintended consequences, and multiplier effects between variables. This means that it challenges a narrow understanding of energy poverty as distinct from income poverty, and as principally caused by energy inefficiency (Middlemiss, 2017).

Attempts to explain energy poverty in a more socially systemic way draw on a range of theoretical traditions, using concepts such as vulnerability (Bouzarovski and Petrova, 2015; Day and Walker, 2013; Middlemiss and Gillard, 2015), precarity (Petrova, 2018), capabilities (Day et al., 2016; Middlemiss et al., 2019), intersectionality (Großmann and Kahlheber, 2017) and assemblages (Day and Walker, 2013). What unites them is their tendency to characterize the problem of energy poverty as having three key facets: (1) energy poverty has to be understood in the context of specific social, political, physical and technological conditions, (2) people's experiences of energy poverty are affected by intersecting challenges (associated with these conditions), and (3) the nature and the progression of these challenges is variable over time. I will explain each of these in more detail in the following three paragraphs.

The "low income, high energy bills, energy inefficiency" characterization of energy poverty, focuses on its technical and economic characteristics. In contrast, a socially systemic account of energy poverty builds a rich picture of a wide range of conditions that shape peoples' daily lives. This includes the social world that they inhabit (politics of energy and beyond, social relations, inequalities), their physical world (climate, health conditions) and their access to technology (energy efficiency and infrastructure). Introducing new conditions to the conversation helps us to see the causal connections between these different elements. To take an example, in her work on energy precarity among young adults in the UK, Petrova shows how policy, young people's lifestyles, demand patterns, and access to resources, shape their persistent inability to access adequate energy services (Petrova, 2018). We can see here that Petrova explains young people's experience of energy poverty, and indeed their vulnerability, in relation to a wide range of conditions that shape these, well beyond the starting point of this paragraph. This amounts to the first facet of socially systemic understandings of energy poverty: the understanding of the phenomenon in the context of a wide range of social, political, physical and technological conditions.

In UK policy, people are characterized as "vulnerable" according to a specific set of demographic characteristics, associated with physical health vulnerabilities (being disabled, older, or very young) (Department of Energy and Climate Change, 2015). This is a "deficit-based" account of vulnerability: an account which defines vulnerability as associated with particular categories of people (Brown et al., 2017). In doing so it tends to associate vulnerability with the individual (of a particular type) rather than the broader conditions that create vulnerability, and risks further individualizing the problem. A socially systemic account understands people's differences as more than this: recognizing that people face challenges as a result of intersecting factors, which can decrease or increase their vulnerability, and recognizing that they also have agency. The work I have led on energy poverty and social relations is an example of this line of argument (Middlemiss et al., 2019). People's access to and influence on powerful others (e.g., MPs, housing officers, social workers), might result in them being able to access more resources (e.g., energy efficiency, bill subsidies) which result in positive outcomes for health (e.g., reduced mental health problems). Here we see causal effects across a web of interlinked factors, as opposed to the direct causal connections made in existing understandings (e.g., disabled people are more vulnerable). Großmann and Kahlheber point at a similar phenomenon in their empirical work in Germany, through an intersectionality lens, showing how multiple forms of inequality intersect, and deepen the experience of energy poverty (Großmann and Kahlheber, 2017). Indeed these authors argue for a conceptual shift: "from energy poverty as a specific form of economic poverty, to energy as a field of inequality in which multiple factors merge to form a state of deprivation" (Großmann and Kahlheber, 2017, p. 14). This amounts to the second facet of socially systemic understandings of energy poverty: understanding that people's experiences are affected by intersecting challenges.

A temporal dimension is rarely referenced in existing understandings of energy poverty. In socially systemic work, time is seen as critical (Day and Walker, 2013). People are able to pass in and out of energy poverty over time, and deeper in and out of energy poverty, for instance, given that their circumstances can change when intersecting challenges mutate (Bouzarovski and Petrova, 2015; Day and Walker, 2013; Kearns et al., 2019; Middlemiss and Gillard, 2015). This suggests a shifting population of the energy poor (as evidenced by Kearns et al., 2019), and makes the concepts of vulnerability and capability especially helpful, since they suggest the need to focus on the presence of a tendency to experience energy poverty, rather than the presence of energy poverty itself. We can also see that both the first (the social, physical and technological conditions) and second (intersecting challenges) facets are subject to change over time, and will therefore shape people's experiences differently depending on a range of factors (e.g., someone's life stage and health status, the presence and design of policies addressing energy poverty, etc.). This amounts to the third facet of socially systemic understandings of energy poverty: a recognition that these challenges change over time.

2.3 Blurring the distinction between energy poverty and income poverty

One of the effects of understanding this problem as a socially systemic phenomenon, is that it blurs the distinction between energy poverty and poverty that is apparent in earlier literature, and in the policy sphere (see for instance the Hills review in the UK (Hills, 2011, 2012), and my analysis of this in Middlemiss, 2017). As we will see below, from the perspective of the lived experience, energy poverty is merely a part of poverty: when people choose between heating, eating, and travelling, there is clearly very little distinction for them between energy, food, and mobility poverty. Some of the solutions to one form of poverty are also likely to have impacts on other forms: having access to a food bank, might mean there is a bit of leftover money for heat.

It is not my intention to undermine the concept of energy poverty here. It clearly has value: particularly as a way of articulating the challenges that income poor people experience in relation to accessing energy services. Certainly without this concept there is a strong likelihood that poverty would be ignored by the energy sector (including energy suppliers, energy networks and governance institutions). However, given the strong association between income poverty and energy poverty, it makes sense to attempt to understand how different forms of poverty intersect, and what the impacts of those intersections are. While Hills finds that people working on fuel poverty tend to defend these boundaries (Hills, 2011), his final recommended measure (*Low Income High Cost*) now used as the principle English indicator of fuel poverty, marks a recognition that income poverty and fuel poverty are strongly intertwined. Income poor people are in many nations more likely to live in poorly insulated homes, to be on higher energy tariffs, to experience cold or heat related ill-health, and to have a disability. When we conceive of energy poverty systemically it forces us to see these connections.

3 Learning from the lived experience

My first engagement with the topic of energy poverty was through encounters with people experiencing the problem in my research on the lived experience, and this has no doubt shaped my understanding of the problem. Many of the authors cited above as taking a socially systemic approach to energy poverty are also working within the qualitative tradition. Here, following an account from one of my interviewees, I will characterize the insights from lived experience research more generally.

3.1 John's story

I went to interview John on a cold winter's day in Leeds in 2016, taking a bus to an affluent dormitory settlement a few miles away from the city boundaries. Situated about 6 miles from the city center, with local amenities in another town about 3 miles away, John's house was typical of the area: a spacious family home. As he welcomed me in, I was glad to have worn some extra layers: he said he had put the heating on, but it was still cold for me. We talked for an hour and a half.

John was a single man in his 50s, and the home belonged to him (with a small mortgage). A few years ago he was married, and earning £60k a year, living with his wife and two children. His wife left with the children while he was still at work, and shortly after that he was made redundant. He had no contact with his wife, and was not formally allowed to see his two daughters, although he did have contact with one of them.

John was just coming out of a very desperate situation. When he was first made redundant 4 years prior, he still had some savings, and access to a car. Over the years he had gradually used up these savings leaving him with no financial resilience, and no means of getting around. Post Austerity, the British social security system does not provide adequate financial help for the long term unemployed, and John's situation is typical of someone who has no additional financial resources. John was barely heating his home (having the heating on for me—a visitor—was a rare treat), regularly not eating for several days between benefit payments, and having to take drastic measures to comply with requirements to receive his benefits. Job-seeking benefit claimants have to "sign on" weekly at a central location: John had to walk or cycle back and forth to central Leeds (6 miles each way) in order to do so, not having money for the bus fare.

John's mental health was very poor. He had attempted suicide a year previously, in connection with the challenges he faced. Luckily for him, his physical health was adequate, given the need to make his own way into central Leeds to sign on. John had relatively few friends and relatives, and as a man in his 50s, was too proud to ask for help from friends and family. Indeed no one knew that he was not heating or eating, or that he was claiming unemployment benefits: in the UK the latter is highly stigmatized in public discourse. He felt his poverty and lack of access to basic services, as an affront to his expectations for himself: his life was not meant to be this way. While John did accept help from charitable foundations (through British Gas) and the government when offered, he also felt torn about this.

John faces a number of intersecting challenges, which make him particularly vulnerable to energy poverty. John's story is from a specific time and place: post 2010 UK, in which austerity policies have had a huge impact on the living conditions of the long-term unemployed. In effect John is living in absolute poverty as a result of the cuts to the incomes of people in his life situation. For John, heating and travelling are activities that amount to flexible costs in his budget: costs that must be cut back if he runs out of money. His geographical location poses extra challenges: the siting of local amenities (3 miles) and government offices (6 miles), and the lack of affordable (for him) transport have major impacts on his life. He faces social challenges too: his previous income and status, as well as his place in the life course (as a middle-aged man), have made him highly averse to asking for help from others, even in desperate times.

3.2 Insights from lived experience research

Qualitative work on the lived experience of energy poverty in the UK began to emerge in earnest in the 2010s, with a number of studies originating in different disciplines offering a characterization of the ways in which the energy poor were living their lives. Key contributions of qualitative work in recent years include: characterizing the impact of energy poverty on households' health, wellbeing and finances (Harrington et al., 2005; Middlemiss and Gillard, 2015; Mould and Baker, 2017); documenting the coping strategies that households engage in (Anderson et al., 2012; Brunner et al., 2012; Tod et al., 2012, 2016); providing insights into how the energy poor experience interventions (Gilbertson et al., 2006); and revealing the challenges that particular types of households face (those containing young children, people with disabilities, older people, young people, tenants, etc.) (Butler and Sherriff, 2017; de Chavez, 2017; Gillard et al., 2017; Hitchings and Day, 2011; Petrova, 2018). Studies have emerged from social policy, environmental social science, transport studies, health studies, and housing studies, and researchers are building links across disciplinary boundaries.

When we use qualitative methods to understand the lives of people experiencing energy poverty, we inevitably add nuance and detail to more top-down conceptions of the problem, which in themselves result in different kinds of recommendation for action. I would argue that the key insights of lived experience research are threefold, with the first two of these mirroring the findings in our socially systemic explanation of energy poverty above. First, we know through qualitative work that vulnerability to energy poverty is complex and multi-faceted: people experience vulnerability due to a number of intersecting reasons. While, in John's account, his income poverty is critical to his experience, he is clearly not only vulnerable because he cannot afford to heat his home, indeed the context of austerity, his location far from services, and his expectations of himself as a middle-aged man prove to be substantial barriers to finding and accessing help. Understanding this complexity makes it much more difficult to offer simple policy solutions. The UK characterization of energy poverty as principally driven by energy inefficiency, and solved by energy efficiency (Middlemiss, 2017), for instance, falls rather flat in the face of this daily reality.

Second, we know through qualitative work that the intersections between a wide range of factors producing vulnerability can further compound the problem. In John's case, we can see a number of factors intersecting to compound his experience of living in a cold home. He experiences energy poverty because he is long-term unemployed and has used up any financial backstop, his home is located far from services, including those that are essential for securing his benefits income and buying food, and his story takes place against the backdrop of austerity and rising energy prices in the UK. These problems are further compounded by his personal life and his health: his relationship breakdown, his unwillingness to ask for help from loved ones, his mental health problems. This complex combination of interlinked drivers gives a very different picture to the three commonly cited drivers of this problem (low income, high bills, energy inefficiency). All three of these drivers are present, but they combine with each other and further factors to create John's particular challenges. Mental health problems, for instance, are exacerbated by John's family and employment situation, but they also make it harder for John to find work, or to ask for help from his close friends. Qualitative work uncovers these interlinked forms of vulnerability, hinting at vicious and virtuous circles (Großmann and Kahlheber, 2017), and shows us how a socially systemic understanding of energy poverty paints a more accurate picture of the lived experience. We have only just begun to investigate the nature of the interlinkages, which are to date rather poorly understood (Großmann and Kahlheber, 2017). The policy and practice solutions stemming from these socially systemic understandings are also in development (Middlemiss et al., 2018).

Third, understanding energy poverty through the lived experience brings to light the range of severity apparent in the UK. John's is a rather an extreme case: given he is regularly not eating, not heating and having to walk 6 miles into Leeds to protect his inadequate income, it would be fair to characterize his experience as one of absolute poverty. Having said this it is not unusual for us to meet people with these experiences in the UK, perhaps especially since post-recession austerity (although a more rigorous investigation into people's experience over time merits further attention). Others in our samples will be coping much more successfully. In previous work we have found factors that make coping more likely: a strong supportive social network, a steady and predictable income, being in good physical and mental health (Middlemiss and Gillard, 2015). To date we have only done this work in the UK, and with a small sample: there is certainly room for further work.

The relative youth of this field of study, and the presence of a number of unanswered questions, suggest an important future for this field. There are certainly unanswered questions about the nature of energy poverty and particularly about the intersecting inequalities that households experience. How precisely do different challenges result in people being able to cope or not, which specific challenges represent critical threats to people's vulnerability, which challenges tend to negatively intersect to intensify vulnerability? We also increasingly understand this problem as dynamic: fluctuating for different households over time (as foreshadowed in our socially systemic explanation of energy poverty above), but the exact nature of these changes is as yet unclear.

One further direction that I hope to develop is through secondary qualitative analysis of existing data. Since 2017, I have led a group of academics in the UK with interests in the lived experience in energy poverty, in a series of projects which attempt to understand the lived experience through secondary analysis of existing qualitative data on this topic (see Middlemiss et al., 2019 for our first output of this work). We have worked with up to 10 qualitative data sets, amounting to around 250 qualitative interviews, gathered for different purposes and over a long time period (2004–18) by a range of primary researchers. Qualitative data is frequently underused by primary researchers: we often gather detailed and complex information about people's lives, experiences and opinions, as well as their responses to different forms of intervention, and then use only a fraction of this data in analysis. I am increasingly minded to believe that this is both a waste of interesting and insightful data, and a failure to give a fair hearing to the people that we interview. Given that being interviewed about poverty of any sort is not likely to be a pleasant experience for our subjects, we have a moral obligation to use the data they give us wisely, and to maximum effect. There are also opportunities through secondary analysis to learn about how this experience has changed over time, and to identify intersections of inequalities across data sets. This method, also being pioneered by secondary qualitative data analysis experts (Edwards et al., 2018), offers the opportunity to combine breadth and depth understandings of energy poverty through the lived experience.

4 The politics of energy poverty

Having characterized the different conceptualizations of the problem of Energy Poverty in play, and explored the possibilities for a socially systemic understanding of energy poverty more extensively, we now turn to the politics of energy poverty. Understanding the ways in which policy conceptualizes energy poverty is a line of enquiry that I have taken in a number of collaborative publications (Kerr et al., 2019; Middlemiss, 2017; Middlemiss et al., 2018). Of course, the way in which the problem is understood impacts on how solutions are conceived of, as well as on the links that are possible to other national agendas (poverty, climate change, health, housing etc.). National and local approaches can be more or less socially systemic in this regard. Policy frequently characterizes energy poverty in ways that allow it to constrain its activities to a strongly boundaried population, experiencing a distinct set of problems, which can be seen to be solved in particular ways. In energy poverty, this takes the form of delineating a specific population, who both "count" as energy poor, are "counted" as such in key indicators and who have access to policy measures. Here I profile two examples of different national approaches to delineating the energy poor, then show how recent research supports an argument for a more socially systemic understanding.

In England, a recent transformation of the policy on fuel poverty has narrowly articulated this problem as one of energy efficiency (see Middlemiss, 2017 for a complete analysis). The new indicator and definition, low income high cost (LIHC), identify households that are both income poor and energy inefficient. In policy terms, the focus is firmly on energy efficiency, with new targets set to improve the efficiency of the housing stock, rather than reducing the numbers of people experiencing fuel poverty (as under the previous definition and indicator). While in many ways this delineation of energy poverty as an energy efficiency issue makes sense, it has narrowing effects on the political agenda. For instance, by characterizing the problem as one of energy efficiency, the effect is to more clearly delineate energy poverty as distinct from poverty itself. This new politics also implies that the governance of energy markets and other policies around poverty and welfare are irrelevant to solving energy poverty: in doing so omitting the privatization and liberalization of energy markets and post-crash austerity as potential explanations for the experiences of the energy poor (see Chapters 1 and 3). Given our characterization of the lived experience of energy poverty above, and our recognition of energy poverty as a socially systemic problem, it is unrealistic to think that energy efficiency on its own will solve this problem.

By contrast, the conceptualization of energy poverty in France is more subtle. French policy on "*précarité énergetique*" characterizes this problem as a complex issue, taking multiple forms, which have overlapping but not mutually exclusive populations (see Kerr et al., 2019 for a complete analysis). Inspired by the British example, the first step towards understanding this problem by French policy-makers was to look closely at an extensive housing survey using a range of indicators (including an adapted LIHC, a subjective measure and a 10% measure). In this initial analysis, they found that the different indicators delineated different, and frequently distinct populations: with many of those individuals captured by the adapted LIHC, for instance, not characterized as energy poor by the other two measures (Devalière and Teissier, 2014). This has led to a much broader, and indeed growing conceptualization of energy poverty: the policy documents openly state that they have not yet got to the bottom of what energy poverty consists of, and there are moves to include measures associated with mobility poverty by the national Observatory on this topic (Observatoire National de la Précarité Energétique, 2014). By anticipating a wide variety of drivers for energy poverty, the French approach accounts for a diversity of experience, and is also likely to enable the continued engagement of a number of policy areas with this problem. A more nuanced measurement and understanding of energy poverty is also apparent in Scotland and Belgium (Baker et al., 2018; Meyer et al., 2018).

Of course France and England are two nations that have recognized the agenda of energy poverty as having worth, other nations around the EU are resisting this, or conceptualizing it differently. In nations with a strong history of social welfare provision, such as the Netherlands, Denmark and Germany (Großmann and Kahlheber, 2017), there is no national policy on energy poverty, and in some cases the agenda is actively resisted or reframed. In Denmark, for instance, where housing is highly efficient, and a relatively strong social security net is in place, the problem is framed as the presence of mould in houses, in contrast to the strong framing of energy poverty as "cold" in houses in the UK (work in progress led by Sirid Bonderup). This is part of a delineation of the problem as a welfare problem rather than an energy poverty problem. Even when a nation does have a policy on energy poverty, there are ways in which boundaries can be drawn around particular people and households that result in them being marginalized. The case of multi-occupancy housing in the UK is a good example of this (Cauvain and Bouzarovski, 2016).

Of course when we see the problem as socially systemic, we also need to think about the impacts of welfare policy, and indeed other policy areas, on households' access to energy services. A number of studies have shown how non-energy policies can have a major impact on energy demand, either reducing or increasing demand irrespective of energy policy (Butler et al., 2018; Royston et al., 2018). Butler et al. (2018) can see direct impacts, as, for example, a result of recent welfare policy in the UK determined by the austerity agenda. This has impacted on the incomes of the energy poor, and resulted in practices such as "sanctioning" that have direct consequences for access to energy services, as well as hitting some demographics (disabled people) harder than others. Impacts can also be indirect, as they put it: "shaping long-term trajectories of social and material change that affect what is possible or not within energy policy, as well as constituting needs for energy" (Butler et al., 2018). So selling off council houses to private ownership in the UK (their example) results in it being challenging for government to enforce or retrofit energy efficiency in much of the housing stock. Given our socially systemic understanding of experiences of energy poverty, we can see that people will be impacted by the functioning of the energy market, the welfare system, housing policy, health policy and practice, and by the distribution of wealth, as well as by the membership of particular social groups that experience intersecting inequalities (Großmann and Kahlheber, 2017).

In previous collaborative research, we offer five principles for integrating understandings of the lived experience into policy and practice (Middlemiss et al., 2018). These principles hold true for a socially systemic approach to addressing energy poverty (see Table 1). Many of these principles are already in evidence in governance approaches. The broader conceptions of energy poverty as a problem being espoused by French, Belgian and Scottish governments are a good example of "measuring progress holistically" (Baker et al., 2018; Devalière and Teissier, 2014; Meyer et al., 2018; Observatoire National de la Précarité Energétique, 2014). We can also see the beginnings of joined-up working in a number of nations by a range of stakeholders. This includes, for instance, water, electricity and gas companies working together to share insights and to coordinate support for vulnerable households under the "Stronger Together Coalition" in Wales (Middlemiss et al., 2018).

5 Conclusions

There is always a risk in research that grows out of qualitative findings, that we merely complicate the understanding of a problem, without offering a way forward. That is absolutely not my intention here. Indeed, the work I have engaged in on the lived experience and in critiques of energy poverty politics, is motivated by the recognition that current approaches to governing this problem are inadequate, and are failing to alleviate people's insufficient access to energy services. As such, by sharing qualitative understandings of the lived experience, critical analyses of the politics

Table 1 Five principles for integrating a systemic understanding of energy poverty into policy and practice

Principle	Description
Create opportunities for joined-up and integrated policy	Socially systemic understandings of energy poverty suggest we should advocate for policy approaches which address this problem collaboratively and from multiple perspectives. This is likely to mean starting conversations which lead to coordinated action on the energy market, the welfare system, housing policy, health policy and practice, and more
Building networks and partnerships beyond the public sector	Given the energy system, and other intersecting systems are governed by a mix of private and public sector actors, integrated action must also reach beyond government, to utility companies, NGOs working on poverty issues, and beyond
Expecting the unexpected	If we view this problem as a systemic one, we also begin to see that action to address it has unexpected consequences, and that these need to be monitored and managed. Keeping an eye on the lived experience is helpful for this, as it allows us to see knock-on effects in people's lives from policy in specific domains
Measuring progress holistically	The English approach of focusing on one key indicator is unhelpful here: if we are to capture the systemic nature of energy poverty we must develop a range of indicators, to allow us to monitor progress and also to pay continued attention to both the lived experience, and to indicators in associated policy areas (e.g., health, housing, welfare).
Just getting on with it	Facing systemic problems can be overwhelming, but there is a strong argument for having a go, and learning from mistakes, rather than waiting for a perfect strategy. There are lots of examples of good practice in this sphere which can offer inspiration and direction.

Adapted from Middlemiss, L., Gillard, R., Pellicer, V., Straver, K., 2018. Plugging the gap between energy policy and the lived experience of energy poverty: five principles for a multi-disciplinary approach. In: Foulds, C., Robison, R. (Eds.), Advancing Energy Policy: Lessons on the Integration of Social Science and Humanities. Springer, Bern.

of energy poverty and an emerging socially systemic understanding of energy poverty, I am very much hoping to help produce more effective responses. Such responses may be more challenging to craft, addressing complexity or shifting a system is never easy, but they are also likely to produce more effective outcomes for those facing these challenges.

There are both academic and practical opportunities to address systemic inequalities more comprehensively in the future. In the academic context, there are opportunities in secondary analysis of qualitative data to learn about how the experience of energy poverty changes over time, and to identify intersections of inequalities across data sets. Policy and practice on energy poverty also has an opportunity to better address energy poverty, by taking into account the systemic inequalities that help to produce this problem. This will involve thinking across the policy silos that impact on energy poverty (energy, housing, health, welfare), but also addressing criticisms of the limits of existing policy. The lived experience is an ideal starting point here: households experiencing poverty are the context in which a range of policy comes together, and these should be prioritized as a key site of interest for ensuring any efforts to address this problem systemically are effective. Monitoring the lived experience as a way of ensuring interventions are effective is a useful starting point.

References

Ambrose, A., McCarthy, L., Pinder, J., 2016. Energy (In)Efficiency: What Tenants Expect and Endure in Private Rented Housing. Eaga Charitable Trust.

Anderson, W., White, V., Finney, A., 2012. Coping with low incomes and cold homes. Energy Policy 49, 40–52.

Baker, K.J., Mould, R., Restrick, S., 2018. Rethink fuel poverty as a complex problem. Nat. Energy 3 (8), 610.

Boardman, B., 1991. Fuel Poverty: From Cold Homes to Affordable Warmth. Belhaven Press, London.

Bouzarovski, S., 2014. Energy poverty in the European Union: landscapes of vulnerability. Wiley Interdiscip. Rev. Energy Environ. 3, 276–289.

Bouzarovski, S., Petrova, S., 2015. A global perspective on domestic energy deprivation: overcoming the energy poverty–fuel poverty binary. Energy Res. Soc. Sci. 10, 31–40 (0).

Brown, K., Ecclestone, K., Emmel, N., 2017. The many faces of vulnerability. Soc. Policy Soc. 16 (3), 497–510.

Brunner, K.-M., Spitzer, M., Christanell, A., 2012. Experiencing fuel poverty. Coping strategies of low-income households in Vienna/Austria. Energy Policy 49, 53–59.

Butler, D., Sherriff, G., 2017. 'It's normal to have damp': using a qualitative psychological approach to analyse the lived experience of energy vulnerability among young adult households. Indoor Built Environ. 26 (7), 964–979.

Butler, C., Parkhill, K.A., Luzecka, P., 2018. Rethinking energy demand governance: exploring impact beyond 'energy' policy. Energy Res. Soc. Sci. 36, 70–78.

Cauvain, J., Bouzarovski, S., 2016. Energy vulnerability in multiple occupancy housing: a problem that policy forgot. People Place Policy 10 (1), 88–106.

Day, R., Walker, G., 2013. Household energy vulnerability as 'assemblage. In: Bickerstaff, K., Walker, G., Bulkeley, H. (Eds.), Energy Justice in a Changing Climate. Zed Books, London.

Day, R., Walker, G., Simcock, N., 2016. Conceptualising energy use and energy poverty using a capabilities framework. Energy Policy 93, 255–264.

de Chavez, A.C., 2017. The triple-hit effect of disability and energy poverty. In: Simcock, N., Thomson, H., Petrova, S., Bouzarovski, S. (Eds.), Energy Poverty and Vulnerability. Routledge.

Department of Energy and Climate Change, 2015. Cutting the Cost of Keeping Warm: A Fuel Poverty Strategy for England, 2015. Available online, https://www.gov.uk/government/publications/cutting-the-cost-of-keeping-warm (Accessed 17 June 2019).

Devalière, I., Teissier, O., 2014. Les indicateurs de la précarité énergétique et l'impact de deux dispositifs nationaux sur le phénomène. Inf. Soc. 184 (4), 115–124.

Edwards, R., Davidson, E., Jamieson, L., Weller, S., 2018. Big data, qualitative style: a breadth-and-depth method for working with large amounts of secondary qualitative data. Qual. Quant. 53, 363–376.

European Energy Poverty Observatory, 2018. European Energy Poverty Observatory Website, 2018. Available online, https://www.energypoverty.eu/ [Accessed].

Gilbertson, J., Stevens, M., Stiell, B., Thorogood, N., 2006. Home is where the hearth is: grant recipients' views of England's home energy efficiency scheme (warm front). Soc. Sci. Med. 63 (4), 946–956.

Gillard, R., Snell, C., Bevan, M., 2017. Advancing an energy justice perspective of fuel poverty: household vulnerability and domestic retrofit policy in the United Kingdom. Energy Res. Soc. Sci. 29, 53–61.

Großmann, K., Kahlheber, A., 2017. Energy poverty in an intersectional perspective: on multiple deprivation, discriminatory systems, and the effects of policies. In: Simcock, N., Thomson, H., Petrova, S., Bouzarovski, S. (Eds.), Energy Poverty and Vulnerability. Routledge, pp. 30–50.

Harrington, B.E., Heyman, B., Merleau-Ponty, N., Stockton, H., Ritchie, N., Heyman, A., 2005. Keeping warm and staying well: findings from the qualitative arm of the warm homes project. Health Soc. Care Community 13 (3), 259–267.

Hills, J., 2011. Fuel Poverty: The Problem and Its Measurement. Her Majesty's Government, London.

Hills, J., 2012. Getting the Measure of Fuel Poverty. Her Majesty's Government, London.

Hitchings, R., Day, R., 2011. How older people relate to the private winter warmth practices of their peers and why we should be interested. Environ. Plan. A 43 (10), 2452.

Jouffe, Y., Massot, M.-H., 2013. Vulnérabilités sociales dans la transition énergétique au croisement de l'habitat et de la mobilité quotidienne. In: 1er Congrès interdisciplinaire du Développement Durable, Quelle transition pour nos sociétéspp. 23–57.

Kearns, A., Whitley, E., Curl, A., 2019. Occupant behaviour as a fourth driver of fuel poverty (aka warmth & energy deprivation). Energy Policy 129, 1143–1155.

Kerr, N., Gillard, R., Middlemiss, L., 2019. Politics, problematisation, and policy: a comparative analysis of energy poverty in England, Ireland and France. Energ. Buildings 194, 191–200.

Liddell, C., Guiney, C., 2015. Living in a cold and damp home: frameworks for understanding impacts on mental well-being. Public Health 129 (3), 191–199.

Liddell, C., Morris, C., 2010. Fuel poverty and human health: a review of recent evidence. Energy Policy 38 (6), 2987–2997.

Meyer, S., Holzemer, L., Delbeke, B., Middlemiss, L., Maréchal, K., 2018. Capturing the multifaceted nature of energy poverty: lessons from Belgium. Energy Res. Soc. Sci. 40, 273–283.

Middlemiss, L., 2017. A critical analysis of the new politics of fuel poverty in England. Crit. Soc. Policy 37 (3), 425–443.

Middlemiss, L., Gillard, R., 2015. Fuel poverty from the bottom-up: characterising household energy vulnerability through the lived experience of the fuel poor. Energy Res. Soc. Sci. 6, 146–154 (0).

Middlemiss, L., Gillard, R., Pellicer, V., Straver, K., 2018. Plugging the gap between energy policy and the lived experience of energy poverty: five principles for a multi-disciplinary approach. In: Foulds, C., Robison, R. (Eds.), Advancing Energy Policy: Lessons on the Integration of Social Science and Humanities. Springer, Bern.

Middlemiss, L., Albala, P.A., Emmel, N., Gillard, R., Gilbertson, J., Hargreaves, T., Mullen, C., Ryan, T., Snell, C., Tod, A., 2019. Energy poverty and social relations: characterising vulnerabilities using a capabilities approach. Energy Res. Soc. Sci. (Accepted).

Mould, R., Baker, K.J., 2017. Documenting fuel poverty from the householders' perspective. Energy Res. Soc. Sci. 31, 21–31.

Mullen, C., Marsden, G., 2018. The car as a safety-net: narrative accounts of the role of energy intensive transport in conditions of housing and employment uncertainty. In: Demanding Energy. Springer, pp. 145–164.

Observatoire National de la Précarité Energétique, 2014. Premier Raport de l'ONPE. Définitions, Indicateurs, Premiers Résultats et Recommandations. Observatoire National de la Précarité Energétique, Paris.

Petrova, S., 2018. Encountering energy precarity: geographies of fuel poverty among young adults in the UK. Trans. Inst. Br. Geogr. 43 (1), 17–30.

Robinson, C., Bouzarovski, S., Lindley, S., 2018. Underrepresenting neighbourhood vulnerabilities? The measurement of fuel poverty in England. Environ. Plan. A 50, 1109–1127 0308518X18764121.

Royston, S., Selby, J., Shove, E., 2018. Invisible energy policies: a new agenda for energy demand reduction. Energy Policy 123, 127–135.

Schwanen, T., Lucas, K., Akyelken, N., Cisternas Solsona, D., Carrasco, J.-A., Neutens, T., 2015. Rethinking the links between social exclusion and transport disadvantage through the lens of social capital. Transp. Res. A Policy Pract. 74, 123–135.

Simcock, N., Walker, G., Day, R., 2016. Fuel poverty in the UK: beyond heating? People Place Policy Online 10 (1), 25–41.

Snell, C., Thomson, H., Bevan, M., 2013. Fuel Poverty and Disability: A Statistical Analysis of the English Housing Survey. Centre for Housing Policy.

Snell, C., Bevan, M., Thomson, H., 2015. Justice, fuel poverty and disabled people in England. Energy Res. Soc. Sci. 10, 123–132.

Thomson, H., Snell, C., 2013. Quantifying the prevalence of fuel poverty across the European Union. Energy Policy 52, 563–572.

Thomson, H., Bouzarovski, S., Snell, C., 2017. Rethinking the measurement of energy poverty in Europe: a critical analysis of indicators and data. Indoor Built Environ. 26, 879–901 1420326X17699260.

Tod, A.M., Lusambili, A., Homer, C., Abbott, J., Cooke, J.M., Stocks, A.J., McDaid, K.A., 2012. Understanding factors influencing vulnerable older people keeping warm and well in winter: a qualitative study using social marketing techniques. BMJ Open. 2. (4)e000922.

Tod, A.M., Nelson, P., de Chavez, A.C., Homer, C., Powell-Hoyland, V., Stocks, A., 2016. Understanding influences and decisions of households with children with asthma regarding temperature and humidity in the home in winter: a qualitative study. BMJ Open. 6. (1) e009636.

Willand, N., Horne, R., 2018. "They are grinding us into the ground"–the lived experience of (in) energy justice amongst low-income older households. Appl. Energy 226, 61–70.

Housing tenure and thermal quality of homes—How home ownership affects access to energy services

Nicola Terry

Qeng Ho Ltd, Cambridge, United Kingdom
Cambridge Architectural Research Ltd, Cambridge, United Kingdom

Chapter outline

1 Introduction 115
2 Housing tenure in the UK and other countries in the EU 117
3 Trends in UK housing tenure 118
4 Who are the private landlords? 120
5 Comparing heating efficiency of privately rented homes with other sectors 121
6 Comparing thermal comfort of private sector homes with other sectors 127
7 Take up of energy efficiency measures 131
8 Reasons for poor take-up 133
    8.1 The tenants' view 134
    8.2 The landlords' view 135
9 Impact of minimum energy efficiency standard 138
10 Summary and conclusions 139
References 140
Further reading 143

1 Introduction

This chapter is about inequalities in home energy efficiency in the UK between owner occupied housing and the two basic types of rented housing: the private rented sector, where rents are at full market rate, and the large (albeit shrinking) stock of reduced rent homes, here referred to as social rented homes. In the UK, in many respects private rented homes are more similar to owner occupied homes than social rent homes.

A key question addressed in this chapter is whether and in what respects dwelling energy efficiency differs between these groups. Because the proportion of private

Inequality and Energy. https://doi.org/10.1016/B978-0-12-817674-0.00006-0

rented housing has been steadily increasing over the past decades in comparison with social rented and owner occupied homes, differences in energy efficiency may bring increasing advantages or disadvantages if these trends continue.

There is a great deal of attention paid to the tenant/landlord split incentive issue in media, political and academic circles. However there is very little data available from monitoring actual temperatures in homes in the private rented sector. This is not surprising, as from personal experience, the hassle involved in obtaining permission to install equipment in homes is bad enough for owner occupied. In the rented sector it is generally much easier to approach landlords in the social rent sector than private landlords, if only because the private market is so fragmented. For example in the Energy Follow Up Survey, one of the most important sources of monitoring data available, only 8% of sample dwellings were in the private rented sector, compared to 29% with reduced rents. This means that the private rented sector is under represented by at a factor of two in this survey while the social rent sector is over represented by 70%. However, the characteristics of the two sectors are very different, as described in Section 5.

This lack of monitoring data is unfortunate, as it makes it difficult to fully understand the apparent paradox whereby private rented homes are similar in energy ratings to owner occupied but tenants are less satisfied with their thermal comfort. All sectors are well represented in surveys and focus groups, but the information from these sources is subjective and potentially affected by a range of other factors. The analysis in this chapter combines information about general housing characteristics, monitoring data and energy bills with survey data in an attempt to explain the comfort paradox.

This chapter is based mainly on the situation in the UK. However, Section 2 begins with some comparisons with other countries in Europe, both for reduced rent and full market rent housing. It also compares housing costs with other countries in Europe.

Sections 3 and 4 explore socio-economic factors for both tenants and landlords in the private rented sector. What sort of people are they and how is this changing?

Sections 5 and 6 explore the thermal efficiency in rented homes of both types compared to privately owned homes, and how residents rate them for thermal comfort. There is clear evidence that private sector tenants are more likely to complain about being cold, and some evidence that private rented homes are not as good quality as other homes. However, the average energy efficiency ratings are actually very similar.

Sections 7 and 8 explore the rate of installation of energy efficiency measures in private rented homes compared to other homes. Comments from tenants and landlords show that they are not entirely in agreement as to where the responsibility for improvements should lie. Also, both are averse to debt; this is a significant barrier to investment in energy efficiency. Finally, in Section 9 some early data on the impact of minimum energy efficiency measures in the rented sector is presented.

2 Housing tenure in the UK and other countries in the EU

Currently, proportions of private renting in the UK are not far from the EU average, although ownership levels vary widely. For example, the market rent sector is very high in Netherlands (30%) and Germany (40%) compared to the UK (18%) and Europe as a whole (20%). However the UK has more reduced rent housing than the EU average: 19% compared to 11% for EU-28. See Table 1 for details.

Residents of rented accommodation are likely to struggle with the costs of housing more than owner occupiers. This is so for households benefiting from reduced rent and even more so for households renting at market rates, as shown in Table 2. The measure of affordability used is the overburden rate; this is the percentage of the population living in households where the total housing costs ("net" of housing allowances) represent >40% of disposable income. For the EU-28, reduced rent tenants are twice as likely to be struggling with costs compared to owner-occupiers, and market rent tenants are twice as likely again—with 28% spending >40% of their disposable income on housing. The UK is worse than the EU average in terms of affordability of rented housing, even with reduced rents.

Table 1 Housing tenure in the UK compared to the EU, 2016.

Country	Market rent (%)	Owner occupied (%)	Reduced rent (%)
EU-28	19.9	69.2	10.8
UK	18.0	63.4	18.6
Netherlands	30.3	69.0	0.7
Germany	39.8	51.7	8.4

Data from Eurostat, 2017. Housing Statistics. http://ec.europa.eu/eurostat/statistics-explained/index.php/Housing_statistics (accessed 21.12.2018).

Table 2 Housing overburden rate in the UK compared to the EU, 2016.

Country	Market rent (%)	Owner occupied (%with mortgage—%no loan)	Reduced rent (%)
EU-28	28.0	5.4–6.4	13.0
UK	35.4	4.8–4.3	16.2
Netherlands	28.0	3.1–3.2	16.4
Germany	23.0	10.3–9.2	19.1

This is the proportion of households where fixed housing costs account for 40% or more of their disposable income.
Data from Eurostat, 2017. Housing Statistics. http://ec.europa.eu/eurostat/statistics-explained/index.php/Housing_statistics (accessed 21.12.2018).

3 Trends in UK housing tenure

In the UK, reduced rent housing is provided either by local government or by housing associations: independent organizations established for the purpose of providing low cost housing to people who cannot afford market rates. Housing associations operate on a non-profit-making basis and any trading surplus is used to help finance new homes. In this chapter the two types of reduced rent housing are grouped together and described as social rent (Table 3).

Both private and social rent sectors include a proportion of houses of multiple occupation (HMO). The definition of an HMO is not entirely clear. Large HMOs need licenses—currently this means dwellings with five or more people from more than one household sharing facilities such as bathrooms and kitchens. As of 2016 there were approximately 230,000 such dwellings in England (HCLG, 2016) out of nine million homes in the whole rented sector (HCLG, 2018a). There are a good deal more HMOs that are smaller and do not need licenses. A survey of 1071 landlords in 2015 found that 11% had at least one shared or HMO property (Shelter, 2016).

Within the UK over the last 20 years the proportion of total rented sector housing has grown slightly, from 32.5% to 34.9%. However there has been a disproportionately large increase in private rents at the expense of social rents, as can be seen in Fig. 1. Over 20 years the proportion of owner occupiers decreased slightly from 67.5% to 65.1% while the proportion in private rented homes more than doubled from 8.6% to 17.3%.

Table 3 Tenure status by age group in 1996 and 2016.

Age group	Tenure	1996 (%)	2016 (%)
Age 16–34	Private rented	21	46
	Owner occupiers	54	34
	Social rent	23	18
Age 35–64	Private rented	6	14
	Owner occupiers	75	67
	Social rent	18	17
Age 65 and over	Private rented	4	5
	Owner occupiers	63	77
	Social rent	31	17
All ages	Private rented	9	17
	Owner occupiers	67	64
	Social rent	23	17

The age is that of the household reference person, usually the person owning the home or responsible for the rent.
Data from Barton C., 2017. Home Ownership and Renting: Demographics. House of Commons Briefing Library. https://researchbriefings.parliament.uk/ResearchBriefing/Summary/CBP-7706 (accessed 21.12.2018).

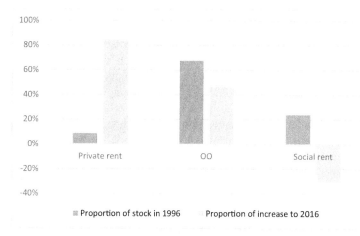

FIG. 1

Changes in the number of households between 1996 and 2016 by tenure. The total increase was 13%. The *dark bars* show the proportion of each tenure in 1996 and the *light bars* show the proportion of the increase. There was a net loss of homes in the social rent sector. The primary source is the Labor Force Survey, a quarterly survey based on a representative sample from the whole of the UK, with 40,000 households in each wave.

Data from Barton C., 2017. Home Ownership and Renting: Demographics. House of Commons Briefing Library. https://researchbriefings.parliament.uk/ResearchBriefing/Summary/CBP-7706 (accessed 21.12.2018).

The decrease in the social rent sector was because many homes were sold under the right to buy initiative. This gives tenants in local government housing the right to buy their home at less than the market price, depending on how long they have been a tenant. The residents who buy may continue to live in the homes they now own but this is not required and sooner or later many are rented out again, this time in the private sector. The right to buy policy was first implemented in 1980; between then and now around 40% of homes sold under the initiative have been converted to private rent status (Barker, 2017).

The demographics of housing tenure in the UK shows an increasing proportion of younger people renting (from 21% to 46%) while people of pensionable age are more likely than before to own their own homes (63–77%). Also ethnic minorities are far more likely to be renting than UK born white people. However, some of this difference is due to the age bias, because in ethnic minorities the proportion of young people is higher (Barton, 2017).

The general effect of these trends in the UK is an increase in the private rented sector as a whole while the social rent sector decreases, and these changes affect younger people the most. Unfortunately, it is also the case that tenants are less likely to be satisfied with the warmth in their home as will be discussed in later sections. This exacerbates the problems for the renting sector who are often already struggling with their housing costs.

When people live in fuel poverty there are impacts on health and other important aspects of wellbeing (Balfour and Allen, 2014). Cold homes can exacerbate respiratory conditions such as asthma and also circulatory problems. Improving thermal comfort helps to relieve mental health conditions such as depression and anxiety—this shows that the increase in mental health problems in people living in cold homes is not just due to people with health problems having less money to heat their homes; causality goes both ways and is very likely therefore self-reinforcing.

Children are particularly vulnerable; living in a cold home can harm their mental and physical health. Cold homes are linked to reduced educational attainment, possibly due to increased absences from school due to illness (Balfour and Allen, 2014).

4 Who are the private landlords?

Most of the data in this section comes from a 2016 survey of landlords in the private rented residential sector for the Council for Mortgage Lenders (Scanlon and Whitehead, 2016). Landlords were recruited by YouGov and Research Now. In total 2517 landlords took part during June 2016. This is the source unless otherwise stated. The English Housing Survey (EHS) includes rented accommodation but the last available landlord survey from the dataset is 2010 (DCLG, 2011). That sample is smaller, just over 1000 landlords and 1100 dwellings.

The private rent sector is characterized by a large number of landlords with only a few properties and a small number with a large portfolio. Fig. 2 shows the distribution both by landlords and dwellings. More than 60% of landlords have only one dwelling but this accounts for only 27% of the stock of privately rented homes. Just under 40% of dwellings belong to landlords with at least five properties. From the EHS survey in 2010, three quarters of rented dwellings are owned by part time landlords.

The main source of funds for most landlords is personal savings, and a significant proportion (17%) were bought using inherited funds. 13% of landlords started out with a home that was initially the main home but became available for rent when they moved out. Landlords generally dislike relying on loans. Almost half (49%) of them have no mortgage at all and say they prefer to avoid debt. Even for landlords with a mortgage the loan to value ratio is low—averaging 60%.

Most landlords are in the business to supplement their income and/or as an investment for capital growth and income. A third are retired. For most landlords, rental income is a small fraction of their overall income—for the majority it is less than a quarter. Only 3% of landlords get >75% of their income from rents.

From the EHS survey, more than half (57%) of landlords manage the properties themselves, rather than employing an agent. However only 8% of landlords are full time in the business and 80% of landlords have no relevant qualifications in property management. Some landlords (17%) have experience in the building trade.

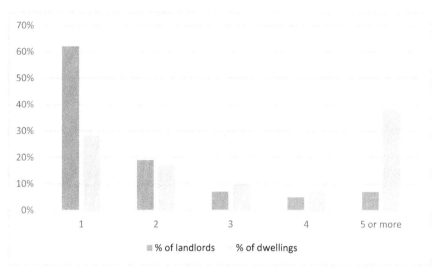

FIG. 2

Distribution of PRS portfolio size: (2016).

Data from Scanlon, K., Whitehead, C., 2016. The Profile of UK Private Landlords. The council for Mortgage Lenders. https://www.cml.org.uk/news/cml-research/the-profile-of-uk-private/the-profile-of-uk-private-landlords-08.05.17.pdf (accessed 21.12.2018).

5 Comparing heating efficiency of privately rented homes with other sectors

All the information in this section is from the English Housing Survey 2015/2016 unless otherwise stated (Box 1).

Private rented housing stock is in poorer condition than the average but not over-whelmingly so. 27% of private rented homes failed the Decent Homes Standard compared to 20% of all stock. This standard requires that the home has no serious health and safety issues, is in a reasonable state of repair, has modern facilities and services and provides a "reasonable" degree of thermal comfort (Box 2).

Box 1 English housing survey (HCLG, 2018c)

The English Housing Survey (EHS) is an annual survey commissioned by the Ministry of Housing, Communities and Local Government (HCLG). It collects information about people's housing circumstances and the condition and energy efficiency of housing in England. Around 13,300 households are interviewed and around 6000 have a physical survey of their dwelling. The sample is deliberately overweight in the rented sector. For aggregate statistics the survey results are weighted to match national statistics for householder age and sex, tenure and region.

Box 2 Decent homes standard (DCLG, 2006)

This standard is designed as guidance for local authorities and social landlords as to the condition of homes that is acceptable. The standard requires:

- No category 1 health and safety issues (such as danger of falls, indoor air pollution, electric shock or collapse of the structure, excessive cold heat or noise, or poor security)
- A reasonable state of repair
- Reasonably modern facilities and services. (For example the kitchen should be no more than 20 years old and the bathroom no more than 30 years.)
- The dwelling must provide a reasonable degree of thermal comfort. This means that there should be an efficient heating system, with appropriate controls, and reasonable insulation: for example at least 50 mm in the loft.

Homes in the private rented sector are generally older - a third were built before 1919 (see Fig. 3). This means they can be hard to upgrade with energy efficiency measures. For example, solid wall insulation costs 11–14 times more than cavity wall insulation (Palmer et al., 2017) and private rented homes are more likely to have solid walls (43%, 28% and 21% solid walls for private rented, private owned and social rent respectively). However even allowing for this, private rented homes are less likely to have wall insulation than owner occupied homes (Fig. 4). Private rented homes with cavity walls are 22% less likely to be insulated and solid wall homes

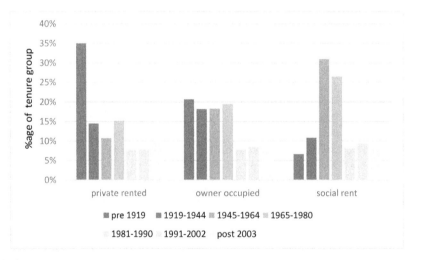

FIG. 3

Dwelling age by tenure.

Data from HCLG, 2018a. English Housing Survey Headline Report Section 1 Household Tables. Ministry of Housing, Communities and Local Government. Available from: https://www.gov.uk/government/statistics/english-housing-survey-2016-to-2017-headline-report (accessed 21.12.2018) (Fig. 2.2).

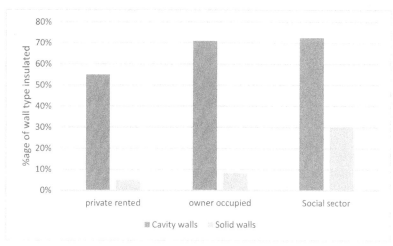

FIG. 4

Wall insulation by tenure.

Data from HCLG, 2018a. English Housing Survey Headline Report Section 1 Household Tables. Ministry of Housing, Communities and Local Government. Available from: https://www.gov.uk/government/statistics/ english-housing-survey-2016-to-2017-headline-report (accessed 21.12.2018) (Fig. 2.12).

are 37% less likely to be insulated than owner occupied homes. Homes for social rent are the most likely to be insulated of any type. For details, see Table 4.

Private rented homes are also more likely to have problems with damp, especially rising damp and penetrating damp which indicate issues with structure or maintenance. (This could be related to the age of the dwellings but the data is not detailed enough to demonstrate this.) Rising and penetrating damp problems are reassuringly rare (see Table 4) but they are three times more common in private rented homes than owner occupied homes. The other main type of damp, condensation and mold, may or may not be a structural issue as it can be due to poor ventilation. It can however be exacerbated by structural issues causing cold bridges. Condensation and mold are more common in both the private and social rent sector, which may indicate that they are due to tenants prioritizing warmth over fresh air, rather than structural or maintenance issues.

Finally, dwellings in the rented sector are more likely to lack central heating. This is partly because the rented sector includes a high proportion of flats where gas is not connected, and partly because gas central heating requires an annual safety check which means arranging for access to the property. This is a hassle as well as an expense for property management. The whole rented sector is affected but private sector homes are even more likely to lack central heating than social sector homes: 16.2% versus 9.1%, compared to only 5.2% for owner occupiers, see Table 4. In terms of energy costs, gas central heating is normally the cheapest and electricity is considerably more expensive. According to a review by the Energy Saving Trust in March 2018, gas is less than half the price of even off peak electricity, on a

Table 4 Comparison of dwelling characteristics (wall construction types, damp issues and central heating and size) between private rented, owner occupied and social rent homes.

Sector	Private rented (%)	Owner occupied (%)	Social rent (%)
Proportion of homes:			
Built before 1919 (see also Fig. 3)	35	21	7
Solid walls	43	28	21
Cavity walls that are insulated (% of cavity walls)	55	71	72
Solid walls that are insulated (% of solid walls)	5	8	30
Rising damp	3.7	1.3	0.8
Penetrating damp	3.2	1.0	1.1
Condensation and/or mold	4.2	1.1	3.7
No central heating (electric storage heaters or other room heaters, usually electric).	16.2	5.2	9.1
Non-decent home	26.8	19.7	12.6
Proportion <50 m^2 total floor area	16.3	2.9	26.9
Mean floor area (m^2)	77	107	66

Data from the HCLG, 2018a. English Housing Survey Headline Report Section 2 Housing Stock Tables. Ministry of Housing, Communities and Local Government. Available from: https://www.gov.uk/ government/statistics/english-housing-survey-2016-to-2017-headline-report (accessed 21.12.2018) (age and size are from Table AT2.1. Also wall type and insulation are AT 2.12. Damp is AT2.4, heating is AT2.8 and non-decent homes is AT2.2).

per kWh basis (EST, 2018). So this is another area where tenants in the private sector may have a disadvantage (Box 3).

Less insulation, more damp and more electric heating suggests that private sector housing is more expensive to keep warm. However, according to the data the energy costs are similar to owner occupied homes. In the UK, the standard measure for energy costs is the SAP rating on a dwelling's Energy Performance Certificate (EPC). This indicates the estimated cost of heating, hot water and lighting, adjusted for floor area. The scale goes from 1 (very poor) to 100 (very good). An SAP rating is calculated for existing homes using a model that requires detailed information about the building geometry, age and envelope constructions, heating and ventilation systems and renewable energy generation if there is any. It does not take into account damp or other defects but it does include insulation and heating fuel. Surprisingly, the mean SAP rating of homes in the private rented sector is about equal to that of the owner occupied sector and has been since about 2012: 60.3 for private rented sector

Box 3 SAP ratings (SAP, 2012)

SAP (which stands for Standard Assessment Procedure) is the method used in the UK to assess energy costs for dwellings that is reported on Energy Performance Certificates. The rating of 1–100 (with 100 being the best) is also converted to an A–G rating (where A is the best). The cost of energy is calculated based on detailed information on the buildings structure, heating and ventilation systems. For existing homes, some of the data is based on assumptions according to the building type and age, using the RdSAP method (Reduced SAP). An air tightness test is not required, with ventilation rates based on the presence of vents and observed draught proofing around windows. The energy costs are based on standard heating patterns and hot water use. Energy assessors can also provide an Occupancy Assessment which is based on the actual heating pattern used by the residents and some other aspects of their behavior.

The SAP rating is not directly related to costs and its adjustment for floor area is also not linear. This adjustment was added in SAP 2001 to ensure that homes that differ in size but are otherwise similar have similar ratings. Previously, larger homes would have better ratings (BRE, 2001). There are several reasons for this. Geometry is a factor, as envelope area increases slower than the floor area, assuming the general shape is the same. Another factor is that some energy use (such as hot water) is closely related to occupancy but the assumed occupancy also increases more slowly than floor area. The final ratings are based on the following equations, where TFA is total floor area:

$$\text{ECF} = \text{deflator} \times \text{totalcost}/(\text{TFA} + 45)$$
$$\text{if ECF} \geq 3.5, \quad \text{SAP} \, 2012 = 117 - 121 \times \log_{10}(\text{ECF})$$
$$\text{if ECF} < 3.5, \quad \text{SAP} \, 2012 = 100 - 13.95 \times \text{ECF}$$

This adjustment increases the energy ratings for smaller homes and decreases the ratings for larger ones, relative to what they would be if based purely on energy use per unit floor area. A small home of $50 \, \text{m}^2$ could have the rating increased by up to 10 points compared to an average size $85 \, \text{m}^2$ with the same cost per unit area.

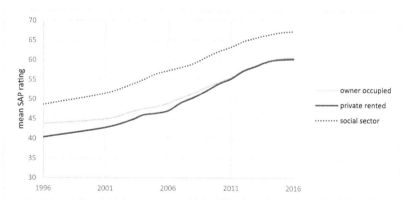

FIG. 5

Mean SAP rating by tenure.

Data from HCLG, 2018a. English Housing Survey Headline Report Section 1 Household Tables. Ministry of Housing, Communities and Local Government. Available from: https://www.gov.uk/government/statistics/ english-housing-survey-2016-to-2017-headline-report (accessed 21.12.2018) (Fig. 2.8).

versus 60.7 for privately owned dwellings. The mean rating for the social rent sector is higher, at 67.3 (see Fig. 5).

There are various possible explanations for this surprising finding. Firstly, the quality of SAP ratings on existing buildings is not as good as it might be. For example "mystery shopper" research in 2014 found that different assessors often rate the same house inconsistently (DECC, 2014). The average range between maximum and minimum ratings on the same house was 11.1 points. Also the range was greater for older properties (built pre 1900) and as we have seen, rented properties tend to be older. Discrepancies can be partly attributed to the fact that the efficiency calculations require input data much of which is hidden from direct view, such as the actual construction of walls. Also, the EPC rating relies on assumptions in build quality and the absence of defects. For example there is no need to perform an air tightness test. This is one of the key issues raised by stakeholder workshops conducted by a team from Leeds Becket University (Fylan et al., 2016).

The UK government has concerns about these and other issues around EPC quality, as related in a call for evidence on the topic (BEIS and HLCG, 2018). For example they are concerned about bias. "…there is now a financial value attached to a higher EPC rating, which could mean that building owners or third parties (e.g., letting agents) try to 'game' the system. This could be by deliberately misleading assessors (for example by making it appear that insulation exists which is not there or does not cover the whole building), or by putting assessors under pressure to generate better EPC scores by tweaking data inputs." An alternative approach would be to measure buildings' actual performance but this is difficult to do, at least while there are people living in the dwellings. The government is currently exploring the possibility of deriving actual dwelling energy efficiency using data from smart meters and smart thermostats.

It is very important that EPC ratings are accurate and trusted. Otherwise there will never be a premium for higher ratings and any policy based on ratings will be flawed. This is not an issue specific to the UK. For example stakeholders in Denmark were very clear on the issue (Ástmarsson et al., 2013).

Size is another possible explanation for the surprisingly high SAP ratings for private rented homes despite the relatively poor insulation. Fig. 6 compares floor area for private rented homes and owner occupied housing. Note that the bars show cumulative percentages. They show that private rented homes are small: nearly half of private sector homes are less than $70 \, m^2$ compared with less than a fifth of owner occupied homes. For such small homes the cost of hot water is a large component of the heating bill and upgrading the heating system can make a substantial improvement in SAP rating because it helps with hot water as well. Private rented homes are slightly more likely to have a condensing boiler (71% of PRS boilers are condensing compared to 67% of owner occupied). However, the social rented sector beats both soundly, with 82% condensing boilers.

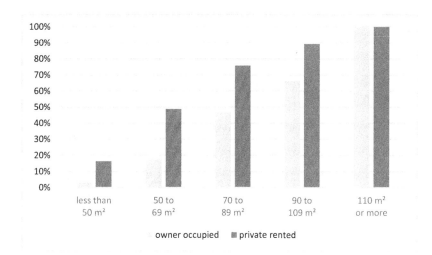

FIG. 6

Dwelling size by tenure, cumulative percentage.

6 Comparing thermal comfort of private sector homes with other sectors

Whether or not the SAP cost rating evidence is a true reflection of reality, private sector tenants seem to be less comfortable in their homes. The Energy Follow Up Survey 2011 (EFUS) collected more detailed survey data from a subsample of the EHS homes from that time and installed temperature loggers in some. One of the questions asked was "During a typical winter (December to February), can you normally keep comfortably warm in your living room." Overall only 6.3% answered this with a "no", a total of 174 respondents, so the sample is rather small for conclusive analysis. However, private tenants were four times more likely than owner occupiers to be uncomfortable: 12% of tenants compared to only 3.5% of owner occupiers. Table 5 gives details and lists segments where the proportion unsatisfied was at least 10%. Notably, poor SAP ratings do not meet this threshold as even in the worst homes only 7% of residents were unsatisfied. However where the rating was 70 or better (A–C) this proportion was down to 4%.

Some of these characteristics are to do with the building and some are to do with the occupants. The most significant characteristic was lack of employment: households with residents of working age but with no-one working were six times as likely to report being cold as households with pensioners. This is hard to explain. Perhaps it is because pensioners are very likely to own their home and hence have lower fixed housing costs, leaving more income to spend on keeping warm.

For occupants who were not warm there was also a follow up question asking why. Approximately a third of unsatisfied occupants claimed this was because of

Table 5 Factors relating to the likelihood of reporting not being comfortable in the living room.

Characteristic	Percentage not comfortably warm	Notes
All homes	6.3	
Owner occupier	3.5	
Private rent	12	
Social rent	11	
Flat	12	In 2016, 37% of private rented dwellings are flats, and 47% of social rents, compared to 8.5% owner occupied. Flats are likely to have electric heating
$<50\,m^2$	11	In 2016, 15% of private rented dwellings are this small and 25% of social rents compared to only 3% of owner occupied
Electric heating	13	Also associated with rented property, see Table 4
Responsible person aged 16–34	11	Older groups were more comfortable. For all age groups of 55+, the proportion who were not was down to 3%
None of the occupants working and none retired	18	The proportion for homes with someone in full time work was only 6% and the least likely were homes with no-one working but one or more retired: 3%
Lowest quintile for income of the responsible person and partner	11	The proportion decreases steadily for wealthier quintiles down to 3% in the highest income quintile.
In fuel poverty (using the indicator of high cost and low income)	13	

Data from the EFUS (2011) reported In Hulme, J., Beaumont, A., Summers, C., 2013b. Energy Follow-Up Survey 2011 Report 7: Thermal Comfort and Overheating. Buildings Research Establishment. https://assets.publishing.service.gov.uk/government/uploads/system/uploads/attachment_data/file/414600/7_Thermal_comfort.pdf (accessed 21.12.2018) (All tenures).

cost and two thirds claimed that the heating system was inadequate. (These proportions include those who reported both problems).

Only a fraction of homes in the EFUS survey had temperature loggers installed, and only 48 of them in households that reported they were not warm. This sample size is rather small for conclusive analysis. However the difference in mean temperature between occupants who were warm and those who were not was very small. Temperature was monitored during 3 months and in two of these the difference in mean living room temperature was <0.1 °C. The largest difference, in February 2011, was still only 0.6 °C. However, this was not statistically significant given the size of the sample.

Nor is there evidence for differences in temperature between private tenures, on average. Other analyses of the EFUS dataset (Hamilton et al., 2013; Hulme et al., 2013a) found no significant difference between mean temperature in the living room or in the bedroom, between owner occupied and private rented homes. Even homes in fuel poverty seemed to have similar temperatures. However social rent homes were significantly warmer as shown in Fig. 7. Research on a different dataset, unfortunately even smaller, with only 13 homes from the private rent sector,[a] found that these homes were warmer than owner occupied homes by 0.97 °C and social rent homes were also warmer (Kelly et al., 2013).

These temperature findings can be explained by factors relating to both the dwelling and the occupants. Firstly, temperature is strongly related to SAP rating, though there are other factors. The link is shown in Fig. 7: homes with poor ratings were considerably colder. This means that similar temperatures between owner occupied

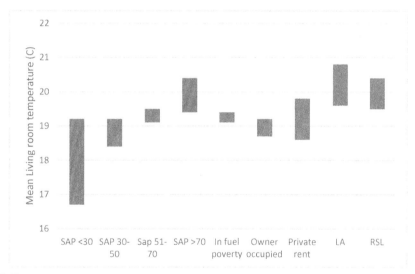

FIG. 7

Mean living room temperature by various factors. The *bars* show 95% confidence ranges. LA (local authority) and RSL (Registered Social Landlord) are forms of social rent.
Data from the EFUS (2011) reported in Hulme, J., Beaumont, A., Summers, C., 2013a. Energy Follow-Up Survey 2011 Report 2: Mean Household Temperatures. Buildings Research Establishment. https://assets.publishing. service.gov.uk/government/uploads/system/uploads/attachment_data/file/274770/2_Mean_Household_ Temperatures.pdf (accessed 21.12.2018) (Table 10).

---

[a]There seems to be a dearth of monitored temperature data in rented homes generally, especially private rented homes. This particular dataset had 71% owner occupied dwellings and only 11% privately rented. In the EFUS sample with monitored temperatures the proportion of homes in private rent was only 8%.

homes and private tenants is consistent with their similar thermal efficiency (as discussed above), and increased temperatures in social housing are consistent with the better efficiency in those homes.

If the rented homes have the same temperature as owner occupied homes, how can it be that their residents are less comfortable? It could be simply that air temperature is not the best measure of thermal comfort. For example draughts and low radiative temperature decrease comfort. However, both of these effects would also reduce air temperature so that seems unlikely.

Alternatively, since there is only a small proportion of tenants saying they are cold (12%), this could be due to more variation in the rented group than in owner occupied, with those tenants suffering low temperatures balanced by other tenants who are warmer than average. This proposition is supported weakly by the SAP data, which shows more private rented dwellings than owner occupied at both ends of the rating spectrum. In 2016 6.6% of private rented dwellings were F or G compared to 5.3% of owner occupied. However, the SAP ratings do not directly reflect damp, and we have seen that more rented homes are damp (Table 4) and therefore more likely to be cold; this is likely to put a small excess proportion of tenants at the cold end of the spectrum.

Another factor noted by Kelly et al. (2013) is that tenants, in particular social tenants, are less likely to be employed and therefore spend more time at home, so are likely to heat their homes for more hours in the day. This means that the mean temperature can be similar in the rented homes while still not reaching the desired temperature during heating hours. This is also consistent with the observation (Table 5) that residents who are not in employment are more likely to say they are cold. However this effect is also weak. Kelly found that each additional hour of heating raised the average temperature by 0.07 °C. An additional 7 h of heating leads to an effective increase of only 0.5 °C. Also, this would only apply to dwellings where the residents are unemployed: around 33% for private tenants and 70% for social tenants compared to 9% of owner occupiers (Kelly et al., 2013).

Another possibility also relates to the occupants: it could be due to general dissatisfaction leading to a negative attitude encompassing thermal comfort that is not justified by actual temperatures. Tenants are more likely to be unhappy about their accommodation than owner occupiers (9% for private tenants and only 3% for owner occupiers). Social tenants were even more likely to be unhappy (13%). There are many possible reasons. For example dissatisfaction seems to be strongly related to overcrowding, with 24% of residents in overcrowded accommodation unhappy in 2016/17 (HCLG, 2018b, table FA5401). This is consistent with the fact that rented dwellings tend to be much smaller than owner occupied homes, particularly in the social sector, as we have already seen. Also, tenants are often unsatisfied with the way that landlords repair and maintain their properties, again more so for social tenants than private tenants: 17% of private tenants and 24% of social tenants were unsatisfied (HCLG, 2018b, table FA5423).

If we accept that mean temperatures are similar in owner occupied and rented homes (even if levels of thermal comfort differ), then it is possible to triangulate the energy efficiency data using actual energy use. The National Energy Efficiency Data-framework is a large dataset correlating energy use from domestic dwellings with a limited range of property characteristics. This shows that rented homes use less gas than owner occupied homes but the difference is driven primarily by property size. Within each size band rented homes still use less gas but the difference between private tenants and owner occupiers is at most 10%. In the smallest size band (less than $50\,m^2$) the difference is <2%. This dataset is large with at least 45,000 dwellings in each group (BEIS, 2018d) so the findings are robust. This suggests that tenants are achieving similar temperatures as owner occupiers while consuming no more heating energy for comparable properties. If anything, rented homes are more efficient.

All the evidence points to the conclusion that owner occupied dwelling stock is no more efficient than private rented housing on average, only it tends to be bigger. In the social rent sector the dwellings are even smaller but the efficiency is better still and the rooms are warmer, at least on average.

---

7 Take up of energy efficiency measures

The private rented sector is under represented in homes in receipt of grants for energy efficiency from the Energy Company Obligation (ECO) which is the main source of grants since 2013 (Box 4).

Overall, the private rented sector is 20.5% of dwellings (HCLG, 2018b). However, only 15% had received ECO grants up to Sep 2017 (BEIS, 2018a). Also, the sector is even more under represented for grants that go primarily to insulation measures. Private rented homes received only 10% of grants for obligations supplying cavity wall insulation and loft insulation. In contrast, they are over represented (26%) for grants in the Affordable Warmth obligation, supplying new or upgraded boilers and other heating measures such as new heating controls (see Fig. 8). This shows that landlords (at the moment) are much more interested in heating systems than insulation. This is consistent with the finding already mentioned that private rented sector boilers are more likely to be efficient condensing boilers. Private rented homes tend to be small so that water heating efficiency is an important factor in overall costs; upgrading a boiler improves both water heating and space heating efficiency.

Also, fitting a new boiler or heating system is generally quicker and less hassle than installing insulation, especially either external or internal insulation for solid walls (the author has personal experience of both). The mess is confined to a few areas and there is no need to empty rooms (for internal insulation) or disrupt access routes or garden plants (external insulation). This disruption is potentially a serious issue for tenants if they are in residence at the time. If they are not then it could prolong the period when the property is vacant, leading to loss of income for the landlord.

Box 4 The Green deal and the energy company obligation (OFGEM, About the ECO Scheme. https://www.ofgem.gov.uk/environmental-programmes/eco/about-eco-scheme)

The Green Deal was a package of related measures introduced by the UK government in 2013. The aim was to roll out energy efficiency improvements in buildings at a large scale. Green Deal assessors were trained to conduct energy efficiency audits and recommend appropriate measures. Green Deal providers and installers were authorized to deliver those measures, funded either privately or through grants or Green Deal loans.

The main source of grants was (and remains) the Energy Company Obligation which requires electricity retail companies to finance energy efficiency measures to households. Targets were set in terms of tonne of carbon emissions saved and companies could meet these targets through auctions where Green Deal providers would offer carbon saving projects to be funded. The price of the carbon savings were thus market driven. However, to ensure that financing was delivered as intended there were three separate targets:

- Affordable Warmth, for households in receipt of benefits (in the case of rented property either the landlord or the tenant can be in receipt of benefits).
- Carbon Savings Community, for regions with a high index of multiple deprivation (and a sub target for rural areas)
- Carbon Saving Target, for measures that are expensive to deliver such as insulation for hard-to-treat cavity walls or solid walls.

The Green Deal loan scheme was a pay-as-you-save plan with payments added to electricity bills. The loan was tied to the electricity supply rather than any individual so would be paid by whoever paid that bill. If the property was sold the new owner would be responsible. In most rented property the tenants pay the bills and the duty to pay would pass from each tenant to the next. Landlords would only be liable while the property was vacant.

One of the main criticisms of the Green Deal loan scheme was the high interest rate. Initially this was 6.9%, much higher than usual for a mortgage or secured loan but this was not a fair comparison since the Green Dean loan had no security. There was also a "golden rule" that the loans would be no longer than the lifetime of the measure (or 25 years) and overall bills would not increase because the loan payments would be less than the predicted energy savings. The high interest rate meant many measures would fail to meet the golden rule.

However, earlier schemes have had a high take-up rate for other measures in the private rented sector. Between 2002 and 2007, 40% of private rented sector homes had a "fabric measure" installed compared to 37% of owner occupied homes. In this study, "fabric measure" includes wall and loft insulation, glazing, and draught-proofing. (Hamilton et al., 2016). The grant schemes available at that time were Warm Front and the Energy Efficiency Commitment. For Warm Front, the tenants had to be receiving social welfare benefits (this is similar to the current Affordable Warmth scheme). The Energy Efficiency Commitment was an obligation on energy companies to meet a target energy saving, similar to the current Carbon Saving Target scheme, however half the savings had to be targeted at priority groups, in receipt of benefits. Most of the installed measures were insulation (cavity wall and loft insulation) and energy saving lighting—at the time this would have been compact fluorescents.

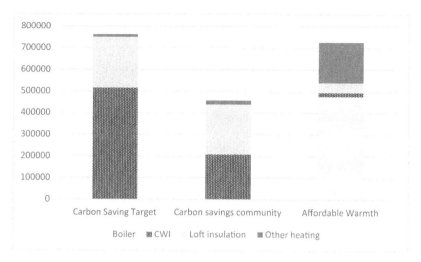

FIG. 8

ECO measures installed by obligation up to March 2017. Private rental sector dwellings are under-represented in Carbon Savings Target and Carbon Savings Community but over-represented in Affordable Warmth.

Data from BEIS, 2018a. Household Energy Efficiency National Statistics. Department for Business, Energy & Industrial Strategy. https://www.gov.uk/government/collections/household-energy-efficiency-national-statistics (accessed 21.12.2018) (Table 2.1.1).

In summary, take-up rates for energy efficiency measures have tailed off recently in the private rented sector except for those measures that are low hassle and reasonably effective in small properties. The rented sector has achieved similar overall efficiency to the owner occupied sector but by a slightly different route, deploying efficient heating in preference to reducing heat loss. However the overall standards of energy efficiency are still low and further improvement will not be possible without more fabric efficiency measures.

8 Reasons for poor take-up

It is widely believed that the main reason for slow take-up of energy efficiency measures in the private rented sector is due to the split incentive issue (Ástmarsson et al., 2013; Ambrose, 2015; Souza, 2018; Trotta, 2018). This problem arises across the world, and the following sub-sections include examples and data from other countries although the focus is on the UK as before. At its simplest, the split incentive issue is that landlords are not motivated to install energy efficiency measures when the tenants get all the benefits from lower heating bills. It is usual in the UK (and in many other countries) for energy bills to be the responsibility of the tenant, except in multiple occupancy homes. The Green Deal loan mechanism was supposed to deal with this by ensuring that tenants pay for the upgrade through their utility bills. However, very few Green Deal loans have been arranged, for a variety of reasons. (For details see Rosenow and Eyre (2016)). As of September 2017, 1.7 million households had benefitted from ECO grants whereas just under 14,000 households (in all tenures) had received Green Deal finance (BEIS, 2018a).

In this section we will look at the issues first from the tenants' point of view and then from the landlords' point of view.

First, however, it is important to note that in the UK there is a severe housing shortage which means that rented accommodation (and housing of all types) is in extreme demand. The requirement for new homes is variously described as between 250,000 and 340,000 per year. However, build rates have not exceeded 200,000 per year for at least 10 years (Wilson and Barton, 2018). Finding a nice place to live at an affordable rent is a huge challenge for tenants.

8.1 The tenants' view

There are two main data sources for this section. The first is a report on research with focus groups involving potential Green Deal consumers (Quadrangle, 2011). There were six focus groups of residential tenants with a range of ages and incomes. The second is an online survey conducted by YouGov for the charity Shelter (Gousy, 2014). There were 4544 respondents who were private tenants. Both reports also included landlords in their research as well as tenants.

Legally, tenants are entitled to ask their landlords for energy efficiency improvements. Under the current UK regulations, the landlord will usually have to comply if the measure can be installed with no upfront cost. (The Green Deal loan scheme was introduced to make this possible. The loan is paid back through energy bills and hence is the tenant's responsibility). In practice, very few tenants will ask for improvements for a variety of reasons. (After consultation, the government has decided to change the no upfront cost rule and landlords will be expected to contribute up to £3500 per dwelling).

Firstly, given the tight housing market, tenants are wary of doing anything to upset their landlord. Asking for energy efficiency measures will at the very least take up time and draw attention to themselves as a potential trouble maker. Even when conditions are very poor, many tenants will not complain. "A key reason why renters do not report poor conditions is because they fear retaliatory eviction—one in eight renters have not asked for repairs to be carried out in their homes or challenged a rent increase in the last year because they fear eviction." (Gousy, 2014).

Secondly, if there is any cost to the measure the tenant usually has to pay and even when the savings on their energy bill are supposed to cover the cost (as with a Green Deal loan under the Golden Rule), there is no guarantee that they will. The consequences of rising bills can be serious: "The main problem is not having a guarantee that this charge won't end up costing you more. Sometimes I don't have enough money at the end of the month, I wouldn't want to risk this." Also, some tenants on very low income rely on being able to reduce their energy bills to zero if they really have to. However with a Green Deal loan they would still have to make the loan payments. (Quadrangle, 2011).

Besides which, if the house is currently cold and the tenants wish for insulation to improve warmth, the bill savings are likely to be very low and not cover the cost—and if the landlord funds the measure there is a fear that they will increase the rent (Eadson et al., 2013).

Some tenants resent being asked to pay the whole cost of improvements on the grounds that the landlord benefits too (in their view). "This is your landlord's property, it's their investment, it's their asset and I think they should be sharing the costs." (Quadrangle, 2011).

Thirdly, there is the hassle factor, which surely applies to tenants as much as home owners, compounded by the expectation that they will move on in a few years anyway. The average length of a tenancy in the UK is less than 2 years (Millman, 2017). Tenants generally prefer to make investments that they can take with them when they move on. Besides which, even among owner occupiers, it usually takes a while before they get around to making energy efficiency improvements. Nearly three quarters of energy efficiency retrofits are done after 2 years of residence (Trotta, 2018).

Long tenancies were related to more efficient appliances in a study of US rented accommodation (Souza, 2018). Their analysis was not able to determine why but they suggest two possibilities. It could be because people who intended to stay a long time were more choosey about where they rented. This would be unlikely in the UK context because of the tight market. Alternatively, it could be that landlords were more likely to give their tenants efficient appliances after they had established good terms and were confident that the tenants would take care of them and stay a long time. This attitude has been observed in Australia (Gabriel and Watson, 2012) and would very likely apply in the UK market too.

8.2 **The landlords' view**

The main data sources used in this section are surveys of landlords in the North of England in 2012: 30 interviews in an unnamed town commissioned by the local authority (Ambrose, 2015) and 20 landlords in Rotherham (Eadson et al., 2013). There was particular interest in this area at that time as a result of the launch of the Green Deal.

The lack of guarantee on bill savings is a concern for landlords too, as this could affect the tenant's ability to pay the rent. Also, between tenants when the property is vacant the landlord is responsible for paying off the loan (Eadson et al., 2013).

Most landlords consider their property as an investment—a long term financial asset (Gousy, 2014), and it would be logical for them to consider spending money on it if this would increase the value. This can be achieved in two ways: through increased capital value and increased income from rent. However, both are uncertain.

A survey of landlords in the north of England (Ambrose, 2015) found that they did not believe tenants would accept higher rents. "Unfortunately, much as we would like to be able to improve energy efficiency in our properties, it's effectively dead money as we'd never see it back. No tenant would be happy with a rent increase to help cover the costs." (quote from a landlord with 40 properties). A survey of landlords in Rotherham found similar views. (Eadson et al., 2013).

Capital value increase is possible but also uncertain. A national survey of the impact of improved SAP ratings on house prices found energy efficiency premiums in some areas but not all. The highest premiums were in the North, but in the South East there were none at all (Fuerst et al., 2013). Rotherham is in Yorkshire, where the premium was reported as 9% for an increase of just one grade from G to F. However, a Rotherham letting agent surveyed in 2013 considered that only houses in the better areas would increase in value (Eadson et al., 2013).

It is sometimes argued that dwellings with better energy efficiency area easier to let. However landlords are under the impression that tenants do not care very much about energy ratings. Neither tenants nor landlords consider that the energy rating is an important factor for the tenants' choice of home to rent (Quadrangle, 2011). Tenants seldom ask for them. "Nobody is interested in EPC ratings, all tenants are concerned with is the condition of the place and how much you are charging." (quote from a letting agent with 200+ properties) (Ambrose, 2015). Landlords believe that tenants are more interested in more cosmetic issues: "I've never known anyone to ask whether the loft is insulated but people do notice how old the kitchen is or if the carpets are stained...those things are the deal breakers" (Eadson et al., 2013).

These attitudes are consistent with those found in a survey of students and staff at Dublin University (Carroll et al., 2016). When asked to rank considerations for rental accommodation, energy ratings came last, below rent, distance to work, safety in the area, condition of the property and size. However their willingness to pay (determined by stated preference using discrete choice experiments) showed that energy ratings were more of an issue at the bottom end than at the top of the energy ratings scale. They suspect this was because low energy ratings are associated with poor comfort as well as higher energy costs. There was no extra willingness to pay for an A rating over a B.

Landlords do expect to spend money on maintaining their property, in order to maintain its value (Gousy, 2014). However they often do not distinguish between general maintenance and energy efficiency improvements—it is all property maintenance. "We renovated the property when we bought it. Extended it, re-roofed it, put in gas central heating, insulated the loft and it's still only an E rating." "We re-wired, put in new kitchens, decorated and put in double glazing, what more can we do?" (Eadson et al., 2013).

It is important to landlords that their property is "lettable". For most that means prioritizing cosmetic improvement. Only a fifth of landlords consider that insulation is even relevant (Ambrose, 2015). Nearly half of landlords in a survey of 53 (47%)

believe tenants are perfectly satisfied with the efficiency of their accommodation (Hope and Booth, 2014). For those that had installed efficiency upgrades the top reasons were to improve thermal comfort for the tenants (47%) or to increase marketability (37%). Reducing running costs came third (33%) (Hope and Booth, 2014). However, upgrades were very rarely instigated by a direct request from the tenant (7%). This suggests that some landlords, at least, think that properties that are more comfortable are easier to let.

Landlords' perceptions of their tenants' experiences are not necessarily accurate. A study in New Zealand including 107 landlords and 126 private tenants found that landlords consistently overestimated the comfort of their tenants. For example 66% of tenants reported that their bedrooms were too cold in winter while the landlords predicted 33% (Phillips, 2012).

Sadly, poor energy efficiency is regarded as the norm in many areas of the UK, especially in older housing. (As mentioned above, there is a high proportion of older properties in this sector). "I know that it's [the property] an E rating from the EPC but I also know that it's no different to the vast majority of houses in [the town]. They were all built at around the same time and are not up to modern standards. So I'm offering the same product as everyone else." (Landlord, 10 properties) (Ambrose, 2015). It is usually possible to achieve an E without resorting to solid wall insulation, especially for terraced homes.

Landlords are averse to any kind of loan finance. Most would only make improvements to their property if they had sufficient cash or equity in the property. "I would only consider it if there was an essential repair that I couldn't fund any other way. Even then I'd rather sell the property on than resort to borrowing. Borrowing's not in my business plan. I only buy properties when I have the cash." This was said by a landlord with 18 properties. Even a Green Deal loan is regarded as dubious, because landlords are concerned it might be harder to let a property with such a loan attached, plus they would be liable if tenants defaulted (Ambrose, 2015). In fact the landlords were not quite correct in this—they would only be responsible during vacant periods.

The fact that most landlords are part time in the business is another factor. Many landlords wish to have as little to do with their tenants as possible, especially if they employ an agent (Quadrangle, 2011). Many are ignorant even of regulations that apply to them: only 15% have heard of the Housing Health and Safety Rating System (HHSRS); for that matter only 50% of agents have (DCLG, 2011). Finding out about energy savings options and evaluating the benefits would take time, and time is valuable. Landlords are less likely to dedicate time to acquiring knowledge regarding the drawbacks, benefits, supply, financing and installation of an innovation than those who pay the energy bills (Ambrose, 2015).

In summary, landlords are averse to investing time and money in energy efficiency because they do not believe that tenants are interested in the issue. They do not believe they will get higher rents and the capital value increase is uncertain too. They regard poor energy efficiency as the norm.

9 Impact of minimum energy efficiency standard

UK regulations for energy efficiency requirements for rented homes have recently been made more strict, albeit only a little. There has been a requirement since 2008 for properties to have a certificate for tenants to be informed of the energy rating. However there was still no required standard at that time. The Minimum Energy Efficiency Standard (MEES), requiring that a rented property must be at least an E rating, was announced in 2015 and took effect in April 2018.

At the moment this only applies to new tenancies but from 2023 it will apply to all properties. Also the UK Government's Clean Growth Strategy 2018 has signaled an intention to ratchet up energy performance over time, perhaps to a minimum C rating by 2030, though the wording in the policy is not definitive (BEIS, 2018c). The current requirement of level E affects relatively few properties. As of 2016, only 6.6% of rented homes were rated below it, down from 25.3% in 2006. Owner occupied homes have improved greatly over the same time frame but from a higher base, as shown in Table 6. Most of these improvements happened before MEES was announced.

There is a long list of reasons for exemption in MEES (BEIS, 2018b). The most serious is probably the rule that says the landlord does not have to make improvements that will have upfront cost. If they have completed all the recommendations on the EPC that can be funded under the Golden Rule (i.e., where bill savings cover the cost) and there is no other finance available (such as the ECO grant scheme), then the property is exempt. As of the end of July 2018 there were < 1900 exemptions registered for domestic rented properties and 59% of these were for lack of funding. Another 20% of exemptions were because all relevant improvements had been made (i.e., all those which would have qualified under the Golden Rule) (PRSRegisterFeedback, 2018).

A smaller but still significant number of exemptions (12%) were due to the negative impacts of insulation. This exemption is allowed "where the landlord has obtained written expert advice which indicates that the measure is not appropriate for the property due to its potential negative impact on the fabric or structure of the property (or the building of which the property forms a part)" (BEIS, 2018b). Some building professionals have concerns that the impacts of insulating older

Table 6 Percentage of homes that are F and G rated (i.e., below the MEES standard)

Tenure	2006 (%)	2016 (%)
Private rented	25.3	6.6
Owner occupied	18.9	5.3
Social rented	17.3	4.8

Data from HCLG, 2018a. English Housing Survey Headline Report Section 2 Housing Stock Tables. Ministry of Housing, Communities and Local Government. Available from: https://www.gov.uk/ government/statistics/english-housing-survey-2016-to-2017-headline-report (accessed 21.12.2018) (the proportions in 5 and G ratings are from AT2.6).

buildings (especially those with solid walls) are poorly understood. They say that the performance improvements are often less than expected, for example if the unimproved fabric is better than models assume. Also there can be a risk of lasting damage to the fabric, for example due to trapped moisture (STBA, 2012).

Finally, these regulations do not apply in the same way to HMOs as to other rented sector housing. The minimum efficiency level E applies, but only if they have an EPC and HMOs will not necessarily have one, unless they have been sold or rented out *as a whole building* within the previous 10 years. Even if they do have an EPC, some professionals have concerns that the assessment is not appropriate for HMOs. For example a house might have reasonable performance as a whole but have a poorly insulated loft conversion. If a single household were resident they could avoid using the loft to some extent but a tenant renting that room has no choice (National Energy Action, 2016).

10   **Summary and conclusions**

The UK is fairly typical of the EU in the proportion of homes with private rent but has more than average reduced rent properties. Unfortunately the latter category is decreasing due to the Right to Buy policy and the private rented sector is increasing. This affects young people most: 46% of people under 35 years old now rent, compared to only 21% 20 years ago.

The UK is worse than the EU average for households spending a high proportion of their income on housing costs.

Most landlords have a small number of properties supplying a fraction of their income and their main business is in other sectors. They regard their properties as an investment as well as a source of income and they are averse to taking out loans to pay for upgrades. Only 8% of landlords are full time in the business and 80% of landlords have no relevant qualifications.

Private rented homes tend to be smaller and older than owner occupied homes. They are less likely to have wall insulation, (especially solid wall insulation) but slightly more likely to have a condensing boiler. The private rented sector has been more eager to take up grant support for heating system upgrades than insulation, at least in recent years. These upgrades are less disruptive than insulation and are effective in small properties as they improve hot water efficiency as well as space heating.

There is no difference between the private rented sector and owner-occupied dwellings for average energy cost rating. This may be due to inaccuracy in the cost ratings, though the similarity is corroborated by similar actual energy use (when adjusted by size) and similar measured living room temperatures. Energy efficiency in the dwelling has a significant effect on temperature, with efficient dwellings being warmer; this applies across tenures.

On the other hand, tenants are more likely than owner occupiers to complain that they are not warm enough in their homes. This finding is not incompatible with the similar average temperature finding; there are a variety of possible explanations. There is some evidence for the following:

- More rented homes appear at both ends of the efficiency spectrum, so that there are more cold homes and also more warm ones.
- Some tenants spend more time at home and therefore run the heating for longer hours, so the mean temperature can be the same as in another home that is heated to a warmer target temperature but for less time.
- Tenants may report dissatisfaction because of negative attitudes due to other issues.

There is a wide range of reasons for slow rates of insulation upgrades in the private rented sector. Tenants expect their landlord to pay for upgrades because it is the landlord's asset and therefore the landlord benefits from the investment. However, landlords doubt that insulation will increase the property value significantly (the evidence for this is mixed) and are also doubtful they will be able to increase rents. Green Deal Finance was supposed to allow for upgrades to be paid for through tenants' energy bills. However, tenants are reluctant to take this on, possibly because they do not believe in the savings, or because they are reluctant to take on debt of any sort. Also, landlords believe that poor energy efficiency is normal, especially for older homes (and it seems that owner occupied homes are just as bad). There is little or no market pressure to upgrade.

Minimum Energy Efficiency Requirements (MEES) came into effect in April 2018. This requires rented property to have an SAP rating of E or better. However this affects relatively few properties (the average is D) and there are many grounds for exemption. The most common reasons given are that finance is not available or that there are no measures that can be made that are cost effective. Also many HMOs are not required to have an EPC and so the MEES regulations do not even apply.

Since MEES relies on the SAP ratings for measuring efficiency, inaccuracy in the assessments could undermine the whole scheme. The government has recognized this and they are exploring alternative ways to measure energy performance of dwellings accurately—but it is a hard problem.

As noted above, the proportion of privately rented housing in the UK has steadily increased, over the past few decades, in comparison to social rented and owner occupied housing. The onus is falling increasingly on private landlords who face all the difficulties and reluctance discussed above, to improve the efficiency of a housing stock that is already poor by western European standards.

References

Ambrose, A., 2015. Improving energy efficiency in private rented housing: what makes landlords act? Indoor Built Environ. 24 (7), 913–924.

Ástmarsson, B., Jensen, P.A., Maslesa, E., 2013. Sustainable renovation of residential buildings and the landlord/tenant dilemma. Energy Policy 63, 355–363.

Balfour, R., Allen, J., 2014. Local Action on Health Inequalities: Fuel Poverty and Cold Home-Related Health Problems. UCL Institute of Public Health. https://fingertips. phe.org.uk/documents/Fuel_poverty_health_inequalities.pdf. (Accessed 21 December 2018).

Barker, N., 2017. Exclusive: 7% Rise in Former Right to Buy Homes Now Rented Privately. Inside Housing. https://www.insidehousing.co.uk/news/news/exclusive-7-rise-in-former-right-to-buy-homes-now-rented-privately-53507. (Accessed 21 December 2018).

Barton, C., 2017. Home Ownership and Renting: Demographics. House of Commons Briefing Library. https://researchbriefings.parliament.uk/ResearchBriefing/Summary/CBP-7706. (Accessed 21 December 2018).

BEIS, 2018a. Household Energy Efficiency National Statistics. Department for Business, Energy & Industrial Strategy. https://www.gov.uk/government/collections/household-energy-efficiency-national-statistics. (Accessed 21 December 2018).

BEIS, 2018b. The Domestic Private Rented Property Minimum Standard. Department for Business, Energy & Industrial Strategy. https://www.gov.uk/government/publications/the-private-rented-property-minimum-standard-landlord-guidance-documents. (Accessed 21 December 2018).

BEIS, 2018c. Clean Growth Strategy: Executive Summary. Department for Business, Energy & Industrial Strategy. https://www.gov.uk/government/publications/clean-growth-strategy/clean-growth-strategy-executive-summary. (Accessed 21 December 2018).

BEIS, 2018d. National Energy Efficiency Data-Framework (NEED) Table Creator. Department for Business, Energy & Industrial Strategy. https://www.gov.uk/government/statistical-data-sets/need-table-creator. (Accessed 21 December 2018).

BEIS, HLCG, 2018. Call For Evidence Energy Performance Certificates for Buildings. Department for Business, Energy & Industrial Strategy and Ministry of Housing, Communities and Local Government. https://www.gov.uk/government/consultations/energy-performance-certificates-in-buildings-call-for-evidence. (Accessed 21 December 2018).

BRE, 2001. Difference in the Rating Scale Between SAP 1998 and SAP 2001. http://projects. bre.co.uk/sap2001/SAP_1998_-_SAP_2001.pdf. (Accessed 7 January 2019).

Carroll, J., Aravena, C., Denny, E., 2016. Low energy efficiency in rental properties: asymmetric information or low willingness-to-pay? Energy Policy 96, 617–629.

DCLG, 2006. A decent home: definition and guidance for implementation. Department for Communities and Local Government. Available from: https://www.gov.uk/government/publications/a-decent-home-definition-and-guidance (Accessed 24 June 2019).

DCLG, 2011. Private Landlords Survey 2010. Department for Communities and Local Government. https://www.gov.uk/government/statistics/private-landlords-survey-2010. (Accessed 21 December 2018).

DECC, 2014. Green Deal Assessment Mystery Shopping Research. Department for Energy and Climate Change. https://www.gov.uk/government/publications/green-deal-assessment-mystery-shopping-research. (Accessed 21 December 2018).

Eadson, W., Gilberson, J., Walsh, A., 2013. Attitudes and Perceptions of the Green Deal Amongst Private Sector Landlords in Rotherham. Sheffield Hallam University, Centre for Regional Economic and Social Research. https://www4.shu.ac.uk/research/cresr/sites/shu.ac.uk/files/green-deal-landlords-rotherham-summary.pdf. (Accessed 21 December 2018).

EST, 2018. Our Calculations. Energy Saving Trust. http://www.energysavingtrust.org.uk/about-us/our-calculations. (Accessed 23 October 2018).

Fuerst, T., McAllister, P., Nanda, A., Wyatt, P., 2013. An investigation into the effect of EPOC ratings on house prices. In: Report for DECC. https://www.gov.uk/government/publications/an-investigation-of-the-effect-of-epc-ratings-on-house-prices. (Accessed 21 December 2018).

Fylan, F., Glew, D., Smith, M., Johnston, D., Brooke-Peat, M., Miles-Shenton, D., Fletcher, M., Aloise-Young, P., Gorse, C., 2016. Reflections on retrofits: overcoming barriers to energy efficiency among the fuel poor in the United Kingdom. Energy Res. Soc. Sci. 21, 190–198.

Gabriel, M., Watson, P., 2012. Supporting sustainable home improvement in the private rental sector: the view of investors. Urban Policy Res. 30, 309–325.

Gousy, H., 2014. Can't Complain: Why Poor Conditions Prevail in Private Rented Homes. Shelter. https://england.shelter.org.uk/__data/assets/pdf_file/0006/892482/6430_04_9_Million_Renters_Policy_Report_Proof_10_opt.pdf%20. (Accessed 21 December 2018).

Hamilton, I.G., O'Sullivan, A., Huebner, G., Oreszczyn, T., Shipworth, D., Summerfield, A., Davies, M., 2013. Old and cold? Findings on the determinants of indoor temperatures in English dwellings during cold conditions. Energ. Buildings 141, 142–157.

Hamilton, I.G., Summerfield, A.J., Shipworth, D., Philip, J.P., Oreszczyn, T., Lowe, R.J., 2016. Energy efficiency uptake and energy savings in English houses: a cohort study. Energ. Buildings 118, 250–276.

HCLG, 2016. New Housing Rental Rules to Protect Thousands of Tenants. Ministry of Housing, Communities and Local Government. https://www.gov.uk/government/news/new-housing-rental-rules-to-protect-thousands-of-tenants. (Accessed 21 December 2018).

HCLG, 2018a. English Housing Survey Headline Report Section 2 Housing Stock tables. Ministry of Housing, Communities and Local Government. https://www.gov.uk/government/statistics/english-housing-survey-2016-to-2017-headline-report. (Accessed 21 December 2018).

HCLG, 2018b. English Housing Survey Attitudes and Satisfaction. Ministry of Housing, Communities and Local Government. https://www.gov.uk/government/statistical-data-sets/attitudes-and-satisfaction. (Accessed 21 December 2018).

HCLG, 2018c. English Housing Survey Headline Report. Ministry of Housing, Communities and Local Government. Available from: https://www.gov.uk/government/statistics/english-housing-survey-2016-to-2017-headline-report (Accessed 21 December 2018).

Hope, A.J., Booth, A., 2014. Attitudes and behaviours of private sector landlords towards the energy efficiency of tenanted homes. Energy Policy 75, 369–378.

Hulme, J., Beaumont, A., Summers, C., 2013a. Energy Follow-Up Survey 2011 Report 2: Mean Household Temperatures. Buildings Research Establishment. https://assets.publishing.service.gov.uk/government/uploads/system/uploads/attachment_data/file/274770/2_Mean_Household_Temperatures.pdf. (Accessed 21 December 2018).

Kelly, S., Shipworth, M., Shipworth, D., Gentry, M., Wright, A., Pollitt, M., Crawford-Brown, D., Lomas, K., 2013. Predicting the diversity of internal temperatures from the English residential sector using panel methods. Appl. Energy 102, 601–621.

Millman, M., 2017. Average Tenancy Lengths Across the UK Revealed. Simple Business. https://www.simplybusiness.co.uk/knowledge/articles/2017/08/letting-agents-your-move-reveal-average-buy-to-let-tenancy-length-across-the-uk. (Accessed 21 December 2018).

National Energy Action, 2016. Fuel Poverty and Houses in Multiple Occupation: Practitioners' Views. https://www.nea.org.uk/research/research-database/fuel-poverty-houses-multiple-occupation-practitioners-views/. (Accessed 21 December 2018).

Palmer, J., Livingston, M., Adams, A., 2017. What Does it Cost to Retrofit Homes? Updating the Cost Assumptions for BEIS's Energy Efficiency Modelling. Cambridge Architectural Research. https://www.gov.uk/government/publications/domestic-cost-assumptions-what-does-it-cost-to-retrofit-homes. (Accessed 21 December 2018).

Phillips, Y., 2012. Landlords versus tenants: information asymmetry and mismatched preferences for home energy efficiency. Energy Policy 45 (C), 112–121.

PRSRegisterFeedback, 2018. Private Correspondence. 24 August 2018.

Quadrangle, 2011. Green Deal and the Private Rented Sector Consumer Research Amongst Tenants and Landlords, Report for DECC. https://assets.publishing.service.gov.uk/government/uploads/system/uploads/attachment_data/file/43019/3506-green-deal-consumer-research-prs.pdf. (Accessed 21 December 2018).

Rosenow, J., Eyre, N., 2016. A post mortem of the green deal: austerity, energy efficiency and failure in British energy policy. Energy Res. Soc. Sci. 21, 141–144.

SAP, 2012. SAP, the current version is BRE, 2014. BRE Group. Available from: https://www.bregroup.com/sap/standard-assessment-procedure-sap-2012/.

Scanlon, K., Whitehead, C., 2016. The Profile of UK Private Landlords. The council for Mortgage Lenders. https://www.cml.org.uk/news/cml-research/the-profile-of-uk-private/the-profile-of-uk-private-landlords-08.05.17.pdf. (Accessed 21 December 2018).

Shelter, 2016. Survey of Private Landlords. Shelter. https://england.shelter.org.uk/__data/assets/pdf_file/0004/1236820/Landlord_survey_18_Feb_publish.pdf. (Accessed 21 December 2018).

Souza, M.N.M., 2018. Why are rented dwellings less energy-efficient? Evidence from representative sample of the U.S. housing stock. Energy Policy 118, 149–159.

STBA, 2012. Responsible Retrofit Guidance Wheel. Sustainable Traditional Buildings Association. http://www.responsible-retrofit.org/wheel/. (Accessed 21 December 2018).

Trotta, G., 2018. The determinants of energy efficient retrofit investments in the English residential sector. Energy Policy 120, 175–182.

Wilson, W., Barton, C., 2018. Briefing Paper 07671: Tackling the Under-Supply of Housing in England. House of Commons Library. http://researchbriefings.files.parliament.uk/documents/CBP-7671/CBP-7671.pdf. (Accessed 21 December 2018).

Further reading

DCLG, 2012. English Housing Survey Homes 2010. Department for Communities and Local Government. https://www.gov.uk/government/statistics/english-housing-survey-homes-report-2010. (Accessed 21 December 2018).

OFGEM, 2005. Energy Efficiency Commitment. https://www.ofgem.gov.uk/ofgem-publications/58749/9520-eecjan05-pdf. (Accessed 21 December 2018).

Eurostat, 2017. Housing Statistics. http://ec.europa.eu/eurostat/statistics-explained/index.php/Housing_statistics. (Accessed 21 December 2018).

Gov.uk, 2012. Eligibility for Warm Front. National Archives. http://webarchive.nationalarchives.gov.uk/20121204221252/https://www.gov.uk/warm-front-scheme/eligibility. (Accessed 21 December 2018).

Hulme, J., Beaumont, A., Summers, C., 2013b. Energy Follow-Up Survey 2011 Report 7: Thermal Comfort and Overheating. Buildings Research Establishment. https://assets.publishing.service.gov.uk/government/uploads/system/uploads/attachment_data/file/414600/7_Thermal_comfort.pdf. (Accessed 21 December 2018).

Cold homes and Gini coefficients in EU countries

7

Ray Galvin

University of Cambridge, Cambridge, United Kingdom
RWTH Aachen University, Aachen, Germany

Chapter outline

1 Introduction 145
2 Measures of fuel poverty 149
3 The variables 151
   3.1 Which variables are relevant? 151
   3.2 Panel data or year by year regressions 154
4 Descriptive statistics 154
5 Results 158
   5.1 Tests for model fit 158
   5.2 Other statistical tests 159
   5.3 The coefficients 163
6 Discussion: The role of the Gini index 165
7 Conclusions 167
References 168
Further reading 171

1 Introduction

Although the EU and UK together make up the wealthiest society on earth (Piketty, 2014), Eurostat surveys indicate that around 10% of EU households say they are unable to keep their homes adequately warm (EU-SILC Survey, 2017). In Bulgaria, Lithuania, Greece and Portugal in 2016 this figure was 39%, 29%, 29% and 25% respectively, and in Britain and Germany 6% and 4%.

In Chapters 5, 6 and 10 of this book we discuss various aspects of fuel poverty and energy poverty, in the UK and more generally. In this chapter I do two things differently. Firstly, I focus on a specific variable: the percentage of households in EU countries who say they are unable to keep their homes adequately warm. This

Inequality and Energy. https://doi.org/10.1016/B978-0-12-817674-0.00007-2

145

is only one among many possible definitions of fuel poverty, and some would not even call it that. To save arguments over terminology, I will call it simply "the percentage of households who say they are unable to keep their homes adequately warm," and to save space I will shorten it to the "percentage of unheatable homes," or simply "UH%," and give it the symbol U. In Section 2 of this chapter I will explain why I think UH% is a useful variable to work with in a study of economic inequality.

Secondly, I will take a country-level view and test a hypothesis about one of the possible drivers of this percentage. Broadly speaking, I will ask, does a country's level of economic inequality influence its percentage of households unable to heat their homes?

The type of economic inequality that has developed over the past 40 years is characterized by increasing numbers of households at the bottom of the income and wealth spectrum and decreasing numbers in the middle and just below the very top (Galvin and Sunikka-Blank, 2018; and see Chapter 1). It has long been known that poverty is one of the three main determinants of fuel poverty, the others being energy-inefficient homes and high fuel prices (Bouzarovski and Simcock, 2017; Sovacool, 2015; Walker and Day, 2012). It has also been shown, quite recently in Europe, that poorer households are less likely to invest in energy efficiency measures for their homes, such as insulation, high-quality windows and boiler upgrades (BRISKEE, 2018).

But economic inequality can have further detrimental effects which could increase the proportion of households unable to heat their homes. Over the last decade Richard Wilkinson and Kate Pickett have mapped many countries' data on economic inequality against these countries' statistics for different measures of human wellbeing, such as longevity, housing affordability and educational achievement (e.g., Wilkinson and Pickett, 2010, 2017; Pickett and Wilkinson, 2015). They consistently report that indicators of human wellbeing are worse in countries with high levels of inequality, even if those countries are richer, on average, than other countries. An interesting aspect of their work is their identification of psychological factors triggered by economic inequality (Wilkinson and Pickett, 2017). They argue this has to do with a balance between humans' evolution-driven competitive and cooperative instincts: as economic inequality increases, competitive instincts dominate over cooperative instincts, putting psychological pressure on poorer people, which can often compromise their health and lead to further disadvantage.

But there are also specific *material* reasons why economic inequality could put some households in the position of finding it harder to heat their homes. To begin with, in a highly unequal society there are some excessively rich people in the market bidding for the goods and services the poor also need. This can increase the price of basic necessities, particularly if certain necessities become prized investment items for the hyper-rich. An example is houses. As Fernandez et al. (2016) have shown, the price of houses at the top end of the market escalates when the hyper-wealthy purchase great numbers of them. This has a cascading effect on houses further down the market, as the merely-wealthy, who can no longer afford top-end houses, purchase great numbers of houses in the next rung down. This effect can ripple down the scale of wealth and housing—often exacerbated by other effects on house prices

(Eaqub and Eaqub, 2015)—so that houses which were previously affordable for poorer people are now way out of their reach. The result is that home ownership rates have been falling for over a decade, both overall and especially among young people, in many western countries, including the US, Australia, New Zealand, the UK and Germany (Galvin and Sunikka-Blank, 2018).

This tends to increase rental prices of homes as far down as the bottom rung of the market, making it more difficult for poorer people to pay for the fuel needed to heat these (often thermally leaky) homes adequately. This dynamic of high rents for poor-quality houses would be less likely in a less rich society where everybody is closer to equal, as there would not be a class of rich rentiers extracting profit from large-scale real estate investments. Hence economic inequality can affect housing affordability differently from mere poverty.

A further material factor is the political power hyper-wealthy people have, to get laws, regulations and the ownership of public goods shifted in their favor, as I discuss in Chapter 3 in relation to the findings of Winters (2011, 2014, 2017) and Stiglitz (2013). We need to do a lot more work investigating the specific actions of the hyper-wealthy that change social structures in ways that benefit themselves and disadvantage and disempower poorer people.

Returning to Wilkinson and Pickett's work, a limitation of their approach is their almost exclusive use of bivariate analysis, i.e., a simple graph with an x-axis and a y-axis. Their x-axis is usually a measure of economic inequality, such as the Gini coefficient, while their y-axis represents a particular parameter of human welfare, such as longevity. Because the data points tend to cluster round a straight line (which appears to have a reasonably high correlation coefficient R^2), Wilkinson and Pickett suggest this indicates causation.

To illustrate the limitations of this reasoning I present a graph, in Fig. 1, showing the Gini coefficient (after tax and welfare transfers) on the x-axis and the percentage of households who are unable to heat their homes adequately on the y-axis, for the EU-26+UK for 2016. By "EU26+UK" I mean all EU countries apart from Croatia, for which insufficient data was available for this study, plus the UK, which was an EU member while this chapter was being written. This graph has the same form as many in Wilkinson and Pickett's work, e.g., Figure 13.1 in Wilkinson and Pickett (2010).

Looking at Fig. 1, there is clearly a correlation between the Gini coefficient and the percentage of households unable to keep their homes adequately warm. These figures are for 2016, but the graphs for all the years 2010–16 are very similar. A major problem with this type of graph is that it does not take into account other factors that might be influencing the percentages of unheatable homes. Both the Gini coefficient and these percentages might be caused by a third factor, or a combination of other factors. Ironically, the Gini coefficient may actually have even greater influence on the percentage of unheatable homes than this graph indicates, because other factors (e.g., wealth or income per person, level of household debt, building regulations) may be working against the effect of the Gini coefficient. These influences, however, are hidden in a simple bivariate graph.

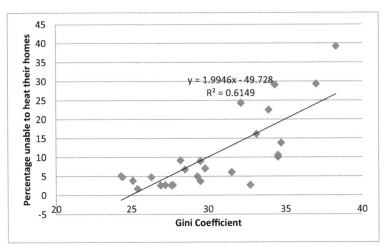

FIG. 1

Graph showing correlation between the Gini coefficient (after tax and welfare transfers) and the percentage of households who are unable to keep their homes adequately warm, EU26+UK, 2016.

From Eurostat, 2018. Your Key to European Statistics. Available from: http://ec.europa.eu/eurostat.

Hence I will use Fig. 1 merely *to form a hypothesis*. My hypothesis is: *that the income inequality represented by the Gini coefficient (after tax and welfare transfers) is a determinant of the percentage of households unable to keep their homes adequately warm in EU countries (and the UK)*. I will test this hypothesis using data for the Gini coefficient *and other factors*—other independent variables—which might also be influencing the percentage of households who are unable to keep their homes warm. I will do this by means of a series of ordinary least squares (OLS) multivariate regression analyses, which test how each of these independent variables correlates with the dependent variable, the percentage unable to keep their homes adequately warm, *while all the other independent variables are held constant* (known technically as "controlling for" the other independent variables). I will thereby investigate whether the data on these other factors enables me to *disprove* my hypothesis. If the hypothesis is disproved, this would imply the economic inequality represented by the Gini coefficient does *not* contribute significantly to the percentage of households unable to keep their homes adequately warm. If my hypothesis is *not* disproved, this would not necessarily mean it is proved. It would mean, however, that it is a strong hypothesis that has stood up to quite rigorous testing and should be taken seriously by policymakers and other relevant actors.

In Section 3 I begin the systematic process of explaining the statistical analysis, and continue this in subsequent sections. First, however, in Section 2, I explain why it is useful to consider the variable "the percentage of households unable to keep their homes adequately warm."

2 Measures of fuel poverty

There are many insightful and informative studies on fuel poverty in the EU using different definitions of fuel poverty and from a range of different perspectives (see reviews in Dubois and Meier, 2016; Moore, 2012; Thomson et al., 2016; Walker and Day, 2012). Most of these are in broad agreement that low income, expensive fuel and thermally inadequate homes are the main contributors to fuel poverty (EC, 2015; Middlemiss, 2017; Pye et al., 2017). Some, such as Bouzarovski (2014) and Thomson et al. (2017), add a fourth determinant, "specific household energy needs." Other studies note the diversity among EU countries, not only of the determinants of fuel poverty (e.g., Dubois and Meier, 2016) but also its within-country geographical diversity (e.g., Bouzarovski and Herrero, 2017).

In studies to date, "fuel poverty" usually refers to home heating while "energy poverty" often refers to all household energy services, sometimes including transport (e.g., Mayer et al., 2014), though other studies use the two terms interchangeably (see review in Thomson et al., 2016). A basic, recurring theme in this literature is the difficulty of defining fuel (or energy) poverty (Moore, 2012). Broadly speaking these definitions can be divided into two types: objective and subjective (Price et al., 2012). Objective definitions consider measurable parameters to do with income, expenditure, household composition, buildings, etc. An example is the UK's relatively new metric, arising from the Hills Review (Hills, 2012, 2011), that a household is in fuel poverty if it has "lower than average income and higher than average fuel costs" (Middlemiss, 2017). Other objective-type definitions among EU countries draw on parameters such as: the ages of household members; an objective measure of the thermal quality of dwellings; and whether or not a household is receiving welfare benefits (Atsalis et al., 2016; Thomson et al., 2017; Price et al., 2012). What these measures have in common is that they seek to indicate whether a household *can be expected* to achieve an acceptable level of energy services, rather than a household's subjective judgments of relevant aspects of their situation, such as whether they find that they are able to heat their home adequately.

These methods generally use a "micro" or "targeting" approach to identify the types and/or locations of households where fuel poverty is most prevalent or households are most vulnerable (see summaries in Pye et al., 2017 and earlier discussion in Thomson et al., 2017). They aim to provide policymakers with detailed information at fine levels of granularity, to enable interventions to be targeted economically efficiently toward vulnerable households (e.g., Liddell and Langdon, 2013). They narrow down the work of identifying specific cases of actual fuel poverty which may need urgent intervention. This is an important endeavor because resources for interventions are always limited and need to reach the people in need.

However, one could argue that "objective" approaches are essentially subjective, since the choice of parameters varies greatly between EU countries (Price et al., 2012). It does not seem possible to establish an objective foundation for parameter selection or the magnitudes of critical thresholds. A further disadvantage is that they are not able to take account of specific households' actual needs in relation to energy use, which may be significantly out of step with the perceived needs embodied in the choice of parameters (Palmer et al., 2008).

A further assumption of such approaches often appears to be that fuel-poor households are passive receptors of targeted help, rather than actors who could be empowered to lift themselves out of fuel poverty if key elements of the macro-economic context were changed. These approaches do not ask what would happen if some of the objectively observable factors—such as poverty itself—were mitigated.

Subjective definitions, on the other hand, ask households one or more pertinent questions on whether they see their energy services as adequate. The best-known questions of this type in EU countries are those used in surveys for the Eurostat Statistics on Income and Living Conditions (EU-SILC). These cover:

- Whether the household is unable to heat their home to a satisfactory level;
- Whether they are in arrears on fuel bills; and
- Whether the building fabric of their dwelling is inadequate (leaking roof; damp walls, floors or foundation; or rot in window frames of floor).

A disadvantage of subjective measures is that households' responses may not be honest. Boardman (2011), cited in Thomson et al. (2017) found people often deny being unable to heat their homes when questioned, often due to pride. Further, different groups or cultures may have different ideas of what thermal comfort is (Chappells and Shove, 2003) or indeed whether there is any need to be comfortably warm (Cupples et al., 2007). Nicola Terry's findings in this book (Chapter 6) indicate that tenant households often complain more readily of cold, than owner-occupier households, for reasons that might not have to do with affordability or building characteristics. The few studies which have used both objective and subjective approaches on the same households find overlaps between the two but also significant differences (Atsalis et al., 2016; Palmer et al., 2008; Papada and Kaliampakos, 2016).

Nevertheless, an important advantage of a subjective approach is that it identifies households who feel cold because they *actually* cannot heat their homes, rather than those who *can merely be expected* to feel cold. These households experience the deprivations of fuel poverty, at least from their own point of view, regardless of what their income and expenditure levels indicate and how sound their dwelling is (Price et al., 2012).

A further advantage of subjective approaches is that they lend themselves to studies on the macro-level, i.e., they yield whole-country results, enabling us to compare countries. Examples of macro-level data are: a country's GDP/capita; its measure of income or wealth inequality; its climate (assuming it is not too geographically disparate); the percentage of its households who feel unable to heat their homes; the percentage who claim to be in arrears on fuel bills; and the percentage who say they are living in damp dwellings. The last three of these are subjective measures and the first three objective, but all are on the macro-level: they give one figure for each country (or country-year) as a single entity, rather than different figures for different types of household within the same country or region.

At least two different types of policy are implicated in discussions of fuel poverty: energy efficiency policy, and social or economic policy. Energy efficiency policy has to do with regulations for building insulation, boiler standards, air-tightness, etc., together with incentive programs to support property owners in bringing their buildings up to standard (Galvin and Sunikka-Blank, 2017). Social and economic policy have to do with setting tax rates and redistributing money for the social good. As I noted above, it has long been established that one of the causes of people's inability to heat their homes adequately is poverty. Poverty is hardly evident in countries with progressive tax and redistribution policies, such as Denmark and Sweden, but much more evident in countries where these policies are weak, such as the US, the UK and New Zealand. This chapter is principally concerned with social and economic policy rather than energy efficiency policy, as it focuses on the post-tax and welfare Gini coefficient. But this should not be taken to imply that energy efficiency policy is any less important than social and economic policy. An important feature of multivariate regression analyses is that they give clues as to how a range of different factors may be influencing the dependent variable. As we will see, the Gini coefficient can explain only part of the differences, between countries, in the percentage of their households who are unable to heat their homes. The results of the analyses may also point to other influences that come under the heading of energy efficiency policy.

3 **The variables**

3.1 **Which variables are relevant?**

The independent variable, "the percentage of households who say they are unable to keep their homes adequately warm" (UH%, symbol U), was given in Eurostat statistics for all years 2008–16 at the time this chapter was written, but I use only the years 2010–16 because other variables are given for those years only. I analyze the data by performing a series of OLS multivariate regressions of the effects of sets of likely independent variables on UH%—or more specifically, on its natural logarithm, for reasons I will explain. All the data for the independent variables comes from Eurostat (2018), except where otherwise stated.

The independent variable which is the focus of this chapter's hypothesis is the level of income inequality (not wealth inequality) *after tax and welfare transfers*. It is therefore based on the differences in *actual* income households can spend. It is represented by the Gini coefficient for this definition of income, calculated by Eurostat and given as a percentage rather than a decimal (see Chapter 1 for an explanation of the Gini coefficient). I abbreviate this as "Gini" (symbol G). I will frequently call it the "Gini index" to avoid confusion with other meanings of the word "coefficient" in regression analysis.

Alongside this inequality measure, the *average* spending power of households is also likely to influence UH%: the more households can spend, on average, the easier it is to thermally upgrade homes and/or to spend on heating whether or not their homes are thermally upgraded. Hence I would hypothesize a negative relationship between average spending power and UH%. It is difficult to capture average household spending power in one variable, but there are three candidates. One is Gross Domestic Product per capita (GDP/capita, symbol D), which enables us to compare country-level incomes. For consistency over time, I inflation-adjust GDP/capita to equivalent 2016 values, using data from the World Bank.

Nevertheless, GDP/capita will not be a good proxy for private income per capita in a country that has large flows of tax haven money, in particular Ireland and Luxembourg. Ireland's GDP/capita in 2016, of €65,207, is about 43% higher than it would be without these international banking flows (Zucman, 2014), so I reduce Ireland's GDP/capita accordingly. Luxembourg's official GDP/capita in 2016 was €91,931, but its banking secrecy is such that there seems to be no way of finding its effective GDP/capita. I therefore leave Luxembourg out of the analysis,[a] leaving just 26 countries, and call these the EU25+UK.

A second candidate for a spending power variable is average net household wealth per person (symbol W), since households' accumulated wealth influences their spending power. Like GDP/capita, data for net wealth/person is inflation-adjusted to 2016 values, using indices from the World Bank.

The third candidate is average household debt to income ratio. On the one hand, interest and principal repayments on poorer households' purchase and credit card debts can seriously reduce their effective income. The average annual credit card interest rate in the UK was 18.35% in September 2018, and interest on debit card overdrafts can be as high as 60%. Research has shown that low-income households often take on these forms of debt to survive from month to month, then descend deeper and deeper into debt due to punitively high interest rates (Achtziger et al., 2015; Soederberg, 2012; Stewart, 2009; The Money Charity, 2018). This factor, which I call "toxic debt," is not captured by other income-related variables such as the Gini coefficient or GDP/capita.

On the other hand, debt can be a positive factor where interest rates are low and the investment accumulates wealth to the borrower, such as a sensibly sized mortgage for buying a house, or a loan for an intelligent stock market investment. Aron et al. (2010) call this type of borrowing a "financial accelerator." In short, we could say that some individuals or households, particularly those of low income, often "borrow their way downwards" to ruin, while others "borrow their way upwards" to riches. The second effect is likely to swamp the first in statistics on debt, because mortgage loans and loans for stock market speculation are often in the hundreds of thousands or millions of euros while the toxic debt of poor households is in the hundreds or thousands. I would therefore hypothesize that a countries' debt to

---

[a]Out of interest I re-ran all the multivariate analyses including Luxembourg among the countries. The results for correlation coefficients and significance levels were only marginally different, but Luxembourg's GDP/capita made for very large outliers in other derived variables.

income ratio is negatively correlated with their percentages if households unable to heat their homes.

Eurostat gives the composite statistic of households' average debt to income ratio (symbol T), but data for Greece, Romania and Malta is missing for some years. I therefore use only the first two spending power variables in regressions on all 26 countries, but perform extra regressions for the 23 for which T is known.

A difficulty with using two or more of these variables in the one regression is that they are closely related to each other: GDP/capita may correlate with net household wealth per person, and both max correlate with the ability to use debt as an income accelerator. They are therefore likely to show multicollinearity with UH%, i.e., they might correlate with each other in such a way as to give false results for each of their correlations with UH%. I therefore apply multicollinearity tests and accept results from only those regressions that do not show this effect.

Another likely independent variable is the number of heating-degree-days (HDDs, symbol H) as this is a measure of how cold, for how long, a country is on average during the winter. There are two different ways this might correlate with UH%. On the one hand, for a particular country over a period of years, we could expect a positive correlation, controlling for other variables: in colder years more people would find it harder to heat their homes. On the other hand, when considering all 26 countries, the correlation is likely to be negative. This is because countries with colder climates, like Finland and Sweden, have far more stringent thermal building regulations and practices than warmer countries like Malta, Portugal and Cyprus (Briggs et al., 2003).

A further likely choice of independent variable is the percentage of inadequate dwellings (ID%, symbol L). Eurostat's household surveys ask households whether the building fabric of their dwelling is inadequate, i.e., has a leaking roof, damp walls, floors or foundation, or rot in window frames or floor. The effects of this variable may not be very strong because different respondents may have different thresholds for what they call rot, damp or a leak. Further, it does not capture differences in thermal quality between dwellings that are *not* regarded as "inadequate" but might still be cold due to poor insulation, inadequate heating systems or poor orientation to the sun. Nevertheless we would expect ID% to be positively correlated with UH%.

Another possible independent variable is the percentage of households who are in arrears on their fuel bills. However, this would act as an alternative dependent variable, since it is likely to be caused by the same factors that cause households not to be able to heat their homes. It therefore seems best to leave it out of the multivariate analysis, but to investigate it as an interesting aside.[b]

---

[b]I performed extra multivariate analyses using this as the dependent variable, and found it gave comparable but far from identical results.

The price of heating fuel is another likely determinant of UH%. However, different countries use different proportions of different fuels for heating, and there does not seem to be a way of bringing these into a composite variable.

The countries in the analysis vary vastly in population size. However, I do not weight the data according to population, for a very important reason: the analyses are on the level of *national jurisdictions*, not individual households. They focus on particular countries in particular years. Each country is treated as an integral unit, with one government, one parliament, one set of decision-making authorities that is able to address the issues raised.

3.2 Panel data or year by year regressions

The regression analyses concern 26 countries over seven specific years. In some circumstances having seven sets of observations for 26 countries enables these to be treated as "pooled panel data," which would give the advantage of having $7 \times 26 = 182$ observations, or $7 \times 23 = 161$ where debt-to-income ratio is included. This would mean, for example, that Bulgaria in 2011, Bulgaria in 2016 and the UK in 2016 would all count as different observations.

There are a number of cautions in using pooled panel data. One is "fixed effects": the possibility that a further, unknown variable is correlating with each country's UH % data by the same magnitude each year. A further caution is the possibility that the within-country variations in the values of variables over the 7 year period are trivial or even random, compared to the between-country differences. In that case, performing a regression of all 182 observations ($n = 182$) might not give us any more useful information than performing a regression of one year's 26 observations ($n = 161$), but the statistical results would fool us because n would be greater, making the standard errors appear smaller and variables appear more statistically significant than they actually are.

Preliminary tests showed that the within-country variations over the 7 year period were random or at least not statistically significant. In other words, between-country differences were the more decisive factors in mapping correlations between variables, rather than within-country differences over a period of a few years. The safe way to proceed was then to simply conduct separate regression analyses for separate years. I conducted regressions for each of the 7 years but, for reasons of space, display results and give comments mostly for 2016. There were no major differences in results from year to year.

4 Descriptive statistics

Table 1 gives the magnitudes of all the variables for the year 2016, to show readers their typical values for each country. I include Luxembourg and the percentage in arrears on fuel bills in this table out of interest, though these are not included in the regressions. The magnitudes of the variables for the other 6 years are comparable

Table 1 Magnitudes of the variables for 2016, by country, including Luxembourg, and the percentage in arrears on their fuel bills, which are not included in the analyses

Country	Percentage unable to heat homes (UH%)	log (UH%)	Gini (%)	GDP/Cap (€)	Net household wealth per person (€)	Debt to income ratio (%)	HDDs	Inadequate dwellings (%)	Percentage in arrears on fuel bills
Belgium	4.8	1.569	26.3	42,046	94,221	103.27	2689	18.0	5.0
Bulgaria	39.2	3.669	38.3	17,460	7335	32.71	2427	30.4	31.7
Czech Rep	3.8	1.335	25.1	31,356	13,598	58.87	3247	13.8	3.0
Denmark	2.7	0.993	27.7	44,184	80,176	244.47	3136	8.7	2.5
Germany	3.7	1.308	29.5	43,827	49,632	82.68	3005	14.0	3.0
Estonia	2.7	0.993	32.7	26,724	11,249	69.65	4208	17.1	7.9
Ireland	9.0	2.197	29.5	45,645	44,414	147.13	2746	11.9	15.1
Greece	29.1	3.371	34.3	24,230	13,002	103.21	1464	18.6	42.2
Spain	10.1	2.313	34.5	32,781	27,913	89.4	1729	16.8	7.8
France	5.0	1.609	29.3	37,057	54,523	61.14	2398	12.8	6.1
Italy	16.1	2.779	33.1	34,563	53,422	178.83	1762	20.6	8.9
Cyprus	24.3	3.190	32.1	29,575	23,595	37.56	680	26.5	15.4
Latvia	10.6	2.361	34.5	23,161	10,150	35.59	4003	25.7	13.2
Lithuania	29.3	3.378	37.0	26,724	8299	40	3827	25.1	9.7
Luxembourg	1.7	0.531	31.0	91,931	12,143	103.21	2967	16.2	4.0
Hungary	9.2	2.219	28.2	23,873	39,953	89.4	2707	30.8	16.2
Malta	6.8	1.917	28.5	34,207	88,026	215.2	322	6.9	9.0
Netherlands	2.6	0.956	26.9	45,609	52,333	84.28	2680	15.6	2.0
Austria	2.7	0.993	27.2	45,609	7254	59.96	3419	13.3	4.2
Poland	7.1	1.960	29.8	24,298	20,967	104.3	3286	22.8	9.5
Portugal	22.5	3.114	33.9	27,591	4312	215.2	1237	18.9	7.3
Romania	13.8	2.625	34.7	20,841	13,848		2919	24.0	18.0
Slovenia	4.8	1.569	24.4	29,907	6154	44.59	2757	30.2	15.9
Slovakia	5.1	1.629	24.3	27,822	29,104	63.11	3172	9.1	5.7
Finland	1.7	0.531	25.4	39,493	96,014	114.24	5338	4.4	7.7
Sweden	2.6	0.956	27.6	44,688	82,701	157.15	5125	8.0	2.6
UK	6.1	1.808	31.5	38,982	94,221	126.42	2976	15.0	5.7

Table 2 Mean, standard deviation, maximum and minimum for the variables used in the analysis

	Mean	Max	Min	Std. dev
Percentage unable to heat home (UH%)	12.14	46.50	0.80	10.92
log (UH%)	2.078	3.839	−0.223	0.961
Gini (%)	29.99	38.30	23.70	3.86
GDP/Cap (€)	31,614	45,645	15,146	8650
HDDs	2871	6058	322	1194
Net household wealth/pers. (€)	30,735	96,014	1613	26,107
Percentage inadequate dwellings (ID)%	16.29	34.70	4.40	6.77
Log (ID%)	2.700	3.547	1.482	0.440
Debt to income ratio (T) %	102	32.71	266.4	60.45

182 observations, EU25+UK, 2010–16. Note that the variable "Debt to income ratio" covers only 161 observations.

to those for 2016, as will be discussed below. Note that UH% is also given in natural logarithmic form.

Table 2 gives the mean, maximum, minimum and standard deviation for each of the variables used in the analysis, covering all 182 observations. Corresponding figures for each year are used to standardize each variable, setting mean $=0$ and standard deviation $=1$, to enable us to make easy comparisons of the impact of each variable on the percentage of households unable to heat their homes adequately.

In order to produce results from an OLS multivariate regression that are relatively straightforward to interpret, it is best if the shapes of the statistical distributions of the variables are reasonably similar, and close to normal—though useful results can still be obtained even if they are not normally distributed. The distributions of most of the variables used here are more or less bell-shaped and more or less symmetrical, i.e., close to normal. However, three of the variables do not fit this pattern. I show these here as pooled data over the entire 7 years, but each year's distribution is similar and the remarks given here also apply to all years separately.

Fig. 2 shows the distribution of the dependent variable UH%. This is sharply right-skewed. Fig. 3 shows the distribution of the natural logarithm of UH%, which is much closer to a bell-shaped, reasonably symmetrical distribution. It makes sense, therefore, to use Log(UH%) as the dependent variable and translate the results back into non-log form in the discussion of their implications.

The distribution of the percentage of inadequate homes is moderately right-skewed, but its log transformation is moderately left-skewed. I therefore tried each of these forms in turn, in alternative multivariate regressions, to see which gave the best model fit. In effect there was hardly any difference, so for simplicity I used the non-log version for all regressions.

Similar to UH%, the distributions of net household wealth per person and debt to income ratio were strongly right skewed and their log transformations close to symmetrical, so I used the log transformations of both of these in all regressions.

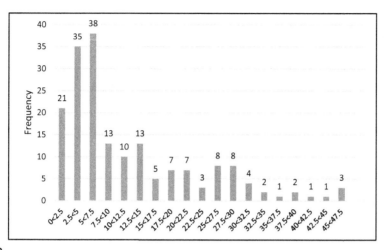

FIG. 2

Distribution of percentage of households unable to heat their homes adequately (UH%) EU25+UK, 2010–16.

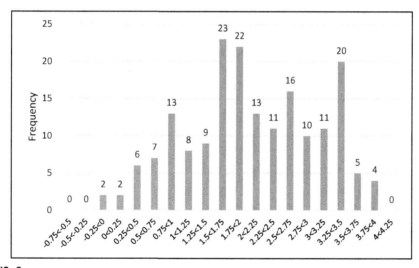

FIG. 3

Distribution of Log (UH%), EU25+UK, 2010–16.

In light of the discussion in Section 3.2 I ran eight multivariate regressions for each of the years 2010–16. Each year gave very similar results, and I present and discuss the results of those for 2016 below, with some comments on 2010 and 2013. Table 3 shows which variables were regressed against Log(UH%) in each model.

Table 3 Variables regressed against log(UH%) for each of 8 models, for each year

Independent variable	Symbol	1	2	3	4	5	6	7	8
Gini	G	✓	✓	✓	✓	✓	✓	✓	✓
GDP/capita	D	✓		✓	✓				✓
HDDs	H	✓	✓	✓	✓	✓	✓	✓	✓
Inadequate dwellings	L	✓	✓	✓	✓	✓	✓	✓	✓
Log(wealth/person)	**Log(W)**		✓			✓	✓		✓
Log(debt/income)	**Log(D)**				✓		✓	✓	✓
Number of countries		26	26	23	23	23	23	23	23

Models 1 and 2 use all 26 countries but do not use Log(Debt–Income-ratio) as this was not available for all 26 countries for all years. Model 1 includes GDP/capita as an independent variable but not Log(Wealth/person), while Model 2 is the reverse of this.

Models 3–8 use the 23 countries for which data on Log(Debt/Income) was available. Model 3 uses the same variables as Model 1, while Model 4 adds Log(Debt/Income) to this. Models 5 and 6 do the same but in relation to Model 2. This enables us to gauge the effect of including Log(Debt/Income) in the model in addition to each of GDP/capita and Log(Wealth/person) in turn.

Model 7 uses Log(Debt/Income) as the only variable related to average spending power. Model 8 uses all three of the spending power variables.

The modeling equation for Model 8 is:

$$Log(U) = A + B_1.G + B_2.D + B_3.H + B_4.L + B_6.Log(W) + B_7.Log(D) + er \qquad (1)$$

where A is the intercept, B_1, B_2 etc. are the coefficients of the independent variables G, D, etc., and er is the error term, i.e., the difference between the theoretical and modeled values of Log(U), which varies for each country's set of independent variables. Modeling equations for Models 1–7 can be written by leaving out the variables not used in these models.

5 Results

5.1 Tests for model fit

Each model was run in conjunction with tests for multicollinearity, randomness of residuals, and omitted variable bias.

Multicollinearity was tested using a VIF (variance inflation factor) test. In this test, each independent variable is regressed against each of the others. The strengths of correlations between variables are given by R-squared values, which are used in the VIF formula.[c] As a rule of thumb, a VIF score under 5 is non-problematic, over 10 is highly problematic, and between 5 and 10 indicates a need for caution. For all of models 1–7 for 2016, 2013 and 2010 (and all other years) VIF scores were very low, ranging from 1.82 to 3.06, but for Model 8 the scores for GDP/capita and Log (Wealth/person) were high, at 10.44 and 8.20 respectively for 2016, and 15.26 and 12.48 for 2010. Hence the regression results of Model 8, which includes all three spending power variables, are not robust and are best ignored.

Considering now the residuals, these are the differences between the observed values of the independent variable and the values the model predicts. Ideally there should be no correlation between the residuals and the observed values. In all eight models for all years the correlations were found to be low to moderate, with R-squared ranging from 0.167 to 0.215 for 2016. This is not high enough to invalidate the models but high enough to be of some concern, as it suggests there may be a missing variable, which could make a difference to the results if identified and used in the regressions.

Finally, tests were performed to check whether the models are correctly specified. The model specified in Eq. (1) is a *linear* model: although some of the variables are in log form, the model is treated as linear in that none of the variables' data values are squared, cubed, or multiplied by each other. The Ramsey "RESET" (Regression equation specification error test) test (Ramsey, 1967; Wooldridge, 2016) is frequently used to check whether this is the best specification or whether there are "omitted variables" in the form of squares, etc., of the variables. Ramsey RESET tests on Models 1–7 for 2016, 2013 and 2010 gave negative results, with P-values[d] ranging between 0.249 and 0.675, indicating there were no omitted variables. The only model that indicated an omitted variable was Model 8 for 2010, which is already rejected due to multicollinearity.

5.2 Other statistical tests

The results for the regressions for 2016 are given in Tables 4 and 5. The coefficients are given with their significance levels if these are <0.1 (*** $P < 0.01$, ** $P < 0.05$, * $P < 0.1$)—but see important note in Box 1 regarding P-values, and reference to Wasserstein et al. (2019). Standard errors are in parentheses. Although all eight models are shown in the table, discussion here is limited to Models 1–7. The results for the eight models for 2010 and 2013 are very similar to those for 2016 and are not shown here due to considerations of space.

---

[c] $VIF = 1/(1 - R_i^2)$, where R^2 is the correlation coefficient of a variable i against the other variables.
[d] The Ramsey RESET test calculates the squares (or other transformation, depending on the user's choice) of the estimated values of the dependent variable and re-runs the regression with these squares as an extra independent variable. Its main output is the P-value for the supposed extra variable within the model. A P-value higher than 0.1 indicates this variable is not a significant predictor of the independent variable.

Table 4 Results for models 1 and 2 for 2016, for 26 countries

Variables	(1) logUH%	(2) logUH%
Gini	0.114***	0.122***
	(0.0248)	(0.0225)
GDP/capita	−2.96e−05**	
	(1.10e−05)	
HDDs	−0.000247***	−0.000277***
	(7.59e−05)	(7.43e−05)
Inad. dwellings	0.0137	0.0187
	(0.0140)	(0.0136)
LogWealthpp		−0.260***
		(0.0873)
LogDebt-IncRatio		
Constant	0.00417	1.401
	(1.047)	(1.354)
Observations	26	26
R-squared	0.830	0.839
Adj R-squared	0.798	0.808
Prob > F	0.000	0.000

Regression of the logarithm of the percentage unable to heat their homes, against selected independent variables. Coefficients are given with their significance levels if <0.1 (***P <0.01, **P <0.05, *P <0.1). Standard errors are in parentheses.
***P <0.01, **P <0.05, *P <0.1.

Looking at Tables 4 and 5, to begin with, all the models (for all the years) show reasonably high adjusted R-squared values of around 0.8, indicating that the model "explains"[e] about 80% of the variance in Log(UH%). Further, the P-values for the F-statistics, given as "Prob > F" in the tables, are consistently extremely low, indicating that each model gives a significant prediction of the sign (plus or minus) of the slope of the dependent variable.

Regarding the form of the numbers, note that −2.96e-05 means -2.96×10^{-5} and all statistics are given to three significant figures.

Looking now at the (asterisks indicating) P-values, we note that the Gini index is statistically significant in all regressions, with very low P-values. This indicates there is virtually zero probability that the sign of the coefficient of the Gini is wrong, i.e., it is always positive: an increase in the Gini is invariably associated with an increase in Log(UH%). Similarly, HDDs has very low P-values in every regression, indicating

---

[e]The word "explains" is standard statistics jargon but is misleading, as it suggests causality. There may be causality but a regression model cannot prove this.

Table 5 Results for models 3–8 2016, for 23 countries

Variables	(3) logUH	(4) logUH	(5) logUH	(6) logUH	(7) logUH	(8) logUH
Gini	0.110***	0.110***	0.120***	0.116***	0.118***	0.118***
	(0.0260)	(0.0258)	(0.0236)	(0.0241)	(0.0244)	(0.0278)
GDPperCap	−3.03e−05**	−1.59e−05				6.84e−06
	(1.17e−05)	(1.74e−05)				(3.24e−05)
HDDs	−0.000294**	−0.000329***	−0.000306***	−0.000331***	−0.000345***	−0.000334***
	(0.000105)	(0.000109)	(0.000103)	(0.000106)	(0.000107)	(0.000110)
BadDwe	0.00697	0.00675	0.0139	0.0110	0.00850	0.0125
	(0.0175)	(0.0174)	(0.0167)	(0.0170)	(0.0172)	(0.0188)
LogWealthpp			−0.271**	−0.175		−0.222
			(0.095)	(0.140)		(0.266)
LogDebtIncRatio		−0.272		−0.220	−0.436**	−0.233
		(0.242)		(0.235)	(0.162)	(0.249)
Constant	0.400	1.199	1.757	1.987	1.185	2.198
	(1.180)	(1.372)	(1.464)	(1.490)	(1.365)	(1.830)
Observations	23	23	23	23	23	23
R-squared	0.823	0.835	0.833	0.842	0.827	0.842
Adj R-squared	0.784	0.787	0.796	0.795	0.789	0.783
Prob > F	0.000	0.000	0.000	0.000	0.000	0.000

Regression of the logarithm of the percentage unable to heat their homes, against selected independent variables. Coefficients are given with their significance levels if <0.1 (***P <0.01, **P <0.05, *P <0.1). Standard errors are in parentheses.

there is virtually zero probability that the models predict its (negative) sign wrongly: a higher number of HDDs is associated with a lower value of Log(UH%).

The only other variables with P-values low enough to indicate statistical significance (in the traditional sense of the term—see Box 1) are the three spending power variables: GDP/capita, Wealth/person and Debt/income ratio. Nevertheless, these only show consistency of significance when in a model without either of the other two. In other words, each model has no more than three statistically significant predictors of Log(UH%): the Gini index; HDDs; and one or other of the three spending power variables. In models 4 and 6, where two of these variables are present in the one model, the statistical significance of each falls away.

The signs of all three of these three income or wealth related variables are consistently negative, which is what we would expect: the higher the average spending power in a country, the lower the percentage unable to heat their homes.

The positive sign of the remaining variable, the percentage of inadequate dwellings, is also what we would expect: the higher the percentage of inadequate dwellings, the higher the percentage unable to heat their homes.

Box 1 How are *P*-values used in this chapter?

P-values are frequently used in statistical analyses and rule-of-thumb test for statistical significance. This approach is increasingly questioned by statisticians. My approach in this chapter is informed by Wasserstein et al.'s (2019) summary of the findings of 43 leading statisticians published in a special issue of The American Statistician (73:51). To begin with, *P*-values are only useful if our data comes from a true random sample of a given population. In this chapter, it could be argued that the sample of 181 country-year observations is the whole population, i.e., all the country-years there are. However, if we want to use the results to make inferences about what might theoretically be likely to happen in subsequent years (all other things being equal) then our 181 observations are a sample. This makes questions of statistical significance, and therefore *P*-values, relevant. For example,—for example we may want to consider what might happen if certain energy or fiscal policies were changed so as to reduce the Gini index or the number of inadequate homes. For this reason issues of statistical significance are relevant here.

But what do *P*-values actually mean? An independent variable's *P*-value in relation to the dependent variable is the probability that its regression coefficient has an indeterminate sign, i.e., that it is not clear whether it is negative or positive. If the *P*-value is 0.1, there is a 10% chance that the sign of the coefficient produced by the analysis is meaningless, and a 90% chance that it does predict the direction of the independent variable's influence on the dependent variable. Given this understanding, in some situations and circumstances we might be justified in tolerating a much higher *P*-value, say up to 0.3, whereas in others (such as the chance of an accidental nuclear missile launch) much lower, say 0.00000000000000001. Hence it is important to understand the context in which *P*-values are being used.

In this analysis I often refer to the traditional *P*-value thresholds of 0.1, 0.05 and 0.01, but also leave open the possibility that variables showing higher *P*-values may still be "significant"—i.e., the sign of the coefficient of a particular independent variable may give a good-enough indication of its direction of influence on the dependent variable. Hence the sign of the coefficients of variables may be relevant even if the *P*-values are considerably higher than 0.1.

5.3 The coefficients

The coefficient of the Gini index is consistently around 0.11 for the 2016 regressions, 0.10 for 2013 and 0.09 for 2010. Because the dependent variable is in log form, care has to be taken in calculating the effect of the Gini on UH% (controlling for the other independent variables). For Model 1, where the coefficient of G is 0.114, Eq. (1) above becomes:

$$\ln(U) = 0.114G + K \tag{2}$$

where K is the intercept in Eq. (1) plus the sum of the products of the coefficients of the other variables and their values for any particular country. The rate of change of U with G is therefore given by:

$$\frac{\partial U}{\partial G} = 0.114e^K \cdot e^{0.114G} \tag{3}$$

For example, for Portugal $\frac{\partial U}{\partial G}$ is 2.550. This means that, controlling for the other three variables, an increase (decrease) of 1% point in the Gini index is associated with an increase (decrease) of 2.55% points in the number of households unable to heat their homes.

We can now compare this with the effect of GDP/capita, symbol D. For D the rate of change is given by:

$$\frac{\partial U}{\partial D} = -2.96 \times 10^{-5} e^J \cdot e^{-2.96 \times 10^{-5}D} \tag{4}$$

where J is the equivalent of K for GDP/capita.

For Portugal $\frac{\partial U}{\partial D}$ is -0.00133. This means that, every thousand euro increase (decrease) in GDP/capita is associated with a decrease (increase) in UH% of 1.33% points. In short, to match the effect on UH% of decreasing the Gini by 1% point, we would need to increase GDP/capita by about €1800 per person.

Portugal is in about the middle of the range of UH%. Comparing this with Sweden, which is toward the low end, we find that $\frac{\partial U}{\partial G}$ falls to 0.231. An increase (decrease) the Gini index of 1% point is associated with an increase (decrease) of 0.231% in the number of households unable to heat their homes.

At the high end of the range of UH% the opposite is the case. For Bulgaria a one percent increase (decrease) in the Gini is associated with a 4.64% point increase (decrease) in the number unable to heat their homes.

This shows that the bivariate graph in Fig. 1 underestimates the statistical association between the Gini and the percentage unable to heat their homes, especially at its high end. The coefficient of the Gini in the bivariate graph is 1.99, indicating that a 1% point increase the Gini is associated with a 1.99 increase in the percentage of unheatable homes. However, when we use multivariate regressions to control for the influence of other variables, this increases to about 2.6 in the mid-range and 4.6 at the high end. Clearly, in the bivariate graph of Fig. 1, the strong dynamic in the association between the two variables is masked by the interactions of other variables—which are properly controlled for in the multivariate regressions.

In all these calculations, the effect of a change in GDP/capita of €1000 is about half that of a change in the Gini of one percentage point, in the opposite direction.

I now consider Log(Wealth/person), which is used in Models 2 and 4 instead of GDP/capita. Its coefficient is −0.260 in Model 2 and not much different, at −0.271, in Model 4. The modeling equation for Model 2 is:

$$\ln(U) = -0.260 \ln(W) + K \tag{5}$$

The rate of change of U with W is given by:

$$\frac{\partial U}{\partial W} = -0.260 \cdot e^K \cdot W^{-1.260} \tag{6}$$

For Portugal, where $K = 5.548$ (see method above), $\frac{\partial U}{\partial W}$ is −0.00176. An increase (decrease) of €1000 in net wealth per household is associated with a decrease (increase) of 1.76% points in the number of Portuguese households unable to heat their homes.

For Sweden, where both HDDs and Wealth per person are much higher while the percentage of inadequate homes is smaller, $K = 3.498$ and $\frac{\partial U}{\partial W}$ is -5.47×10^{-6}. Here, an increase (decrease) of €1000 in net wealth per household is associated with a decrease (increase) of a mere 0.00547% points in the number of Swedish households unable to heat their homes. Because Swedes are so wealthy, a kind of saturation appears to have been reached in the effect on thermal comfort of increasing wealth per person even higher.

For Bulgaria the result is not much different from that of Portugal, with $K = 3.498$ and $\frac{\partial U}{\partial W} = -0.00137$. Although Bulgaria has a much higher percentage of households unable to heat their homes, it also has higher net household wealth per person than Portugal. This may be partly because Bulgaria has a higher home ownership rate than Portugal, at 83% compared to 75%.

Considering the variable Log(Debt-to-income-ratio), here the coefficient in Model 7 is −0.436. This results in a function with a similar shape to that of Wealth/person.

Turning now to heating degree-days (HDDs), the coefficient of H is also negative: the colder the winters, the lower the percentage who cannot heat their homes. This indicates that the second effect noted in Section 3 dominates: countries with colder climates tend to have more stringent thermal building regulations than countries with warmer climates and therefore, over time, the higher the number of HDDs, the better the thermal quality of homes constructed. The coefficients of H are around −0.0003 for 2016 and 2010 and somewhat larger, at −0.00045, for 2013. Although policy cannot change the number of HDDs, policymakers in warmer countries can consider whether their countries' thermal building regulations are adequate for the cold spells that do occur there.

Finally, the coefficient of the percentage of inadequate dwellings is consistently positive, ranging from about 0.01 to 0,015 in the more reliable models, namely those which include only one spending power variable. The P-value of this variable is quite high, at 0.337, in Model 1 and even higher in Models 3, 5 and 7 (see Table 6), so we have to be cautious about drawing implications for it that go beyond the actual observations.

Table 6 Regression results using standardized variables and showing *t*-statistics, models 3, 4 and 7, 2016

logUH	Coef.	Std Err.	t	P-value
Model 3 2016				
Gini	0.438262	0.103875	4.22	0.001
GDP/Cap	−0.271863	0.105077	−2.59	0.019
HDDs	−0.34132	0.122539	−2.79	0.012
Inad. Dwe	0.05124	0.128516	0.40	0.695
_constant	−0.07717	0.092024	−0.84	0.413
Model 4 2016				
Gini	0.478122	0.094476	5.06	0.000
HDDs	−0.355808	0.119295	−2.98	0.008
Inad. Dwe	0.102206	0.122743	0.83	0.416
LogWealthpp	−0.264656	0.092189	−2.87	0.010
_constant	−0.124588	0.087284	−1.43	0.171
Model 7 2016				
Gini	0.470421	0.097545	4.82	0.000
HDDs	−0.400958	0.124753	−3.21	0.005
Inad. Dwe	0.062452	0.126277	0.49	0.627
LogDeb-IncRatio	−0.255242	0.094649	−2.70	0.015
_constant	−0.146955	0.089239	−1.65	0.117

It is also worth considering the t-statistic of each variable. The t-statistic (coefficient divided by standard error) gives an indication of the relative magnitude of each independent variable's impact on the dependent variable: the larger the t-statistic (i.e., the higher its absolute value), the larger the changes in the variance of the dependent variable that are associated with changes in its variance. A similar effect can be seen by regressing the standardized values of the variables (subtract each variable's mean from its value and divide the result by its standard deviation). These values are given in Table 6 for Models 3, 5 and 7 for 2016 (each of which uses only one spending power variable). Clearly both the standardized coefficients and the t-statistics for the Gini are significantly larger than those of the spending power variables.

6 Discussion: The role of the Gini index

The Gini index was a significant predictor of Log(UH%) in all the regressions in 2016 and for all other years 2011–15, with very low *P*-values, a high t-statistic and a large positive regression coefficient. Though not recorded in this chapter, results for regressions for all 7 years 2010–16 were similar. The size of the coefficients indicated that moderate to small changes in the Gini are associated with large changes in UH%, and that these changes in UH% are larger for higher initial values of UH%: the higher the magnitudes of UH% and the Gini, the more difference it makes to UH% if the Gini is changed, at least in theory.

Further, at least in Models 1, 2, 3, 5 and 7 (where only one average spending power variable was used), these findings were robust against tests for multicollinearity and the Ramsey RESET test for missing derived variables. The only weakness in the results is that the residuals showed moderate correlation with the estimated values of Log(UH%), indicating that a further, unidentified independent variable might be influencing results. However, the very low P-values and high correlation coefficient for the Gini indicate that a further variable would be unlikely to prove it a false predictor.

The results therefore indicate strong correlation between the Gini and Log(UH%), controlling for other variables. It must be emphasized, of course, that this does not prove causation, since a regression analysis can only prove correlations. It would always be possible, even in the best and most robust regression analysis, that some other factor may be influencing between-country variation in both the Gini and Log(UH%) in tandem.

Nevertheless, the regressions have not been able to disprove our hypothesis that the Gini index is a predictor of Log(UH) and therefore of UH%, the percentage of households unable to heat their homes. There are good intuitive reasons for believing this hypothesis. A high Gini coefficient means there are a lot of very poor people, who are unlikely to have the means to heat their homes adequately or to renovate them to be made more energy efficient. In countries like Bulgaria, with a high Gini coefficient but high rates of owner occupied homes, many homeowners may be stuck with thermally poor dwellings which they cannot afford to upgrade. A high Gini coefficient also means there is a minority of extremely rich people. In countries with modest rates of home ownership, real estate is a prime target of the very-rich. This not only inflates its price, but also leads to a larger private rental sector, and private sector tenants can find it very hard to persuade landlords to thermally upgrade their properties, as Nicola Terry's contribution in this book indicates (Chapter 6).

It should also be noted that, of all the independent variables in this study, the Gini coefficient (after taxes and transfers) is by far the easiest for governments to influence. This can be done simply by Act of Parliament and associated administrative changes, for example, by increasing marginal tax rates on the highest incomes while providing more adequate universal welfare. Fig. 4 shows the ratio of pre- and post-redistribution Gini coefficients in EU+UK countries in 2016. It indicates, for example, that the post-tax and transfer Gini in Sweden was 0.48 (48%) of its pre-tax and distribution level in 2016, indicating a radical redistribution of income. The country with the least radical redistribution was Bulgaria, at 0.62. The eight countries who redistribute income most radically include three from Eastern Europe, while the eight least radical include three from Western Europe and the Mediterranean. Hence it would be difficult to argue that (poorer) Eastern European countries find it inherently more difficult to redistribute income.

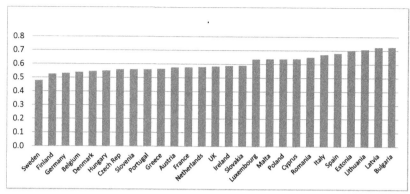

FIG. 4

Ratio of post-redistribution Gini % to raw Gini %, 2016, EU countries and UK.

Countries need to consider, then, that having a high percentage of households unable to heat their homes may well be a symptom of poor social and economic policy resulting in high levels of economic inequality. However, this does not suggest countries can solve the problem of households unable to heat their homes *solely* by social and economic policy. Energy efficiency policy and interventions are also needed. Reducing the Gini coefficient would enable some households to replace inefficient boilers, insulate lofts and walls, replace the worst windows and afford adequate energy services. But a combination of social, economic and energy efficiency policy interventions would be needed to address some of the deep, underlying issues of inefficient building stock in many European countries.

7 Conclusions

In this chapter I tested the hypothesis that income inequality, represented by the Gini coefficient after tax and welfare transfers, is a determinant of the percentage of households unable to keep their homes adequately warm in EU25+UK countries (EU excluding Luxembourg and Croatia but including the UK), which I abbreviated to UH%. I proposed this hypothesis because, on the face of it, there is a clear, positive relationship between these two variables, for these countries, for each of the years 2010–16.

I tested this hypothesis by performing a series of multivariate regression analyses, mapping the logarithm of the percentage of households unable to heat their homes (Log(UH%)) as the dependent variable, against the joint effects of a number of relevant independent variables. These included the Gini coefficient, three different variables serving as proxies for average household spending power, plus the number of heating degree-days (HDDs) and the percentage of "inadequate" homes, i.e., where occupants reported a leaking roof, damp walls, floors or foundation, or rot in window frames of floor. The spending power variables were GDP/capita and the logarithms of average wealth per person and average household debt to savings ratio. These three variables could not be used together because of multicollinearity, and the best fits to the data were obtained by using each of them separately in different models. Further, where debt-savings-ratio was used, the number of countries had to be reduced since data was incomplete for Romania, Malta and Greece.

I avoided using the variable "percentage of households in arrears on their energy bills" as an independent variable, since this was most likely an effect of the (other) independent variables.

I used the natural logarithmic form of UH%, wealth per person and debt-income-ratio because their distributions were highly right-skewed whereas the distributions of their logarithms were close to normal, as were that for the Gini, HDDs and the percentage of inadequate homes. I ran tests for multicollinearity, omitted variable bias, and the randomness of the residuals.

I avoided using the observations for all 7 years together as pooled panel data, since there was little variation within each country over the 7 year period, and the regressions would have given false significance results. Instead I regressed the data separately for each year, focused on the regressions for 2016, and compared the results with those of other years to check for anomalies.

I performed each analysis twice, once with the actual values of the independent variables and once with their standardized values, i.e., setting mean $=0$ and standard deviation $=1$, so as to observe the relative strengths of the coefficients of different independent variables. The F-statistics for all the analyses showed the models to be significant predictors of the independent variable, and the adjusted R^2 values showed very high correlation.

The regressions showed the Gini, HDDs and one or other of the spending power variables to be statistically significant predictors of Log(UH%), while the coefficients for the percentage of inadequate dwellings were consistently positive, as would be expected.

I began with the hypothesis that the Gini coefficient is a significant predictor of the percentage of households unable to heat their homes. The magnitude of the coefficients of the Gini in all the valid models, together with its extremely low P-values and large t-statistics, prevented the hypothesis being disproved.

References

Achtziger, A., Hubert, M., Kenning, P., Raab, G., Reisch, L., 2015. Debt out of control: the links between self-control, compulsive buying, and real debts. J. Econ. Psychol. 49, 141–149.

Aron, J., Duca, J., Muellbauer, J., Murata, K., Murphy, A., 2010. Credit, Housing Collateral and Consumption: Evidence From the UK, Japan and the US. University of Oxford, Department of Economics. Discussion Paper Series, No. 487.

Atsalis, Mirasgedis, S., Tourkolias, C., Diakoulaki, D., 2016. Fuel poverty in Greece: quantitative analysis and implications for policy. Energ. Buildings 131, 87–98.

Boardman, B., 2011. Participant benefits and quality of life: the challenge of hard to measure benefits. In: IEA Workshop Evaluating the Co-Benefits of Low-Income Weatherisation Programmes, Dublin. 27–28 January 2011.

Bouzarovski, S., 2014. Energy poverty in the European Union: landscapes of vulnerability. WIREs Energy Environ. 3, 276–289.

Bouzarovski, S., Herrero, S., 2017. Geographies of injustice: the socio-spatial determinants of energy poverty in Poland, the Czech Republic and Hungary. Post-Communist Econ. 29 (1), 27–50.

Bouzarovski, S., Simcock, N., 2017. Spatializing energy justice. Energy Policy 107, 640–648.

Briggs, R., Lucas, R., Taylor, G., Todd, Z., 2003. Climate classification for building energy codes and standards: part 1-development process. ASHRAE Trans. 109, 4610–4619.

BRISKEE, 2018. Two H2020 research projects that provide empirical evidence to support energy efficiency policies. Findings available fromhttps://www.briskee-cheetah.eu/.

Chappells, H., Shove, E., 2003. An Annotated Bibliography of Comfort Research. Available from, http://www.lancaster.ac.uk/fass/projects/futcom/comfort_biblio.pdf.

Cupples, J., Guyatt, V., Pearce, J., 2007. "Put on a jacket, you wuss": cultural identities, home heating, and air pollution in Christchurch, New Zealand. Environ. Plan. A 39 (12), 2883–2898.

Dubois, U., Meier, H., 2016. Energy affordability and energy inequality in Europe: implications for policymaking. Energy Res. Soc. Sci. 18, 21–35.

Eaqub, S., Eaqub, S., 2015. Generation Rent: Rethinking New Zealand's Priorities. BWB Texts, Wellington.

EC, 2015. Communication from the Commission to the European Parliament, the Council, the European Economic and Social Committee and the Committee of the Regions—A Framework Strategy for a Resilient Energy Union With a Forward-Looking Climate Change Policy: COM(2015) 80 Final, Brussels, 2015. .

Eurostat, 2018. Your Key to European Statistics. http://ec.europa.eu/eurostat.

EU-SILC Survey, 2017. European Union Statistics on Income and Living Conditions. Available from: http://ec.europa.eu/eurostat/web/microdata/european-union-statistics-on-income-and-living-conditions.

Fernandez, R., Hofman, R., Aalbers, M., 2016. London and New York as a safe deposit box for the transnational wealth elite. Environ. Plan. A. 48, 2443–2461 in press.

Galvin, R., Sunikka-Blank, M., 2017. Ten questions concerning sustainable domestic thermal retrofit policy research. Build. Environ. 118, 377–388.

Galvin, R., Sunikka-Blank, M., 2018. Economic inequality and household energy consumption in high-income countries: a challenge for social science based energy research. Ecol. Econ. 153, 78–88.

Hills, J., 2011. Fuel poverty: the problem and its measurement. In: Interim Report of the Fuel Poverty Review. Centre for Analysis of Social Exclusion, LSE, London.

Hills, J., 2012. Getting the measure of fuel poverty. In: Final Report of the Fuel Poverty Review. Centre for Analysis of Social Exclusion, LSE, London.

Liddell, C., Langdon, S., 2013. Tackling Fuel Poverty in Northern Ireland-an Area-Based Approach to Finding Households Most in Need. http://www.ofmdfmni.gov.uk/de/tackling-fuel-poverty-in-ni-liddell-lagdon.pdf.

Mayer, I., Nimal, E., Nogue, P., Sevenet, M., 2014. The two faces of energy poverty: a case study of households' energy burden in the residential and mobility sectors at the city level. Transp. Res. Procedia 4, 228–240.

Middlemiss, L., 2017. A critical analysis of the new politics of fuel poverty in England. Crit. Soc. Policy 37 (3), 425–443.

Moore, S., 2012. Definitions of fuel poverty: implications for policy. Energy Policy 49, 19–26.

Palmer, G., MacInnes, T., Kenway, P., 2008. Cold and Poor: An Analysis of the Link Between Fuel Poverty and Low Income. New Policy Institute, London.

Papada, L., Kaliampakos, D., 2016. Measuring energy poverty in Greece. Energy Policy 94, 157–165.

Pickett, K., Wilkinson, R., 2015. Income inequality and health: a causal review. Soc. Sci. Med. 128, 316–326.

Piketty, T., 2014. Capital in the Twenty-First Century (Translated From the French by Arthur Goldhammer)Belknapp-Harvard University Press, Cambridge, MA.

Price, C., Brazier, K., Wand, W., 2012. Objective and subjective measures of fuel poverty. Energy Policy 49, 33–39.

Pye, S., Dobbins, A., Baffert, C., Brajković, J., Deane, P., De Miglio, R., 2017. Energy poverty across the EU: analysis of policies and measures. In: Europe's Energy Transition—Insights for Policy Making: Findings Informing the European Commission. pp. 261–280. Available from: https://www.sciencedirect.com/science/article/pii/B9780128098066000304.

Ramsey, J., 1967. Tests for specification errors in classical linear least-squares regression analysis. J. R. Stat. Soc. B. Methodol. 32 (2), 350–371.

Soederberg, S., 2012. The US debtfare state and the credit card industry: forging spaces of dispossession. Antipode 45 (2), 493–512.

Sovacool, B., 2015. Fuel poverty, affordability, and energy justice in England: policy insights from the warm front program. Energy 93, 361–371.

Stewart, N., 2009. The cost of anchoring on credit-card minimum repayments. Psychol. Sci. 20 (1), 39–41.

Stiglitz, J., 2013. The Price of Inequality. Penguin, London.

The Money Charity, 2018. Available from: https://themoneycharity.org.uk/.

Thomson, H., Snell, C., Liddell, C., 2016. Fuel poverty in the European Union: a concept in need of definition? People Place Policy, 5–24. https://doi.org/10.3351/ppp.0010.0001.0002 University of York.

Thomson, H., Bouzarovski, S., Snell, C., 2017. Rethinking the measurement of energy poverty in Europe: a critical analysis of indicators and data. Indoor Built Environ. 26 (7), 879–901.

Walker, G., Day, R., 2012. Fuel poverty as injustice: integrating distribution, recognition and procedure in the struggle for affordable warmth. Energy Policy 49, 69–75.

Wasserstein, R., Schirm, A., Lazar, N., 2019. Moving to a world beyond "$p < 0.05$". Am. Stat. 73 (Suppl 1), 1–19. https://doi.org/10.1080/00031305.2019.1583913.

Wilkinson, R., Pickett, K., 2010. The Spirit Level: Why Equality is Better for Everyone. Penguin, London.

Wilkinson, R., Pickett, K., 2017. The enemy between us: the psychological and social costs of inequality. Eur. J. Soc. Psychol. 4 (7), 11–24.

Winters, J., 2011. Oligarchy. Cambridge University Press, New York.

Winters, J., 2014. Wealth defense and the limits of liberal democracy. In: Paper for Annual Conference of the American Society of Political and Legal Philosophy, Washington, DC, August 28–31, 2014. Revised 29 April 2015.

Winters, J., 2017. Wealth defense and the complicity of liberal democracy. In: Knight, J., Schwartzberg, M. (Eds.), Wealth, NOMOS LVIII. In: A Special Issue of the American Society for Political and Legal Philosophy, NYU Press, pp. 158–225 2017.

Wooldridge, J., 2016. Introductory Econometrics—A Modern Approach, sixth ed. Cengage Learning, Mason (OH), pp. 273–278.

Zucman, G., 2014. Taxing across borders: tracking personal wealth and corporate profits. J. Econ. Perspect. 28 (4), 121–148.

Further reading

ONS (Office for National Statistics), 2018. Statistical Bulletin: Effects of Taxes and Benefits on UK Household Income: Financial Year Ending 2017. Office for National Statistics, UK. Available from: https://www.ons.gov.uk/peoplepopulationandcommunity/personalandhouseholdfinances/incomeandwealth/bulletins/theeffectsoftaxesandbenefitsonhouseholdincome/previousReleases.

Sunikka-Blank, M., Galvin, R., 2012. Introducing the prebound effect: the gap between performance and actual energy consumption. Build. Res. Inf. 40 (3), 260–273.

Why are women always cold? Gendered realities of energy injustice

8

Minna Sunikka-Blank

Department of Architecture, University of Cambridge, Cambridge, United Kingdom

Chapter outline

1 Introduction 173
2 Gendered household practices and the feminization of demand side response 175
3 Fuel poverty, single parents and intergenerational immobility 179
4 Conclusions 184
References 185
Further reading 188

1 Introduction

If energy justice literature is "an attempt to provide an ethical framework for discussing issues of fairness between people" as outlined by Galvin in Chapter 4 of this book, it seems necessary to also include fairness between different groups of the population in this discussion. The practices of these groups are not always determined by socio-economic characteristics such as wealth or class but also by other factors such as gender or ethnicity. The ethnicity dimension of fuel poverty has been studied previously (e.g., Reames, 2016) but a gender perspective on energy justice is an underexplored area, especially in urban and Western contexts.

Most of the discussion on economic inequality in recent years—such as that outlined in Chapter 1—uses generic statistics that do not distinguish between the economic situation of women compared to men. A figure for the Gini index, the Robin Hood index or the income share of the poorest 1% of a country's population is blind to gender and may mask some very significant differences. Hence in this chapter I seek to go beyond the generic statistics and findings on inequality and energy—worrying as they are—and ask whether and how women might be affected in specific ways within these.

Inequality and Energy. https://doi.org/10.1016/B978-0-12-817674-0.00008-4

My energy research of the past decades has been largely Euro-centric, focused particularly on the UK, but also includes empirical work among women consumers in non-western countries, particularly India. My findings from work among women in India help inform my outlook, and also shed more light on the economic implications of the ways many women in high-income, European countries are involved in energy consumption. Hence this chapter moves freely between these different countries as the comparisons become apparent.

Current policies on energy and climate change have been criticized for focusing too much on deterministic technological solutions and economic impacts rather than on the human dimension (Skinner, 2011; Wang, 2016). As Sovacool (2014) found in his analysis of 15 years of energy research, this may be related to the fact that most research on climate change and energy has been dominated by white males with a background in Engineering or Sciences and 87.4% of authors coming from European or North American institutions. Meanwhile, studies drawing from social sciences, arts and humanities have been treated as secondary and only 12.6% of the surveyed energy papers used qualitative methods. Yet it is increasingly acknowledged, even in the popular media (e.g., The Guardian, 2018a), that climate change is ultimately caused by people and therefore efforts from all disciplines, including arts and humanities, are required to tackle it.

Sanne (2002) argues that current climate change policies rely too much on existing theories that see consumption and energy use as an individual choice but ignore the fact that in reality a consumer's choices are always influenced by wider structural factors in her or his society. The application of social practice theory in energy studies has been popular in recent years, and despite its limitations (see Galvin and Sunikka-Blank, 2016), it has brought invaluable insights into energy use as heavily socially influenced practice and why it is so hard for households to change their practices. Domestic practices are not only utilitarian actions. They are constrained and enabled by the social structures in which consumers find themselves, and also carry meanings, sometimes even from childhood, and are heavily shaped by cultural norms and routine practices, especially in developing countries (Khalid and Sunikka-Blank, 2017). Shove (2010, 2014) argues that entrenched practices form the basis for everyday domestic energy demand and that many attempts by policy to reduce consumption are in fact up against very ordinary features of everyday life.

Some scholars have argued that addressing climate change can never be considered to be completely gender-neutral (Dankelman, 2002). Further, if climate change policies remain gender-neutral, they will be both unjust and insufficient (MacGregor 2009). Yet current climate change and energy policies have been observed to fall short of consideration of gender (Vinz, 2009; Hemmati and Rohr, 2009; Skinner, 2011), even in Scandinavian countries (Magnusdottir and Kronsell, 2014). However, gender has been identified to be a significant factor in residential energy consumption, for example in the US (Elnakat and Gomez, 2015; Elnakat et al., 2016). In developing countries policies have addressed cleaner cooking fuels and energy infrastructure in improving women's situations in rural settings, but in general women's involvement in energy policies, or the achievement of the UN Sustainable

Development Goal (SDG) 7 (energy) remains an oddly under-researched area (see Sovacool, 2014; Sunikka-Blank et al., 2018).

Improving gender equality can also contribute to more equal wealth distribution. According to World Economic Forum (2015), greater gender equity and increasing female economic participation are associated with lower income inequality in a society. If indexed, an increase from zero ("perfect gender equality") to one ("perfect gender inequality") is associated with an increase in net income inequality of almost 10 points as measured by the Gini index. Gender inequalities can be further worsened by poverty, especially in low-income countries. Gender gaps in education and health are a barrier to a more equal income distribution (IMF (International Monetary Fund), 2015). However, general improvements in economic conditions by means of increasing GDP and elimination of absolute poverty will not necessarily equally benefit all groups of a given society: "horizontal inequalities" which can be based on gender or ethnicity can persist (Kabeer, 2015). In India, despite economic growth and increase in GDP, women's participation in the workforce is actually on the decline (Klasen and Pieters, 2015; Majumdar and Buckley, 2019).

Equally, gender has been observed to be a key factor in wealth and income distribution in Europe (e.g., Botti et al., 2012). This chapter adopts a gendered view towards energy justice. It asks if women are equally affected by energy policies as men or whether there are factors that advantage or disadvantage them. It also questions whether the division of domestic labor in households leaves women and men in different positions in terms of time and money and suggests that policies should ask whether the responsibility of demand side management is introduced as one extra task for women.

It is worth mentioning the UK government's poverty line that considers a household to be in "relative poverty" if its income is below 60% of the median household income, which in 2016 was £413 per week or £21,600 per year (Department for Work and Pensions, 2017). This would make the poverty line £13,000 in annual household income or £248 per week.

The rest of this chapter is structured as follows. Section 2 discusses the gendered nature of household practices, women's unpaid domestic work and the wealth distribution between men and women. Section 3 describes the link between fuel poverty and gender. It challenges energy poverty as an individual comfort issue and discusses it as a demonstration of an accumulation of disadvantages that should be addressed as a part of wider social policy. Section 4 concludes.

2 Gendered household practices and the feminization of demand side response

Gender inequality in wealth is rooted in asymmetries in the amount of time that men and women have at their disposal, how they use their time and whether their labor is being paid. Further, even when in employment outside the home, women still do a disproportionate share of household practices and they also spend more time at home than men, especially if they have children. Gender equality in terms of who does the bulk of the domestic work, and who is primarily responsible for looking after the children, has made very little progress in the UK (Park et al., 2013). According to the Office of National Statistics, in 2016 women on average were doing 40% more household tasks than men (ONS (Office of National Statistics UK), 2016). In the UK, men's contributions to household practices averaged 16h per week compared to 26h by women. Women who were on maternity leave were doing up to 60h of household work per week.

The situation is similar in the United States (Bianchi et al., 2012). In Finland, women do household tasks on average for 221 min per day, compared to 153 min by men (Tilastokeskus (Statistics Finland), 2009). In Catalonia, women dedicate twice as many hours per week to household tasks than men: men on average did 15h of household tasks per week compared to 28h by women and this changed very little in the 10 years from 2002 to 2012 (Gonzalez Pijuan, 2018).

Some household practices remain more gendered than others. Laundry is done by women in 70% of households in Catalonia and repairs around the house by men in 75% of households, with almost no change in the past 20 years. Doing grocery shopping or looking after sick family members are more equally balanced tasks and equally done by both women and men in 43% of Catalonian households in 2012 (Gonzalez Pijuan, 2018).

There are also new demands on women's time at home. Schultz (1993) argues that the consequences of new environmental policies, such as waste recycling or even healthy eating, are borne by women in the household in Germany. These expectations increase women's workload at home and feminize the responsibility for the environment. They can also increase women's sense of guilt if they feel they do not correctly perform these duties in their household. Johnson (2018) argues that the rollout of smart meters in the UK calls for the mode of 'flexibility woman' who may need to shift her household practices. This is not only a European issue: Wang (2016) argues that existing eco-consumption policies in Taiwan unintentionally increase women's burden of domestic labor and that domestic "green practices" are highly gendered. In fact recycling, saving energy, saving water, driving less, buying organic, and avoiding products for environmental reasons have been identified as "women's tasks" in 30 countries (Sandilands, 1993; Dzialo, 2017). Dzialo (2017) suggests that environmentalism is more likely to be feminized in neoliberal countries, where there is a greater focus on individual citizen-consumers rather than changes in the system.

It can be that women are likely to take up these new practices not only because they are in charge of the household but also because women on average have been observed to have more environmentally friendly attitudes than men (Zelezny et al., 2002; Dzialo, 2017) and to consume less: single women households, on average, have been found to consume less energy than their male counterparts both in Sweden and the US (Räty and Carlsson-Kanyama, 2010).

Why is this distribution of domestic tasks relevant for energy use? Firstly, most household practices like cooking or cleaning are closely tied to energy use. This makes women the main stakeholders of energy policy in terms of load shifting or absorbing the shock of increasing energy prices, and it applies to reducing comfort standards as well as reliance on electrical appliances. Second, household work is unpaid and the hours that women put into domestic tasks will not increase their economic wealth.

In developing countries, women's traditional role at home leads to them having even more responsibility in the sphere of domestic energy consumption. Women act as shock absorbers in the household in case of increased energy expenditure, either by searching for secondary employment to pay for their bills or by compromising their own comfort standards (see Sunikka-Blank et al., 2018). In urban India, women's primary role is seen to be in domestic labor, and despite women getting more involved in income generating work, this has not been accompanied by men's participation in care work (McDowell et al., 2006; Klasen and Pieters, 2015.) Even if women work outside the home their main identity lies in being a mother. As in Western countries, lack of childcare limits women's prospects: despite having university degrees women may not be able to keep "proper" jobs as they have to fit their working hours within school hours, locking them in a trap of lowly part-time paid jobs, the informal economy and no time to develop their skills (Sunikka-Blank et al., 2018). Energy is a part of this equitation: poor housing design increases appliance ownership and energy costs, thereby increasing the pressure to find flexible employment and reducing the time for self-development. Debnath (2019) describes children's key role in choosing new household appliances in slum rehabilitation in urban India and their aspirations as a determinant of which appliances are chosen.

Ha and Williams (2013) suggest that motherhood can be an opportunity for behavioral change, but they conclude that new mothers see energy use as necessary for keeping the baby warm and comfortable, and while they consider costs they do not see energy in relation to climate change. Motherhood has been considered to be a factor in greening women's consumption practices, but according to the Australian Bureau of Statistics (ABS) (2012), household's energy use increases after having a child.

Yet the notion of motherhood, both in India and the UK, is important in understanding the socially constrained and determined nature of household practices. It can be argued that rather than looking at energy use based in binary consideration of gender, the true difference in household practices and who performs them comes from becoming a mother. The average mother on maternity leave in the UK does 37 h of active household work and childcare per week: 8 h of cooking, 7 h of housework, 2 h of laundry and 6 h of transporting themselves or their family (ONS (Office of National Statistics UK), 2016). In the US, women in middle class households took on a disproportionate share of the household work and caring tasks after their first child was born. This was further reflected in leisure practices: 35% of the time when women were doing housework, the men were engaged in leisure practices, whereas women engaged in leisure activities for 19% of the time that men spent engaged in housework or childcare (Kamp Dush et al., 2018).

The demands of childrearing practices on women's time are increased and intensified by the fact that childcare practices in many Western countries have become more intensive and the notion of parenthood more demanding, especially in middle class households. Children tend to spend more time indoors and supervised. They also use more electrical appliances, making childrearing practices more energy intensive. It has been concluded that children in the UK spend less time outdoors than prison inmates (The Guardian, 2016). Moreover, the costs of childcare in countries like the UK have radically increased in the past decades, overtaking increases in wages: UK childcare costs increased by 52% per week in 2008–18 for families with a full-time and a part-time working parent, whereas the parents' wages increased only by 17% (The Guardian, 2018b).

As a consequence of becoming a mother, women are likely to start working part time. In 72% of UK households with children, both parents were working but in most cases the father was in full-time work and the mother in part-time work, especially if the children were less than 12 years old (ONS (Office of National Statistics UK), 2017). There are nearly four times as many women in part-time work as men: they are likely to have lower hourly rates than full-time workers and have fewer opportunities for self-development. Having children leaves fathers relatively unaffected. Yet working part time is a deliberate choice for most women: 9 out of 10 mothers working part time said they did not want a full time job while 3 of 5 mothers looking for work said they prefer to work part time (ONS (Office of National Statistics UK), 2017). In any case the situation has consequences for women's economic situation and wealth. If we look at UK households without children where both partners are employed, men are paid on average 6% more than women, but in families with children, by the time the child is around 12 years old men are earning around a third more than women (Barnard, 2016).

Unequal division of unpaid domestic labor, childcare duties and working part-time all contribute to the unequal division of wealth between men and women. On average, women have a lower average income than men (Tunstall, 2018); in Catalonia, women earn 22% less than men in the age group of 30–44 and 36% less in the age group 45–64, while 22% of women work in Catalonia work part-time compared to 7% of men (Gonzalez Pijuan, 2018).

Discussion on the nexus between energy and gender is not complete without considering housing as an asset and women's access to it. In the UK, women comprise 42% of those owning a house outright and 31% of those with a mortgage (Tunstall, 2018). Home-ownership is reflected in the discrepancy in wealth: in 2016, the mean net wealth among women in the UK was £176,000, compared to £248,000 among men (D'Arcy and Gardiner, 2017). This economic disparity starts to develop after the age of 40, peaking in one's 60s when the median wealth of men is £100,000 higher than for women (Tunstall, 2018). At the other end of the spectrum, in the UK nearly 63% of adults in households claiming housing benefit are women (Department for Work and Pensions, 2017). 61% of all low paid workers are women in the UK and 25% women in employment are low paid, compared to 16% of men (Barnard, 2016). It has been estimated that 20% of UK women live in poverty: 5.1 million compared to 4.4 million men (Barnard, 2016). It is fair to say that women are significantly more affected by poverty than men in the UK, and this of course has implications for energy consumption since energy costs money.

Time poverty, unpaid domestic work and unequal wealth distribution make women more vulnerable to energy poverty. There are risk moments in women's lives that make them especially vulnerable to poverty such as becoming a mother or separating from your partner. The link between gender and energy poverty is further examined in the next section.

3 Fuel poverty, single parents and intergenerational immobility

Boardman (2010) has identified low-income single parent families as one of the most vulnerable groups exposed to fuel poverty in the UK. In the UK, 90% of UK single parents are women. It should also be considered that the number of single parent households is increasing in most Western countries (Härkönen, 2017) and that most single parents in Europe are women (Jalovaara and Andersson, 2017).

Table 1 shows the percentage of European households who were unable to heat their homes to an adequate level in 2017 in single parent households and in all households (figures for Ireland are for 2016). The notion of being unable to heat one's home adequately is one definition of fuel poverty (Dubois and Meier, 2016; Price et al., 2012; and see discussion in Chapter 7 of this book).

The figures show that in the UK, 14% of single parent households are in fuel poverty, by this definition, compared to 5.9% of all households. In the UK, 42% of all single parents live in relative poverty when housing costs are taken into account (Bennett, 2015). Single parenting is a common factor among households that are long-term poor in the UK, even where being a single parent is only a transitory phase (Brewer et al., 2012). Women are more likely to be in relative poverty if they are single parents: 46% compared to 22% of couple parents (JRF (Joseph Rowntree Foundation), 2017). There can be underlying cultural and class attitudes and pre-judgments in the UK towards single parents: single parenting can be associated with young mothers but in fact fewer than 1% of single parents are teenagers (Rabindrakumar, 2018).

In Germany, 9.9% of single parent households are in fuel poverty compared to 3% of all households. As in the UK, single parent households with children have been identified as the poorest German households, at 44% risk of poverty compared to 11% for a two-adult household with children (Destatis, 2016). Similarly, single-parent households are the most vulnerable group exposed to poverty in Catalonia (34%)—and in 80% of the single-parent households the adult is a woman (Gonzalez Pijuan, 2018).

It is worth noting that in some countries (Belgium, Latvia, Lithuania, Cyprus, the Netherlands, Finland) a household is more than twice as likely to be fuel poor if it is led by a single parent, whereas in a number of countries this is not necessarily the case (e.g., Italy, Luxembourg). This can be due to adequate housing, subsidies or support from family networks.

Table 1 Percentage of households unable to heat their homes in EU countries in 2017

Country	Percentage of single parent households unable to heat their home	Percentage of all households unable to heat their home
Bulgaria	45.2	36.5
Cyprus	41.6	22.9
Lithuania	40.5	28.9
Greece	34.0	25.7
Portugal	23.8	20.4
Ireland[a]	15.5	5.8
Italy	16.3	15.2
Latvia	15.2	9.7
Malta	14.6	6.6
Spain	14.1	8.0
United Kingdom	14.0	5.9
Croatia	12.8	7.4
Belgium	12.3	5.7
Poland	11.9	6.0
Germany	9.9	3.3
Denmark	9.1	2.7
Romania	9.0	11.3
Hungary	8.8	6.8
Austria	8.0	2.4
France	7.9	4.9
Slovenia	7.8	3.9
Netherlands	7.4	2.4
Slovakia	7.0	4.3
Sweden	6.2	2.1
Czechia	5.6	3.1
Finland	3.9	2.0
Estonia	3.4	2.9
Luxembourg	2.0	1.9

[a]Ireland (2016).
From Eurostat, 2019. Your Key to European Statistics. Available from: http://ec.europa.eu/eurostat (accessed 13 May 2019).

Fig. 1 shows the ratio between the percentage of single parent households unable to heat their homes and the percentage of all households unable to heat their homes, in EU countries in 2017. Countries with developed welfare systems (Denmark, Sweden, Germany or the Netherlands) still have a high percentage of fuel poverty in single parent households compared to their fuel poverty levels overall. It should be considered however, that in these countries the absolute number of fuel poor

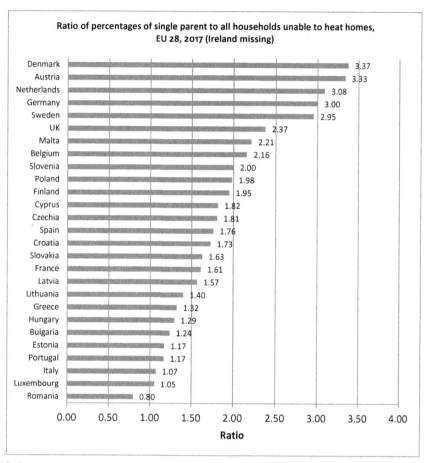

FIG. 1

Percentages of single parent households compared to all homes who are unable to heat their homes in EU countries in 2017.

Author's calculations based on Eurostat, 2019. Your Key to European Statistics. Available from: http://ec.europa.eu/eurostat (accessed 13 May 2019).

households can be relatively small compared to the UK for example. In the UK, one quarter of families with dependent children are single-parent families. It could also be that the share of single parent households in countries like the Netherlands is disproportionately higher in lower socio-economic groups or among immigrant groups.

Why are single parent households so vulnerable to energy poverty? In the UK, 67% of single parents are employed and their employment rate has been on the increase, but most single parents are limited by commuting distance, working hours and flexibility. Further, only 38% of single parents in the UK received regular child maintenance payments from their ex-partners. Consequently, 27% of working single parents in the UK are in poverty (Rabindrakumar, 2018). Even among single parents there is a gender payment gap: the median gross weekly pay for male single parents is £346, while for female single parents it is £194.

Due to their disadvantaged economic situation, single parents in the UK are less likely to own their own property and therefore have little control over improving its condition. In 2017, 27.1% of UK single parent households reported poor conditions in their homes: a leaking roof, mold, damp walls, floors or foundations, or rot in window frames, compared to 17.0% of all households (see Table 2). Table 2 shows that in all but five EU countries the housing conditions in single parent households are poorer compared to the whole population (the exceptions are Latvia, Greece, Spain, Romania and Sweden). In EU countries, on average, 18.3% of single parent households reported problems with their housing such as leaking roofs or mold compared to 13.3% of the whole population. In Belgium this was 24.8% and 18.5% respectively and in Germany 18.7% and 12.5% (for comparison with 2015 see Dubois, 2017).

When 47% of children in UK single parent families live in relative poverty, which is twice the risk of relative poverty faced by children in couple families, it should be considered that fuel poverty affects children in different ways. A household may choose self-imposed restrictions on energy consumption, such as cooking or heating (*"heat or eat"*) that can have health implications such as asthma on child's health and wellbeing, or cutting other general spending at the expense of children's other needs (see discussion by Middlemiss in Chapter 5 of this book).

Research suggests that in families in lower socio-economic groups the risk of single-parenthood is higher than in higher educated families. Jalovaara and Andersson (2017) compared the children of tertiary-educated mothers to those of basic-educated mothers in Finland, and concluded that children born to basic educated mothers are almost twice as likely to be born to an unmarried couple and four times as likely to be born to a single mother. They are also much more likely to experience parental separation. On average, children of basic-educated mothers spent half of their childhood years living with both parents. This suggests an "accumulation of disadvantages" that can limit child's prospects over the life course and across generations (TA and Eirich, 2006).

In fact, some social scientists argue that family structure has become an important mechanism in the reproduction of class, race, and gender inequalities, and in increasing inequality and reducing intergenerational mobility in the US (see e.g., McLanahan, 2004; Massey, 2008; Western et al., 2008; McLanahan and Percheski, 2008). Children in lower socio-economic classes are more likely to experience parental separation and father absence, and there is evidence that recent demographic changes in families (i.e., increased single parenthood in low income groups) are contributing to the intergenerational persistence of inequality. Similar social transitions have been observed in Europe where mothers with lower education levels, who already have had less opportunities and resources, are more likely to be single mothers, for example in Spain (Garriga et al., 2015) and in Finland (Härkönen, 2017). In previous decades it was more common among women with affluent backgrounds and higher education to be able to separate form their partner (Garriga et al., 2015).

Table 2 Percentages of single parent dwellings, compared with all dwellings, that are 'inadequate', i.e., with a leaking roof, damp walls, floors or foundation, or rot in window frames of floor

	Percentage of single parent homes that are inadequate, 2017	Percentage of all homes that are inadequate, 2017
European Union	18.3	13.3
Belgium	24.8	18.5
Bulgaria	15.4	12.2
Czechia	10.8	8.0
Denmark	24.0	14.9
Germany	18.7	12.5
Estonia	17.4	13.9
Ireland	24.8	12.6
Greece	12.8	13.5
Spain	10.8	11.5
France	17.3	11.1
Croatia	18.2	11.4
Italy	16.4	16.1
Cyprus	37.5	29.3
Latvia	21.1	22.8
Lithuania	19.6	15.7
Luxembourg	18.5	17.4
Hungary	41.1	24.8
Malta	13.7	8.4
Netherlands	23.7	13.5
Austria	16.5	11.9
Poland	16.3	11.9
Portugal	25.7	25.5
Romania	5.0	11.1
Slovenia	27.6	22.0
Slovakia	12.5	6.7
Finland	8.5	4.2
Sweden	5.0	7.0
United Kingdom	27.1	17.0

From Eurostat, 2019. Your Key to European Statistics. Available from: http:/ec.europa.eu/eurostat (accessed 13 May 2019) (EU-SILC survey [ilc_mdho01]).

However, this has changed, with separation and single parenthood common across all classes. As single parenthood makes women even more vulnerable to poverty, this development can contribute to the increasing inequality among mothers and children from different socio-economic backgrounds both in terms of their well-being and future opportunities (McLanahan and Percheski, 2008; Härkönen, 2017).

As the UK has one of the highest Gini coefficients in Europe and has recently been warned that it is heading for US levels of inequality (The Guardian, 2019), it should be asked what impact this development of increasing economic inequality and the demographic trend of more single parent households in low income cohorts will have on the most vulnerable household type and their intergenerational social mobility.

Changes in family structure can also contribute to increases in disparities between different ethnic groups. Single parents are more likely to come from a black or minority ethnic background in the UK: 19% of single parents had a black or minority ethnic background, compared with 14% of people in the UK (Rabindrakumar, 2018).

The statistics given in Tables 1 and 2 and Fig. 1 show that the risk of fuel poverty is significantly higher in single parent households, which in 90% of cases are led by women. This makes tackling energy poverty (with subsidies, elimination of pre-paid meters, provision of social and low cost housing and more progressive income and welfare redistribution) an urgent issue. Fuel poverty is not only a question of compromised comfort standards in a family but a wider structural issue. Energy should not accelerate gender inequality in term of making vulnerable women even worse off, or reduce social mobility in terms of disproportionately disadvantaging children who come from single parent households.

In addition to adequate housing provisions, energy distribution is an important part of ensuring energy justice. Single parent households are currently the largest group to be on prepaid meters in the UK. Along with Belgium and Australia, the UK is one of the few countries where prepaid meters are required in households that have credit problems and may not get a proper contract with a provider. The tariffs are higher than in a regular contract, and this means that vulnerable households pay the highest rate for their energy—a situation that again adds to the energy poverty burden of single mothers. Other countries are more supportive in this regard. In Finland, one energy provider in each area is obliged to make a contract with a household that may not have credit. This ensures that at least energy is not a driver or facilitator that pushes a household into deeper poverty.

4 Conclusions

This chapter has argued that women are in a different position from men as stakeholders of energy policies. Firstly, women perform most of the domestic practices that are closely tied to energy and water use. Therefore any domestic energy policy change is likely to have more impact on women's practices and comfort standards, including the unintended effect of the feminization of pro-environmental practices at the expense of women's time.

Second, if energy infrastructure is considered as gender neutral, it is ignored how women and men mostly in very different positions in terms of wealth and income, and also in time as a resource. Women already do most unpaid household work and childcare in families, which leaves them in lower paid part time jobs and economically worse off than men, especially once they become mothers. As a result, women are more exposed to poverty if their relationship breaks down. Single parent households are predominantly women-led in all EU countries, and face double the risk of energy poverty compared to families with couples. Due to the accumulation of disadvantages in single parent families, women from lower socio-economic groups and ethnic minorities have a higher probability of separating from their partner: fuel poverty can persist across generations and reduce intergenerational mobility.

A review of keywords used in energy justice literature shows that most studies are linked to poverty, economics or the environment, whereas housing features as a more secondary issue (Debnath, 2019). Yet statistics show how single parent households in the EU are disproportionally affected by poor housing conditions. Increasing inequality in wealth, which sets the context for the housing market, will put further pressure on single parent households.

Finally, rather than focusing on energy use by two binary genders, energy research should consider key moments in women's lives that can expose them to poverty such as becoming a mother or separating from their partner. The concept of motherhood, rather than gender, affects how households perform and share practices, work outside the home, accumulate wealth and use energy. This needs to be explored further.

References

Australian Bureau of Statistics (ABS), 2012. Household Energy Consumption Survey. Australia. Available at http://www.abs.gov.au/ausstats/abs@.nsf/Lookup/4670.0main +features100062012 (Accessed 13 May 2019).

Barnard, H., 2016. Women Carry the Burden of Poverty—We Should End that Injustice. Joseph Rowntree Foundation, London. Available at https://www.huffingtonpost.co.uk/helen-bar nard/women-in-poverty_b_13428572.html?guccounter=1 (Accessed 13 May 2019).

Bennett, F., 2015. Poverty in the UK: The Need for Gender Perspective. UK Women's Budget Group (WBG, London.

Bianchi, S.M., Sayer, L.C., Milkie, M.A., Robinson, J.P., 2012. Housework: who did, does or will do it, and how much does it matter? Soc. Forces 91 (1), 55–63.

Boardman, B., 2010. Fixing Fuel Poverty: Challenges and Solutions. Earthscan, London.

Botti, F., Corsi, M., D'Ippoliti, C., 2012. The Gendered Nature of Multidimensional Poverty in the European Union, CEB Working Paper no 12/026. Universite Libre de Bruxelles, Brussels.

Brewer, M., Costa Dias, M., Shaw, J., 2012. Lifetime Inequality and Redistribution, Working Paper W12/23. Institute for Fiscal Studies (IFS), London.

Dankelman, I., 2002. Climate change: learning from gender analysis and women's experiences of organising for sustainable development. Gend. Dev. 10 (2), 21–29.

D'Arcy, C., Gardiner, L., 2017. The Generation of Wealth, Asset Accumulation across and within Cohorts. Resolution Trust, London.

Debnath, R., 2019. Invisible Drivers of Electricity Consumption in Slum Rehabilitation in India. First year PhD report. University of Cambridge, Cambridge.

Department for Work and Pensions, 2017. Households below Average Income: An Analysis of the UK Income Distribution: 1994/95–2015/16. Available at: https://assets.publishing.

service.gov.uk/government/uploads/system/uploads/attachment_data/file/600091/house holds-below-average-income-1994-1995-2015-2016.pdf (Accessed 13 May 2019).

Destatis, 2016. Einkommen, Einnahmen und Ausgaben privater Haushalte. Statisches Bundesamt, Germany. Available at https://www.destatis.de/DE/Publikationen/ Thematisch/EinkommenKonsumLebensbedingungen/EinkommenVerbrauch/Einnahmen AusgabenprivaterHaushalte.html.

Dubois, U., 2017. Energy Poverty across the EU: Main Characteristics and the Urgency of the Issue. ISG Business School, Paris.

Dubois, U., Meier, H., 2016. Energy affordability and energy inequality in Europe: implications for policymaking. Energy Res. Soc. Sci. 18, 21–35.

Dzialo, L., 2017. The feminization of environmental responsibility: a quantitative, cross-national analysis. Environ. Sociol. 3 (4), 427–437.

Elnakat, A., Gomez, J., 2015. Energy engenderment: an industrialized perspective assessing the importance of engaging women in residential energy consumption management. Energy Policy 82, 166–177.

Elnakat, A., Gomez, J.D., Booth, N., 2016. A zip code study of socioeconomic, demographic, and household gendered influence on the residential energy sector. Energy Rep. 2, 21–27.

Galvin, R., Sunikka-Blank, M., 2016. Schatzkian practice theory and energy consumption research: time for some philosophical spring cleaning? Energy Res. Soc. Sci. 7, 55–65.

Garriga, A., Sarasa, S., Berta, P., 2015. Mother's educational level and single motherhood: comparing Spain and Italy. Demogr. Res. 33, 1165–1210.

Gonzalez Pijuan, I., 2018. Gender Inequality and Energy Poverty, a Forgotten Risk Factor. ESF/Association of Engineering without Borders. Available at: https://esf-cat.org/ wp-content/uploads/2018/02/ESFeres17-PobresaEnergeticaiDesigualtatdeGenere-ENG. pdf (Accessed 13 May 2019).

Guardian, T., 2016. Three-Quarters of UK Children Spend Less Time Outdoors Than Prison Inmates—Survey. Available at https://www.theguardian.com/environment/2016/mar/ 25/three-quarters-of-uk-children-spend-less-time-outdoors-than-prison-inmates-survey (Accessed 13 May 2019).

Guardian, T., 2018a. Securing a Future for Humanities: The Clue Is in the Name. Available at https://www.theguardian.com/education/commentisfree/2019/mar/18/securing-a-future-for-humanities-the-clue-is-in-the-name (Accessed 13 May 2019).

Guardian, T., 2018b. Working families' Childcare Costs Outstrip Wage Rises, Says TUC. Available at https://www.theguardian.com/money/2018/sep/03/childcare-costs-working-families-outstrip-wage-rises-tuc (Accessed 13 May 2019).

Guardian, T., 2019. Britain Risks Heading to US Levels of Inequality, Warns Top Economist. Available at https://www.theguardian.com/inequality/2019/may/14/britain-risks-heading-to-us-levels-of-inequality-warns-top-economist (Accessed 14 May 2019).

Ha, T., Williams, K., 2013. Does Becoming a Mother Make Women 'Greener'? The Conversation/Environment+Energy. Available at http://theconversation.com/does-becoming-a-mother-make-women-greener-19390 (Accessed 13 May 2019).

Härkönen, J., 2017. Single-Mother Poverty: How Much Do Educational Differences in Single Motherhood Matter? Research Reports in Demography 02. University Department of Sociology Demography Unit, Stockholm.

Hemmati, M., Rohr, U., 2009. Engendering the climate change negotiations: experiences, challenges and steps forward. Gend. Dev. 17 (1), 19–32.

IMF (International Monetary Fund), 2015. Empowering Women, Tackling Income Inequality. Available at https://blogs.imf.org/2015/10/22/empowering-women-tackling-income-inequality (Accessed 13 May 2019).

Jalovaara, M., Andersson, G., 2017. Disparities in Children's Family Experiences by Mother's Socioeconomic Status: The Case of Finland. Stockholm Research Reports in Demography 22. Stockholm University Demography Unit, Stockholm.

Johnson, C., 2018. Is Demand Side Response a Woman's Work? Gender Dynamics. Presentation at: "In Search of a 'Good' Energy Policy" seriesCentre for Research in Arts, Social Sciences and Humanities (CRASSH), University of Cambridge. 13 February 2018. Available at http://www.crassh.cam.ac.uk/events/27521 (Accessed 13 May 2019).

JRF (Joseph Rowntree Foundation), 2017. UK Poverty 2017, A Comprehensive Analysis of Poverty Trends and Figures. Joseph Rowntree Foundation, London. Available at https://www.jrf.org.uk/sites/default/files/jrf/files-research/uk_poverty_2017a.pdf (Accessed 13 May 2019).

Kabeer, N., 2015. Gender, poverty, and inequality: a brief history of feminist contributions in the field of international development. Gend. Dev. 23 (2), 189–205.

Kamp Dush, C.M., Yavorsky, J.E., Schoppe-Sullivan, S.J., 2018. What are men doing while women perform extra unpaid labor? Leisure and specialization at the transition to parenthood. Sex Roles 78, 715–730.

Khalid, R., Sunikka-Blank, M., 2017. Homely social practices, uncanny electricity demands: class, culture and material dynamics in Pakistan. Energy Res. Soc. Sci. 34, 122–131.

Klasen, S., Pieters, J., 2015. What explains the stagnation of female labor force participation in urban India? World Bank Econ. Rev. 29 (3), 449–478.

MacGregor, S., 2009. A stranger silence still: the need for feminist social research on climate change. Sociol. Rev. 57, 124–140.

Magnusdottir, G.L., Kronsell, A., 2014. The (in) visibility of gender in Scandinavian climate policy-making. Int. Fem. J. Polit. 17 (2), 308–326.

Majumdar, R., Buckley, P., 2019. How the financial crisis affected the world work force. Deloitte Rev. 24, 151–162.

Massey, D.S., 2008. Categorically Unequal, the American Stratification System. Russell Sage Foundation, New York.

McDowell, L., Ward, K., Fagan, C., Perrons, D., Ray, K., 2006. Connecting time and space: the significance of transformations in women's work in the city. Int. J. Urban Reg. Res. 30 (1), 141–158.

McLanahan, S., 2004. Diverging destinies: how children are faring under the second demographic transition. Demography 41 (4), 607–627.

McLanahan, S., Percheski, C., 2008. Family structure and the reproduction of inequalities. Annu. Rev. Sociol. 34 (1), 257–276.

ONS (Office of National Statistics UK), 2016. Women Shoulder the Responsibility of 'Unpaid Work', UK: 2015. Available at https://www.ons.gov.uk/employmentandlabourmarket/peopleinwork/earningsandworkinghours/articles/womenshouldertheresponsibilityofunpaidwork/2016-11-10 (Accessed 13 May 2019).

ONS (Office of National Statistics UK), 2017. Families and Labour Market, England: 2017. Available at https://www.ons.gov.uk/releases/familiesandthelabourmarketengland2017b (Accessed 13 May 2019).

Park, A., Bryson, C., Clery, E., Curtice, J., Phillips, M. (Eds.), 2013. British Social Attitudes: The 30th Report. NatCen Social Research, London.

Price, C., Brazier, K., Wand, W., 2012. Objective and subjective measures of fuel poverty. Energy Policy 49, 33–39.

Rabindrakumar, S., 2018. One if Four, A Profile of Single Parents in the UK. Guingerbread Foundations, London.

Räty, R., Carlsson-Kanyama, A., 2010. Energy consumption by gender in some European countries. Energy Policy 38 (1), 646–649.

Reames, T., 2016. Targeting energy justice: exploring spatial, racial/ethnic and socioeconomic disparities in urban residential heating energy efficiency. Energy Policy 97, 549–558.

Sandilands, C., 1993. On "green" consumerism: environmental privatization and "family values". Can. Woman Stud. 13 (3), 45–47.

Sanne, C., 2002. Willing consumers or locked in? Policies for a sustainable consumption. Ecol. Econ. 42, 273–287.

Schultz, I., 1993. Women and waster. Capital. Nat. Social. 4 (2), 51–63.

Shove, E., 2010. Beyond the ABC: climate change policy and theories of social change. Environ Plan A 42, 1273–1285.

Shove, E., 2014. Putting practice into policy: reconfiguring questions of consumption and climate change. Contemp. Soc. Sci. 9 (4), 415–429.

Skinner, E., 2011. Gender and Climate Change. Institute of Development Studies UK, London.

Sovacool, B., 2014. What are we doing here? Analyzing fifteen years of energy scholarship and proposing a social science research agenda. Energy Res. Soc. Sci. 1, 1–29.

Sunikka-Blank, M., Bardhan, R., Nasra Haque, A., 2018. Gender, domestic energy and design of inclusive low-income habitats: a case of slum rehabilitation housing in India. Energy Res. Soc. Sci. 49, 53–67.

TA, D.P., Eirich, G.M., 2006. Cumulative advantage as a mechanism for inequality: a review of theoretical and empirical developments. Annu. Rev. Sociol. 32, 271–297.

Tilastokeskus (Statistics Finland), 2009. Ajankäyttötutkimus 2009, Ajankäytön muutokset 2000-luvulla. Tilastokeskus, Helsinki.

Tunstall, B., 2018. Housing and Gender. UK Women's Budget Group (WBG), London.

Vinz, D., 2009. Gender and sustainable consumption: a German environmental perspective. Eur. J. Womens Stud. 16 (2), 159–179.

Wang, S., 2016. Green practices are gendered: exploring gender inequality caused by sustainable consumption policies in Taiwan. Energy Res. Soc. Sci. 18, 88–95.

Western, B., Bloome, D., Percheski, C., 2008. Inequality among American families with children: 1975–2005. Am. Sociol. Rev. 73 (6), 903–920.

World Economic Forum, 2015. Why Gender and Income Inequality Are Linked. Available at https://www.weforum.org/agenda/2015/10/why-gender-and-income-inequality-are-linked (Accessed 13 May 2019).

Zelezny, L.C., Chua, P.P., Aldrich, C., 2002. New ways of thinking about environmentalism: elaborating on gender differences in environmentalism. J. Soc. Issues 56 (3), 443–457.

Further reading

Eurostat, 2019. Your Key to European Statistics. Available at: http://ec.europa.eu/eurostat. (Accessed 13 May 2019).

Lee, A.N., Ho, R.Y., Ko, L.C., 2010. Gender differences in energy consumption and energy-saving consciousness. In: Proceedings of the Energy and Economy Conference, Taipei, November 2010.

Inequality and renewable electricity support in the European Union

Lawrence Haar

Oxford Brooks Business School, Oxford Brookes University, Oxford, United Kingdom

Chapter outline

1 Introduction and background 189
2 Investment and costing issues with renewable electricity 192
3 Renewable energy support 194
   3.1 Background 194
   3.2 Growth in RE capacity 195
   3.3 Expenditure on renewable electricity 195
4 Pricing of electricity 197
   4.1 Cost components 197
   4.2 Pricing of electricity by household size 200
   4.3 Relating household electricity consumption to household income 209
5 Renewable energy and electricity pricing 211
6 Conclusions 214
Acknowledgment 217
References 218
Websites consulted 219
Further reading 220

1 Introduction and background

In this chapter I argue that the rapid increase in renewable electricity in most European Union (EU) countries over the past two decades has led to regressive pricing structures for electricity. "Regressive" in this context means electricity is more expensive per kilowatt-hour (kWh) for poorer households than for wealthier households; that it is less expensive still for many large firms; and even less expensive for the largest firms.

This is partly because of the way subsidies are granted for renewable energy installations such as photovoltaics and wind turbines, but that is not the whole story.

Inequality and Energy. https://doi.org/10.1016/B978-0-12-817674-0.00009-6

More generally, the pricing structure for electricity was developed, in former decades, to suit a market structure where most electricity came from burning fossil fuels (coal, oil, gas) or from reacting nuclear materials (plutonium, uranium). In these methods of generating electricity there are three main costs to consumers: the fixed costs of building and maintaining power plants and distribution networks; the cost per kWh of consuming fuel (fossil or nuclear) for each extra kWh produced: and government-induced taxes and levies. Each consumer paid a relatively small fixed price per month to compensate power companies and distributors for their fixed costs. On top of this, customers paid for each kWh of electricity they consumed, to compensate power companies for their extra cost per kWh of burning fossil fuel. Because hydrocarbon fuels were expensive making the variable cost component of total costs large, prices in markets were set by the marginal producer, typically a combined cycle gas turbine, household electricity bills were more or less proportionate to the volume of electricity the household consumed. Further, taxes and levies were also roughly proportionate to the volume consumed, and did not necessarily favor one type of household over another.

Renewable electricity production from wind turbines and photovoltaic panels is technologically, however, different from fossil and nuclear electricity production in two fundamental ways. Firstly, with wind turbines and photovoltaic panels there is often a very big difference between the capacity rating of a generator—expressed as "kilowatt-peak" power (kWp)—and its actual moment-by moment output—expressed as kilowatt-hours of energy (kWh). A photovoltaic panel rated at 1 kWp can produce a maximum power output of 1 kW in ideal, sunny conditions, and if this persists for an hour it will have produced 1 kWh of electrical energy. However, in the sunniest regions of Germany a 1 kWp photovoltaic panel produces, on average, about 950 kWh per year—despite there being 8760 h in a year and (theoretically) about 4380 h of sunlight. Further, with both wind and photovoltaics the output varies enormously from day to day and hour to hour as wind and sun ebb and surge. This leads to extra cost burdens, such as for supply and distribution that have to adapt to very changeable supply. Hence it is very important not to confuse capacity with output, when considering the economic merits renewable electricity.

Secondly, with wind turbines and photovoltaic panels there is no extra cost to the producer for each extra kWh of electricity produced (the "marginal cost" per kWh is zero). If the wind blows harder or the sun shines brighter today than yesterday, each additional kWh costs nothing to produce. In fact, almost all the producer's costs are in the building of the power plants—the wind turbines or photovoltaic panels—and the distribution lines, transformers, regulators, etc., associated with these. So, for example, if E.ON's huge offshore wind farm in the English Channel produces 1000 GWh of electricity on a windy day, this costs E.ON no more than the 200 GWh it produced on calmer day. Nevertheless, E.ON has to somehow recoup this investment along with its investment in distribution lines and transformers. It does this by increasing its fixed charges. When all the major electricity companies are doing this together, customers pay higher fixed charges. The result is that the fewer kWh a household consumes, the more it pays per kWh. Most of this chapter describes and comments on the details of how this is working itself out within EU countries. First, however, I comment on several ways in which it relates to the themes we have covered in this book.

Firstly, and most clearly, it illustrates one way in which the current and emerging electricity market exacerbates economic inequality. Although poorer households consume less electricity than wealthier, larger ones, because the price of electricity is discounted according to volume, poorer households pay more for each unit of energy consumed. By analogy, this would be like a motorist with a larger, expensive vehicle paying less per liter at the petrol pump than a motorist with small, efficient vehicle.

Secondly, it is in large part a *socio-technical* phenomenon and not primarily (or solely) the result of the shifts in economic policy which are outlined in Chapter 1 of this book. The liberalization of markets is certainly part of the picture. But at a more basic level, the pricing structure issue has to do with the shift in technologies from fuel-based electricity generation to renewable forms. In the 1980s a group of scholars developed an understanding of how social phenomena are deeply inter-twined with the types and functions of technology that are used in different contexts and periods (Bijker et al., 1987). One of the key works in this literature is Thomas Hughes' *Networks of Power: Electrification in western society* 1880–1930 (Hughes, 1993). A review of this social science literature is beyond the scope of this chapter, but the key point is that a new technology or set of technologies can have unexpected consequences within society. Trying to align renewable elec-tricity technology with the historic systems and structures that have arisen under the technology of fossil fuel generated electricity, creates issues and problems, from a socio-economic perspective. Future researchers may find it interesting to see how the socio-technical systems literature could enrich the findings of the present chapter.

Thirdly, the regressive pricing structure that has evolved in tandem with renew-able energy is to a large extent enabled by the liberalization of markets. The liber-alization of markets was a central tenant in the growth of neoliberalism, a theme touched in in Chapters 1–3 of this book. It is not inevitable that renewable energy—with its high capital costs and near-zero cost per kWh—should produce a highly regressive pricing structure. For example, in New Zealand in the 1940s, 1950s, 1960s and 1970s, most electricity came from renewables: hydro and geother-mal. But in the very non-liberalized, egalitarian society of New Zealand in those decades, the electricity market was government controlled and anything but regres-sive (Bertram, 2006). Hence, within this chapter I will occasionally refer to today's liberalized electricity markets as one of the factors that favors markets leading to regressive pricing when they have a high share of renewables.

Fourthly, direct subsidies for renewables have mostly come in the form of feed-in tariffs (FITs). These are direct payments to producers per kWh of renew-able electricity produced, and the money comes from a per kWh levy on con-sumers' electricity bills. As many researchers have already shown (e.g., Andor et al., 2015; Frondel et al., 2015) this form of levy is inherently regressive: it transfers money from all households, including the poorest, to the relatively wealthy households who have invested in photovoltaics on their roofs or in wind farms via local consortia or other investment structures.

It must also be stated at the outset that this chapter is in no way intended as a criticism of renewable electricity and nor is it criticism of liberalized electricity markets which for two decades delivered investment in conventional generation and low-cost electricity. Clearly a much higher, not lower, proportion of our electricity will need to be generated from renewables if we are to mitigate damaging climate change. Rather, my purpose in this chapter is to show how the integration of renewables into electricity markets is happening largely at the expense of poorer consumers. It does not have to be this way. If we can better understand how, why and in what ways the increase in renewables is leading to regressive electricity pricing, we will be in a much better position to change our pricing systems so that the burden of climate change mitigation is more equitably shared.

In Section 2 I offer an overview of the investment and costing issues that have arisen due to the advent of increasing volumes of renewable electricity in energy markets in EU countries. In Section 3 I quantify the level of support for renewable electricity. In Section 4 I explore how electricity is priced for retail-household consumers. In Section 5, comparing EU countries over time, I investigate the proposition that the observed electricity pricing structure is related to the growth in and reliance upon renewable electricity and consider how this may be explained from the perspective of regulatory theory. I conclude in Section 6 with policy observations and recommendations.

2 Investment and costing issues with renewable electricity

In liberalized power markets, electricity generated by renewable energy in the form of solar photovoltaics and wind turbines requires various support mechanisms to induce private investment, because of its high cost on a "levelized" basis (the "levelized" cost of electricity generation is the average price the owner of the generator must receive, to break even over its lifetime). Unlike conventional methods of generating electricity, the annual output from a renewable electricity facility may be one-quarter or less of the expected output of facilities using fossil fuel, making its Levelized Cost of Electricity Output (LCOE), i.e., Capital Costs ÷ Output, very high. With conventional forms of electricity generation, the mixture of fixed and variable costs means that, even with lower utilization, the impact upon LCOE remains small. But with renewable electricity generation, because there are fewer hours of output and the LCOE is driven largely by capital costs, potential owners must receive incentive prices rather than market prices, to be motivated to invest. Even with incentive prices, however, in liberalized electricity markets like those of Europe, the UK and North America, a further challenge arises: The intermittent nature of RE generation means that conventional, quickly "dispatchable"[a] fossil-fuel plants need to be available for back-up. But they are likely to be under-utilized or only used when renewable output is not generating. Hence a lot of capital resides in under-utilized conventional plants. For such

---

[a]"Dispatchable" in this context means able to be ramped up and down quickly in response to demand.

plants to remain *on-the-system*, they must be rewarded typically through *capacity payments* and other such mechanisms.

Recognizing the challenges that deregulated/privatized electricity markets pose for renewable electricity, various incentives and mechanisms to encourage investment have been introduced. They include incentive tariffs, price premiums and supports as well as tax credits and allowances (see, European Commission: Guidance... Support Schemes, 2013). In addition to having incentive prices to compensate for the low level of capacity utilization, to address the stochastic nature of renewable electricity output, renewables are given "dispatch priority" through power purchase agreements (PPA): while renewable plants are generating, other plants with flexible and controllable output must reduce their output. Dispatch priority creates short-term balancing costs for the grid and fossil fuel plants. For this reason, in several European countries, capacity payments were introduced to ensure plant availability when renewable electricity is not being generated. In the Irish market, conventional fossil fuel generators are paid for the energy they produce, but they also receive a capacity payment for being available to produce when needed, with peaking plants receiving the largest amounts. In the UK, with the Act of 2012, capacity payments were introduced for dispatchable plants to ensure availability.[b] In Germany, although not described as a capacity payment, a payment was introduced in 2015 to reward plant availability (https://energytransition.org/2015/01/did-germany-reject-or-just-postpone-capacity-payments/).

Across support schemes, in general, we see a transference of risk from private investors to third parties (Ayoub and Yuji, 2012; Haar and Haar, 2017). In transferring risk, such support mechanisms may have both a direct cost in additional expenditure over what would have been incurred had fossil fuel generation been utilized, and indirect costs in managing the random nature of renewable electricity output. Further, the distributed nature of renewable electricity generation generally requires additional investment in connections and transmission. Although both direct and indirect costs fall initially upon integrated utilities and aggregators—the *Licensees*—ultimately such costs reside with stakeholders: retail customers or ratepayers, businesses, industry and share-holders. Given widespread exemptions for industry and businesses from green levies and surcharges, it appears that support for renewable electricity is largely left to retail consumers[c] (Reuters, 2016). The question then arises: are they all treated equally?

Across the EU, retail prices for electricity paid by households, comprising fixed and variable cost components, differ widely. Furthermore, within individual countries, how electricity is priced to households varies according to the level of annual

---

[b]In November 2018, the European General Court found that the UK payment scheme to ensure back-up capacity was ruled to be a violation of State Aid rules and a "standstill period" was imposed (McCormick, FT, 15-11-2018)

[c]According to Der Spiegel (04–9-12), in Germany over 2300 businesses have managed to exempt themselves from the green energy surcharge through claiming they face international competition.

consumption or customer categories. In the remainder of this chapter I examine the structure of pricing and seek to explain such differences. The potential for energy policy having distributional effects has been raised in the literature (Chawla and Pollitt, 2013 and Brazilian et al., 2014). In this light, in the following sections I investigate whether such differences are related to how renewable electricity is supported.

3 Renewable energy support

3.1 Background

The ultimate effects of renewable electricity on the pricing structure would not be important if the expenditure on supporting it were small relative to total expenditure upon electricity, but this is not the case. Below I will look at how renewable electricity has been supported and review the growth in renewable electricity capacity and what this means for expenditure on this form of electricity generation. The combination of generous support schemes and growth in renewable electricity output has together resulted in large increases in total expenditure upon this form of electricity generation. To understand this trend, I begin with the various incentive schemes introduced in the EU from 2010. Across the EU, feed-in tariffs (FiT) were the most widely utilized support schemes for renewable electricity (Yurchenko and Thomas, 2015). These schemes offered a fixed price per MWh of electricity produced. The FiT support scheme involved private generators of renewable electricity selling output to private purchasers or Licensees, which were typically integrated utilities and aggregators. Under the scheme, private generators received a tariff price from the Licensee according to the scale and type of technology (RES-Legal[d]). The incentive schemes provided attractive returns over many years with little risk: minimum prices were guaranteed while dispatch priority ensured a market for their output. In addition to the FiT support schemes, some countries including Denmark, Germany, Spain and the United Kingdom introduced "contracts-for-differences" (CfD) for larger facilities to support RE. A contract-for-differences is an agreement between a buyer and a seller, where the seller will pay to the buyer the difference between the current value of an asset, and its value at the time the contract was agreed. This means the renewable energy producer gets the same (high) price for each kWh she produces, even when its actual market value is low. In the UK, under the Energy Market Reform Act of 2012, facilities exceeding 300 MW of capacity were offered a CfDs to provide stable and predictable incentive prices for investment in low-carbon generation such as solar photovoltaic and wind turbines (Brinckerhoff, 2012; Couture et al., 2010). So, at times when the spot-price of electricity fell below a certain level due to excess generation from renewables, the owners of the larger renewable electricity plants still received the agreed price for each MWh they generated.

---

[d]http://www.res-legal.eu/compare-support-schemes/

At a time of stable to declining prices for wholesale electricity, where prices have averaged in the mid €40s/MWh across the EU, the incentive prices awarded to green generation have been exceptionally attractive (Platts PowerVision, 2018). As examples, in 2010 Belgium, Italy and Spain all offered around €300 per MWh for solar photovoltaic electricity. Sunny Greece in 2010, was offering nearly €500 per MWh for photovoltaic electricity. Up until 2014, the levelized cost of photovoltaic electricity in Germany exceeded €400/MWh until it was reduced to €90/MWh, though this is still roughly double the average wholesale price for that year (Erbach, 2016). In 2012, across the EU member states, the average support for renewable energy through FiTs was €110.65/MWh. It was highest in the Czech Republic (€194.51/MWh), and lowest in Estonia (€10.56/MWh). Measured on a per MWh basis, hundreds of Euros were spent for each unit of renewable electricity produced, while wholesale prices of electricity, as set by the marginal dispatchable gas-fired plant, were stable to declining, averaging below €45.00/MWh.

3.2 Growth in RE capacity

The combination of attractive incentive prices described above, together with guaranteed markets for renewable electricity output, has resulted in significant investment in renewable electricity capacity, a growing contribution of renewable electricity to total electricity consumption and unprecedented expenditure on this form of electricity generation. The growth of investment in renewable forms of electricity generation along with its growing contribution are widely regarded as a policy achievement of the European Union (Yurchenko and Thomas, 2015; EEA Report, 2016). Examining the largest 10 economies of the European Union, we see in Fig. 1 that the percentage contribution to total energy consumption coming from renewable electricity has grown steadily since 2007.

For the European Union, the contribution of renewable electricity to total electricity consumption increased nearly threefold between 2007 and 2016. Some notable milestones in the contribution of renewable electricity were in countries such as Romania, where it increased from nearly zero to 20%; Italy from zero to 14%; and the United Kingdom from 1.5% to 16%. Summarizing the results, we see in Table 1 that between 2010 and 2016, for all 28 members of the European Union, the percentage of electricity consumed which was generated from wind turbines or solar photovoltaics increased from less than 5% to over 12%. In 2016 the minimum was just under 2% for Latvia, while the maximum was just over 43% for Denmark.

3.3 Expenditure on renewable electricity

The surge in investment in renewable electricity generating technology, earning incentive prices as described above, has necessitated a high level of expenditure in support. Through incentive mechanisms, direct expenditure in support of renewable electricity output has been significant. According to work for the EU Commission by Ecofys and KPMG, in 2012 alone over €27 billion euros was spent supporting

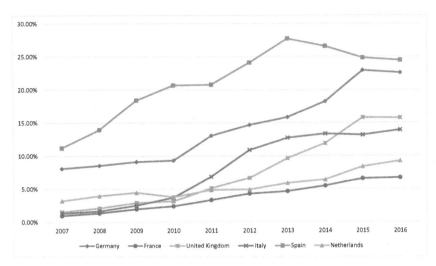

FIG. 1

Percentage contribution from wind and solar to total electricity consumption for 10 WU countries.

From http://appsso.eurostat.ec.europa.eu/nui/submitViewTableAction.do.

Table 1 Percentage of electricity consumption from wind and solar energy

Stats/year	2010 (%)	2016 (%)
Minimum	0.06	1.97
Maximum	24.34	43.42
Average	4.64	12.10

From http://appsso.eurostat.ec.europa.eu/nui/submitViewTableAction.do.

solar and wind generation (Ecofys, 2014). From other research we see that Germany's support scheme for renewable technologies is likely to exceed €400 billion over the next two decades (Andor et al., 2017). In 2017 Ofgem, the UK Regulator, stated in their Annual Report that in the previous year £1.25 billion was spent on just the UK FiT support Scheme, an increase of £170 million from 2015 (Ofgem, Annual Report, 2017: 11). Using 2016 data we can make a top-down estimate of the incremental *direct* expenditure in support of renewable electricity across the EU + Norway, as shown in Table 2.[e] This estimate *excludes* additional network costs, balancing costs and other effects of managing intermittent renewable output under dispatch priority which fall upon stakeholders and have been estimated to be very large (Haar and Haar, 2017).

---

[e]These sources are consistent with the various *Status Reviews of Renewable and Energy Efficient Support Schemes in Europe* in 2012, 2013 and 2015 published by the Council of European Energy Regulators.

Table 2 EU 28 + Norway: Estimated additional expenditure on renewable electricity in 2016 (assuming wholesale fossil fuel generation price of €45.00 per MWh and a weighted average price of solar photovoltaic (4%) and wind (11%) of €172.77)

Electricity consumption by sectors (2016) (MWh)		Green energy contribution— 15% (MWh)	Additional expenditure on wind and solar (Euros, Billions)
Industry	1,013,148,000	151,972,200	€ 26.26 bn
Transport	63,828,000	9,574,200	€ 1.65 bn
Households	1,707,301,000	256,095,150	€ 44.25 bn

Authors' calculations using EuroStat data and CEER Annual Reports.

Reflecting various concerns, Feed-in Tariffs for *new* investments have been reduced across the EU since their introduction in 2010, but efforts by various governments to apply the revised tariff prices *retrospectively* have been largely unsuccessful.[f] Thus, even with reduced tariffs, the legacy of renewable electricity investments enjoying historic prices has ensured that the total expenditure on this mode of generation remains considerable. This raises the issue of how such costs are recovered from various stakeholders. To the extent that industry, business and transport *do not* absorb the costs of supporting renewable electricity, an additional €28 billion in cost beyond the €44 billion already faced by retail consumers (mostly households) must be met, directly through the pricing structure or indirectly through burden shifting. The magnitude of renewable electricity support across the EU raises the question of whether retail consumers, mostly households, within a given country are treated equally and how such pricing compares between countries. Further, we ask, in addition to energy costs, if taxes and network charges have been applied to retail consumers in an equitable manner. Critically, we investigate if pricing differences are indeed related to investment in renewable electricity.

4 Pricing of electricity

4.1 Cost components

At retail consumer level, the price of electricity as paid by households varies greatly by country and by the contribution of the various components, namely (i) Energy or Wholesale Costs, (ii) Network costs and (iii) Taxes and Levies. As we see in Table 3

---

[f]In the case of the United Kingdom, the High Court ruled *against* the Department of Energy and Climate Change (DECC) and in favour of the Solar Industry which had petitioned against retrospective changes to tariffs. The High Court ruled that retrospective changes to FiT prices were unlawful. The Government lost on Appeal and the Secretary of State in 2012 refused DECC permission to challenge the ruling. The Judge also denied the request for DECC to go to the Supreme Court to challenge the lower court decision.

Table 3 Components of retail electricity prices in the UK 2016

Component	Percentage of total cost
Wholesale costs	33.52
Network costs	25.46
Environmental and social obligation costs	17.45
Other direct costs	1.26
Operating costs	17.15
Supplier pre-tax margin	0.4
Value added tax	4.76

From https://www.ofgem.gov.uk/consumers/household-gas-and-electricity-guide/understand-your-gas-and-electricity-bills.

below, according to data from the UK regulator the Office of Gas and Electricity Markets (OFGEM), in 2016 the UK price of retail electricity for the average household comprised the following components:

Explaining the various components, wholesale costs refer to how much a supplier must pay for electricity supply to the retail customer. Such electricity may be purchased on an exchange or through a contract with an electricity generator. Some suppliers are also part of companies that generate their own electricity. Network costs concern the wires for transmission and distribution carrying electricity across the network and into homes and businesses, and are paid for through "Use of System Charges". Suppliers are charged for the costs of maintaining and using these networks, and these charges are then passed on to customers through bills. Taxes and involve all manner of policy initiatives. For example, environmental and social obligation costs—the third row in Table 3—are government programs to save energy, reduce emissions and encourage use of renewable energy.[g] Taxation varies greatly by member states, and may be adjusted to budgetary considerations. Levies represent distributed costs of specific policies, of which support for renewable energy is the most prominent. In the UK in 2016 network charges plus taxes and levies were nearly 43% and exceeded the costs of the actual electricity per kWh. Further, programs like the FiT Scheme described above, as supported through taxes and levies, may affect some customers' bills indirectly as well: households participating in the scheme will see a reduction in their electricity prices while those *not* participating, will pay more, through the transference of such costs. Suppliers incur costs relating to sales, metering and billing of renewable electricity on top

[g]All EU households pay value added tax on their electricity bill. Rates vary between 5% in the UK and 27% in Hungary. The tax is applied to the total electricity price, including energy and network component and taxes and levies. The VAT falls under the heading taxes and levies, but the nominal value depends on the total price. Consumers in Denmark pay VAT at a rate of 25% which in 2015 was €6.2 ct/kWh. In 2015, Germans paid the second highest VAT at a rate of 19% or €4.7 ct/kWh. Portugal increased the VAT rate from 6% to 23% in 2016. In Belgium, the VAT rate was reduced to 6% in 2014, but was then restored to 21% in 2015.

of normal expenditures. For the UK in 2016, environmental costs plus suppliers' pre-tax profit margin represented about 18% of the total retail price of electricity as paid by households.

Across the EU, wholesale electricity exchange markets have been established to provide day-ahead, forward and intra-day trading. Flexible and liquid markets promote an efficient matching of supply and demand that lowers generation costs and therefore prices. Such exchanges provide guidance to "over-the-counter" contract prices in the most mature markets. Gradually, separate national wholesale markets (through grid-interconnections and harmonization of network codes) have become integrated, encouraging market convergence (European Commission, 2016). Notwithstanding the integration trend in wholesale markets across the EU, at country level retail electricity prices paid by households vary greatly because of different tariff structures and the respective contributions from network charges, taxes and levies. In retail markets at country level, we see big differences in how electricity is priced, reflecting the relative importance of the above described components. For example, as shown in Fig. 2, in 2017, using data from the European Commission, medium-size households with an annual consumption between 2500 kWh and 5000 kWh faced the highest electricity prices in Germany (EUR ***0.305 per kWh), Denmark (EUR 0.301 per kWh) and Belgium (EUR 0.288 per kWh), while the lowest electricity prices were in Bulgaria (EUR 0.098 per kWh), Lithuania (EUR 0.111 per kWh) and Hungary (EUR 0.113 per kWh).

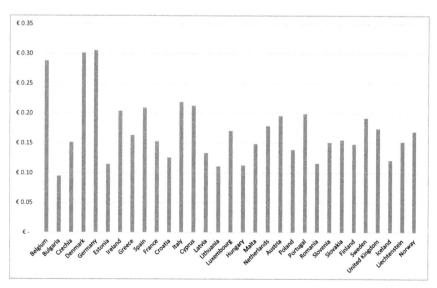

FIG. 2

Retail electricity prices in €/kWh including network charges, levies and taxes; (averages 2013–2018, 2500–5000 KWh).

From https://ec.europa.eu/eurostat/statisticsexplained/index.php/Electricity_price_statistics#Electricity_ prices_for_household_consumers.

Table 4 Average EU electricity prices (euros/kWh) for consumption band 2500–5000 kWh per annum

Retail prices with and without taxes and levies	Electricity prices excluding taxes and levies	Electricity prices including taxes and levies
Minimum	€ 0.08	€ 0.09
Maximum	€ 0.19	€ 0.30
Average	€ 0.12	€ 0.15
Percent difference Minimum to maximum	92%	119%

From Eurostat: https://ec.europa.eu/eurostat/statistics-explained/index.php/Electricity_price_statistics#Electricity_prices_for_household_consumers.

Comparing retail prices as paid by households across Europe, *excluding* levies and taxes we still see large differences between countries, as shown in Table 4. We see a 92% difference between the highest and lowest retail electricity prices for average households excluding taxes and a 119% difference between the minimum and maximum electricity price when taxes and levies are included. Given the convergence of European wholesale power markets and its contribution which wholesale prices make to household electricity prices, it is not easy to explain the observed differences between countries. Comparing countries, even with taxes and levies excluded we still see large differences in what households pay for electricity. These observations are for retail consumers who consume between 2500 and 5000 KWh per annum, which is the most common consumption band. As I will indicate below, country differences increase when different sized customers are considered.

4.2 Pricing of electricity by household size

As explained in Section 2 above, the retail price as paid by households for electricity comprises the energy component, the network charges and taxes plus levies. As we saw in Table 3, such prices vary considerably by country. I now turn to how consumers of different sizes are treated. In the EU, household consumers are categorized by their annual consumption using five levels of consumption, as shown in Table 5.

As mentioned in the Introduction, in almost all countries, the prices paid per kWh for electricity are a *decreasing* function of annual consumption: The more one takes, the less one pays per kWh. Comparing retail electricity prices, as we see in Fig. 3, exclusive of taxes, levies and network charges, in the 10 largest countries of the European Union retail consumers with the lowest annual consumption (hereinafter called the "smallest" consumers) paid more than those with highest consumption (hereinafter called the "largest" consumers).

Purely on energy costs, we see from Fig. 3 that among Spanish households, the smallest retail consumers pay over three times as much per kWh as the largest retail

Table 5 Annual household electricity consumption by Eurostat categories (kWh)

Household consumption levels	Minimum	Maximum
First	0	1000
Second	1000	2499
Third	2500	4999
Fourth	5000	14,999
Fifth	15,000	No limit

From Eurostat.

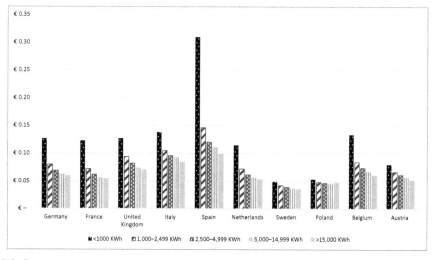

FIG. 3

Retail electricity prices in the EU's 10 largest economies, excluding taxes, levies and network charges, by consumer category, 2017.

From Eurostat.

consumers, those consuming more than 15,000 KWh per annum. In Poland, in contrast, we see very low prices and a flat retail pricing structure, resembling that of Sweden. In Germany, considering the energy cost component, the smallest customers pay roughly double what is paid by the largest residential consumers per KWh. Expanding to examine the entire European Union and adding Norway, we see that the regressive price pattern is common, as shown in Fig. 4, with a few exceptions.

Apart from the few countries where the pricing structure is invariant to how much one consumes, such as in the Baltic countries and Poland, in all other countries what one pays for electricity per kWh at retail level *declines with the quantity purchased.*

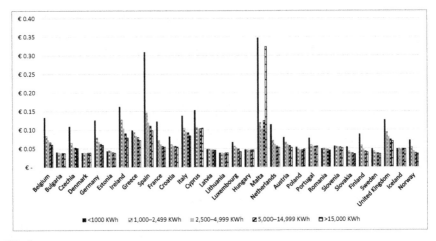

FIG. 4

Retail electricity price by consumer category, excluding taxes, levies and network charges, 2017.

From Eurostat.

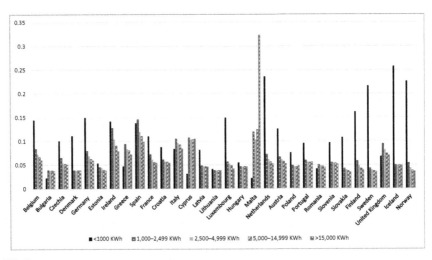

FIG. 5

Network charges by consumer category 2017.

From Eurostat.

For network charges and for taxes and levies, as shown in Figs. 5 and 6 respectively, we again see the same pattern. Looking first at Fig. 5, network charges per unit of electricity consumed decline with annual consumption, with the exceptions of Cyprus and Lithuania. Looking at taxes and levies, in Fig. 6, we again see that in

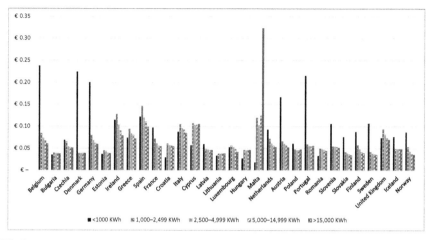

FIG. 6

Retail taxes and levies on retail electricity, 2017.

From Eurostat.

general, expenditure per kWh declines with volume of annual consumption, with Estonia and Hungary as exceptions. In Romania, the very smallest consumers are given a break on network charges, paying € 0.03 cents per KWh while those consuming around the median level pay € 0.05, and the very highest—those consuming more than 15,000 kWh per annum—pay € 0.04 per kWh.

Looking at the above data over time, comparing the experience of the smallest versus largest consumers is also revealing. For the smallest consumers, those taking less than 1000 KWh per annum, we see in Fig. 7, that between 2008 and 2017, taxes and levies nearly doubled while energy costs were roughly stable and network charges increased slightly. By comparison, in Fig. 8, for the largest consumers, those households taking more than 15,000 KWh per annum, network charges fell while taxes plus levies increased only slightly. In addition to the observed comparative trends, we further observe large differences in absolute expenditure between the smallest consumers and the largest consumers: All three components are greater for the smallest consumers than they are for the largest consumers.

We further note that network charges since 2011 have been falling for the largest consuming households while rising for the smallest consumers, thereby increasing the absolute difference. We observe that taxes and levies have grown from €0.05 to €0.9 per kWh for the smallest consuming households while for the largest consuming households these have grown from €0.03 to €0.05 per kWh. Critically, we note that the energy cost component of total expenditure has been falling over time, which may be related to the penetration of zero marginal cost renewable energy. As explained in Section 1, with renewable electricity each extra kWh produced on a particular day costs nothing, whereas with fossil and nuclear-fueled generation, each extra kWh costs the producer the fuel used to

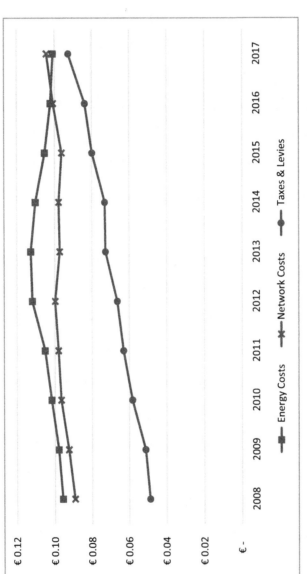

From Eurostat.

FIG. 7

EU average electricity expenditure per kWh by components—smallest consumers.

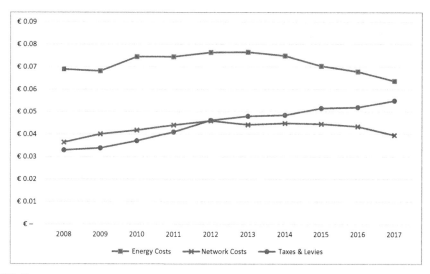

FIG. 8

EU average electricity expenditure per kWh by components—largest consumers.

From Eurostat.

generate it. The more renewable electricity there is in the mix, the lower the cost component per kWh and the higher the fixed costs.

Regarding taxes and levies, Fig. 9 compares categories of customers for Europe's largest economies, looking at total expenditure on taxes and levies by retail customers. This shows some large differences in the burden of taxation. Per kWh, the largest consumers, those taking more than 15,000 KWh per annum, enjoy the lowest tax burden while in seven countries the highest rate of tax is imposed upon the smallest consumers, those taking less than 1000 KWh per annum.

For the smallest consumers, network charges now exceed actual energy costs, while taxes and levies are close. For Europe's largest economies, we see in Fig. 10 that, with the exception of Spain, in 2017 the sum of taxes and levies plus network charges was greater than the energy cost component, with the smallest consumers paying the most per kWh. In Austria, for the smallest consumers, who take 1000 KWh or less per annum, taxes, levies plus network charges were over 3.5 times as high as actual energy costs, while in the Netherlands this was nearly three times as high and in Germany 2.5 times as high. Apart from the smallest households, taxes and levies plus network charges were double what was paid for the energy component of the total electricity price. This point holds across all EU countries. But the burden on the smallest consumers, as shown in Fig. 11, varies greatly.

Looking across the EU for 2017 at retail electricity prices in the aggregate (i.e., the sum of energy charges, taxes plus levies, and network charges), we see in Fig. 11 that Belgium leads in regressiveness, with the largest absolute

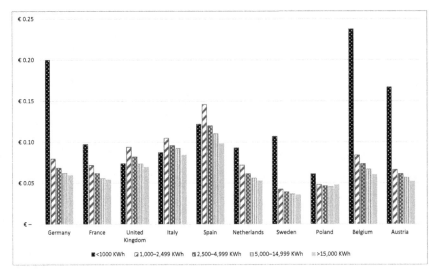

FIG. 9

Taxes and levies per kWh by customer category, selected EU countries, 2017.

From Eurostat.

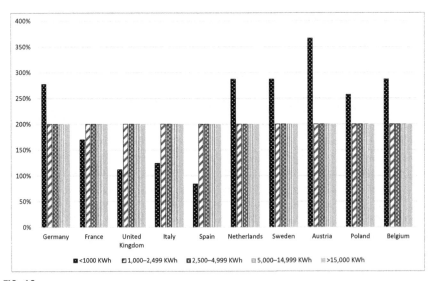

FIG. 10

Percentage difference between taxes, levies plus network charges and energy charges, for electricity per kWh consumed, for different sized customers, 2017.

From Eurostat.

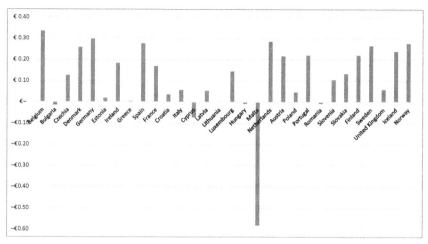

FIG. 11

Absolute difference between total electricity price per kWh for smallest verses largest consumers, 2017.

From Eurostat, 2018.

difference between what the smallest consumers versus the largest consumers pay, a difference of €0.34 per kWh.

The difference between what the smallest and the largest consumers pay per kWh derives from all three components comprising the retail price of electricity: the energy charge; the network charges; and taxes plus levies. Above we saw in Figs. 5 and 6 that what one pays per kWh generally falls linearly as annual consumption increases. Summarising the above data, we see in Table 6 that across the European Union, the smallest consumers pay on average 42% more per KWh for the energy component than the largest consumers, with a maximum of 115% and a minimum 0%, which is for Lithuania where all customers pay the same per unit of consumption regardless of annual consumption. Network charges are on average 84% greater across the EU for the smallest consumers than for the largest, with a maximum of 203%. For both Network charges and taxes, Malta is the exception: it is one of

Table 6 Retail electricity pricing structure: EU + Norway comparisons of energy price components, percentage difference in costs for smallest vs. largest consumers, 2017

	Energy costs (%)	Network charges (%)	Taxes (%)
Maximum	115	203	166
Average	42	84	49
Minimum	0	−2	−10

From Eurostat (2018).

the few countries where taxes on electricity per unit of consumption are greater for large consumers, and taxes are levied in a progressive manner. All three components of electricity prices explain the greater burden upon the smallest consumers: on average across the EU, the smallest consumers pay approximately twice as much per kWh as the largest consumers.

If one excludes the very small countries Cyprus and Malta from the data set with their progressive pricing policies, the average difference in prices paid per kWh between the smallest and largest households rises to over 150%.

We can also examine how the prices of electricity have changed over time. Comparing EU averages for the smallest consumers and the largest consumers, and adding the three components (energy costs, network charges and taxes plus levies), we see in Fig. 12 that, on average, in 2017 the consumers with the smallest annual consumption paid roughly double what was paid by consumers with the largest annual consumption. Tracking this over time since 2008, while expenditure per kWh has held steady or even declined for the largest consumers, for the smallest consumers the trend has been gradually upward. Of course, volumetric discounts are common to many commodities and consumer goods. The difference with electricity is that the volumetric discount *results* from spreading standing charges over a larger volume. Given the growing reliance upon zero marginal cost renewable energy, we ask whether this approach equitable.

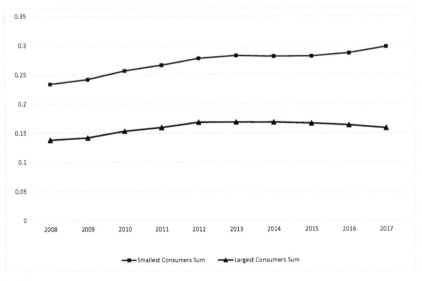

FIG. 12

Electricity cost (energy costs + network costs + taxes and levies) per kWh, EU average, smallest vs. largest consumers.

From Eurostat, 2018.

Table 7 Home ownership and financing

Household wealth (quintiles)	Own with mortgage (%)	Own outright (%)	Rent (%)
1 (Least wealthy)	7.80	0.20	92.00
2	43.20	11.40	45.40
3	50.40	38.00	11.60
4	44.60	49.80	5.60
5 (Most wealthy)	40.40	56.40	3.20

From UK ONS.

4.3 Relating household electricity consumption to household income

While it may seem intuitive to associate higher electricity consumption with higher income, comparing for example a family in a five-bedroom suburban home with a family living in a one-bedroom flat, I now examine this premise critically. Can we relate annual electricity consumption and its above described incidence, to income per capita? Although we have observed that the more one consumes, the less one pays per kWh, can we conclude that the structure of retail electricity pricing is regressive from the standpoint of household income? Do those with lower incomes pay more per kWh than those with higher incomes? Can we further observe whether the phenomenon of regressive electricity pricing at retail level has grown with time?

Data on electricity consumption by socio-demographic statistics is not available for all countries of the EU.[h] The European Union collects retail electricity price data according to the five annual household consumption levels I have used above. For Germany there are a number of household expenditure surveys.[i] The UK regulator, Ofgem, as we see below in Table 7, collects consumption data by size of dwelling and other demographics parameters such as age and number of occupants. However, data is not collected for electricity consumption according to income cohorts.

One possible way forward is to propose using size of dwelling *as a proxy* for income, arguing that the larger one's home, the greater one's income. Naturally, there may be exceptions: the banker with a pied-a-terre in town in addition to a country house, or a retired, wealthy couple who have down-sized their dwelling. But in

---

[h]Eurostat provides data on the Structure of consumption by income quintiles but this aggregated to the classification of Housing, Water, Electricity, Gas and other fuels (https://ec.europa.eu/eurostat/web/products-datasets/-/hbs_str_t223).

[i]For Germany, electricity consumption at household level, is available in the Income and Expenditure Survey (EVS) 2008: 80% Scientific-Use File of the Research Data Centres of the Statistical Offices. The SOEP data are from the 2010 survey year (wave 27, Data Distribution 1984–2010, V.27). For tenants, there was a survey of monthly advance payments for electricity while the average expenditures for electricity in the previous calendar year were surveyed for homeowners. Source: TNS Infratest (2011). See also Wagner et al. (2008).

general, the proposition that size of domestic residence increases with income tends to follow from existing data.

Electricity consumption increases with the size of dwelling, and the size of dwelling is correlated to income. As a percentage of total income, expenditure upon electricity at household level declines with income, as noted in other research (Neuhoff et al., 2013). For the United Kingdom as an example, according to data collected in 2011, the average household consumes 71 kWh of electricity per annum per square meter, with an end-terraced house as high as 81 kWh per square meter and a medium terrace house as low as 52 kWh per square meter (Statista, 2011). Smaller homes tend to rely upon electricity for heating (UK- Office of National Statistics and have less cavity insulation (ONS), 2018),[j] making them particularly vulnerable to high electricity prices. Although this might place them in a higher consumption band with lower electricity price per kWh, it also means that if they reduce their heating to economize, they end up paying more per unit of heating. The effect of size upon electricity consumption falls in medium to larger homes because of the greater reliance, in these larger homes, on natural gas for space heating.

According to the UK Fuel Poverty Report of 2016, in the lowest and second lowest income deciles, fuel poverty is observed in over 42% of households. In Europe 79% of residential energy consumption is for space and water heating of which 36% comes from the burning of natural gas (Eurostat, 2018). In Germany, the encouragement of heat-pumps in homes has led to greater demand for electricity (https://www.bmwi-energiewende.de/EWD/Redaktion/ EN/Newsletter/2015/ 09/ Meldung/infografik-heizsysteme.html).

House size and house price are also related. In the United Kingdom, the ratio of median house price to median gross annual earnings by country and region in England and Wales, is quite stable and has averaged 7.30 between 2010 and 2017 (ONS, 2018). The reliance upon mortgage financing means that the ability to amortize debt is naturally a function of income, popularly known as the house to income ratio. In the United Kingdom, even for the most-wealthy, as shown in Table 7, mortgage finance remains important for ownership. Although the house price to income ratio has increased somewhat, reflecting many factors including the 2008–2009 financial crisis, the correlation between the size of one's home and of one's income is supported empirically.

Despite the decreasing reliance upon electric heating as the number of bedrooms per dwelling increases, overall electricity consumption rises with size of residence and the size of one's residence reflects household income and family wealth. Given the inverse relationship described above between retail electricity prices and annual consumption, we can deduce that across the EU, with a few exceptions,

---

[j]The data from the UK Office of National Statistics shows the much greater reliance upon electric heating in smaller dwellings. Source: ONS, 2018 release using 2011 Census Data. Officially, fuel poverty is defined as domestic situations in which expenditure upon fuel would leave the family below the official poverty line. Accordingly, the following data is observed for domestic residence size and using electric heating: 1 Bedroom 36%, 2 Bedrooms 15%, 3 Bedrooms 5%, 4 Bedrooms 2% and 5 Bedrooms 3%

electricity is priced in a regressive manner: in energy cost, in the cost of network charges and in levies and taxes. The price per KWh falls with the amount consumed; thus, the lowest *income* families pay more than those with greater income and resources. Altogether, given the greater reliance upon electricity heating in the smallest dwellings and the lower use of cavity insulation (Department for Business, Energy and Industry 2016) this amplifies the impact of regressive pricing of electricity.

5 Renewable energy and electricity pricing

There is no necessary reason why the regressive pricing structure observed above should correlate to reliance upon renewable energy, but insight is provided by using Newberry's (1999) "normative perspective" on electricity pricing.[k] As a natural monopoly, electricity generation presents a problem: According to a normative theory of regulation, to correct the inefficiency of a natural monopolist, prices may be set through regulation at "marginal costs", i.e., the cost of each extra unit produced. This would make customer expenditure a linear function of price and quantity. However, this would lead to a loss, requiring a subsidy to continue operating, as shown in Fig. 13.

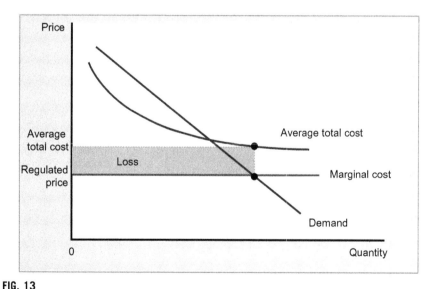

FIG. 13

Marginal cost pricing for a natural monopoly.

[k]Newbery's Normative theory of Regulation is in contrast to a Positivist approach (Newbery, 2009)

To cover the loss, a two-part non-linear tariff consisting of a fixed amount or fee, invariant to consumption, plus a price per unit is levied. This approach to pricing electricity addresses the shortfall but raises the question of how the fixed portion should be shared. To cover the loss shown in Fig. 13, an amount Loss/N, where N is the number of customers, could be charged. Under this approach all customers, regardless of consumption, pay an equal portion of fixed costs. Although EU electricity markets are deregulated and privatized, the technology of renewable electricity, which is almost entirely driven by fixed costs but supported per unit of output, creates the problem of how such cost should be shared across customers. If supply were entirely dependent upon renewable electricity generation, then sharing fixed costs equally by customers regardless of the volume consumed would imply that the effective price per kWh falls linearly with the volume. A household using 2000 kWh per year would pay five times as much per kWh as a household consuming 10,000 kWh per year.[1] Paying for renewable electricity therefore presents challenges. But how does this relate to our observations on the regressive pricing of electricity?

As I show below, three observations may be made with respect to renewable energy and regressive pricing of electricity:

a. Over time, across the EU, there is a positive relationship between the contribution of renewable energy and the regressive retail electricity pricing structure;
b. Across the EU, over time, there is an inverse relationship between electricity prices and the size of retail consumer, measured in annual consumption; and
c. For certain countries, the observed results are acutely greater than the observed EU averages.

Beginning at the aggregate level, we observe in Table 8 that there is a positive correlation between the contribution of renewable energy and regressive pricing,

Table 8 Correlations between green energy consumption and regressiveness: Smallest customers vs. largest customers, EU 28 + Norway 2008 to 2017

Correlations	Smallest vs. largest consumers (%)
Correlation between renewable electricity consumption and regressiveness in energy costs	28
Correlation between renewable electricity consumption and regressiveness of network charges	2
Correlations between renewable electricity consumption and regressiveness of taxes and levies	44

Author's calculations using EuroStat Data.

---

[1] Under such circumstances, some customers may be driven from the market if the two-part tariff exceeded their consumer surplus, although smaller households may have limited scope for such avoidance.

Table 9 Correlation between regressiveness and renewable electricity contribution, top 10 EU economies 2016

Regressiveness	Correlation (%)
Regressiveness in prices	37
Regressiveness in prices excluding Belgium and Poland	47
Regressiveness in taxes and levies	37
Regressiveness in taxes and levies excluding Belgium and Poland	40

Author's calculations using Eurostat data.

measured as the difference between what the smallest electricity consumers pay and what is paid by the largest consumers.

Although regressiveness in *network* charges is not related to renewable growth, for both energy costs and especially for taxes and levies there is a positive relation between network charges and growth in renewable electricity consumption. Narrowing the analysis, I examine results for the largest economies of the EU, as shown in Table 9, where we see stronger relationships:

Table 9 indicates that for the EU's 10 largest economies in 2016, there are strong correlations between the growth of renewable energy consumption and regressive retail pricing of electricity. Overall there was a 37% correlation between renewable energy consumption and regressiveness in retail electricity pricing, as measured by how much more the smallest consumers pay per kWh relative to the largest consumers. As a matter of interest, if we exclude Poland, where electricity generation is 86% from coal and follows a flat pricing structure, and Belgium, where the regressiveness of electricity pricing fell between 2008 and 2017 from a 98% difference to a 80% difference, as we saw in Fig. 4, we then find that the correlation between regressiveness in prices and the contribution from renewable electricity increases to 47%. The correlation between regressiveness in taxes and levies with growth in renewable electricity consumption also increases, as shown in the third and fourth rows of Table 9.

Highlighting the correlation between regressive pricing and renewable electricity contribution, I look at some individual countries in 2008–2017 in Table 10. In Finland, Austria and the Netherlands there is nearly perfect correlation (R^2 almost 100%) between regressive pricing and renewable electricity contribution. We see that the relationship between regressive electricity pricing and growth in renewable electricity contribution holds across many countries of the European Union: As the contribution of renewable electricity has grown, so has the regressiveness in electricity pricing as measured by the difference between what the smallest consumers pay and what is paid by the largest consumers. For some countries, as posed in our third proposition, the correlations between regressiveness and renewable electricity consumption are much greater than the EU average. Noticeably, France has not been included in the list. Although France has a regressive pricing structure, over three quarters of its electricity comes from nuclear energy, while renewable

Table 10 Ranking of correlations between electricity price regressiveness and renewable electricity contribution, 2008–2017, in 16 EU countries

Country	Correlation coefficient (%)	Country	Correlation coefficient (%)
Finland	98	Ireland	68
Austria	95	United Kingdom	67
The Netherlands	95	Denmark	63
Czech Republic	79	Bulgaria	54
German	74	Hungary	45
Slovenia	72	Romania	44
Greece	71	Sweden	44
Italy	70	Slovakia	30

Author's calculations using Eurostat data.

electricity makes a small contribution, averaging less than 4% between 2008 and 2017 (Eurostat, 2018). From the above results we can see the three propositions presented above are empirically supported. There is a strong correlation between the growth and reliance upon renewable energy and a regressive pricing structure, and much of this can be explained by network charges and levies-taxes.

Looking at energy costs and the taxes and levies imposed upon energy, the countries in which renewable electricity has made the greatest contribution also have the most regressive pricing structures. Over time and across countries, regressiveness in both pricing of energy and in taxing of energy are correlated with the growth in renewable energy consumption. Denmark is an interesting case. In Denmark, renewable electricity is now contributing to almost half the electricity consumed but the energy cost component does *not* change according to annual requirement: In 2017, regardless of annual consumption, the same price was paid for the energy component of €0.0338 pence per kWh. But in this Nordic country, both taxes and levies as well as network charges are highly regressive, as shown above in Figs. 6 and 7. In my conclusions below, I reflect on why the growth and use of renewable energy is associated with regressive pricing of electricity to consumers. From the above we see that the growing reliance upon renewable electricity lends itself to a highly regressive pricing structure. Spreading the fixed costs of electricity across greater levels of consumption means that the more one takes, the less one pays per unit. The smaller the household, the more one carries the fixed cost of renewable electricity.

6 Conclusions

In this chapter I have shown the strong correlation over time and across the countries of the EU, between the growth in renewable electricity consumption and the regressive pricing of electricity. The findings of this chapter go further than the observations of the ECOFYS—Fraunhofer ISI (2016) study, which found that from 2008 onwards, the taxes and levies component of electricity pricing has increased largely because of the support for renewable electricity. According to this study, these payments increase the energy and environmental taxes paid on the use of energy carriers in Europe. The impact is amplified by value-added tax which is applied as a percentage to the total retail price including environmental taxes and levies. In the research reported here I have examined how the costs of renewable energy are shared across retail consumers. Rather than looking at the general effects or the burden upon a typical household, I have assessed the impact at different levels of annual consumption as a proxy for income distribution percentiles. I have found, across the EU, the burden of supporting renewable electricity falls disproportionately upon smaller retail consumers. These European wide results go further than current findings on the German photo-voltaic market, where it was shown that the FiT support scheme in effect redistributes income from lower income groups to upper income groups (Grosche and Schroder, 2014) and other work based upon expenditure surveys (Neuhoff et al., 2013). It is not just the FiT that causes regressiveness, but also the fact that in liberalized markets, the ratio between producers' fixed and per kWh costs increases as the proportion of renewable electricity increases. With a few exceptions, in most EU countries electricity is priced in a regressive manner: the more one purchases, the less one pays per KWh, because the fixed costs of renewable energy are spread over greater consumption. Looking at energy costs, taxes and levies and network charges, the differences between what the smallest consumers pay versus the largest pay per KWh can be up to fivefold. Upper income groups, through making renewable energy investments, become in effect the beneficiaries of the prices imposed upon lower income cohorts. The effects of the regressive pricing structure are made worse because poorer households spend a larger proportion of their disposable income on electricity and have less scope for reducing consumption through investing in efficiency. Estimates of price elasticity for electricity vary, but according to most studies, especially in the short-run, it is very low (i.e., consumers do not easily reduce consumption of this good when the price increases) (Cialani and Mortazavi, 2018 and Labandeira et al., 2012), implying that consumers have few choices with regard to paying for any additional costs of renewable electricity. In practice, support mechanisms for renewable electricity under the existing electricity market structure both facilitate and exacerbate the regressive pricing structure, transferring money from poorer to richer households.

How did the present circumstances arise? The current design and structure of electricity in Europe was premised upon a dominant role for dispatchable fossil fuel electricity generation. Regulators set price caps using the factors discussed in Section 3, giving energy companies an incentive to maximize profits through reducing costs, including in the wholesale power market. Vertical unbundling was also encouraged to promote competition. With dispatchable generation like gas turbines or modern coal-fired plants, energy costs contribute about one-third to household bills, making the volume of electricity consumed critical. But with increasing reliance upon renewable electricity, the variable component of bills falls and the fixed cost component increases. With renewable electricity, network charges and "policy costs", as described by Ofgem, become dominant. In the UK for example, as reported by British Gas, charges due to government energy policies are now a bigger share of household electricity bills than wholesale prices (Thomas and Ward, 2017). In the United States it has been argued that regressive prices and increased reliance upon taxes and levies are a consequence of the growth in distributed renewable electricity. According to this research, utilities increasingly seek to recover their costs using fixed monthly charges instead of relying upon a per-unit of energy consumed tariff (Synapse 2016). Further, in research on the Irish electricity market it was shown that a flat-rate Public Service Obligation Levy was the most regressive approach to funding of renewable electricity (Farrell and Lyons, 2015). Altogether we see that the trend towards fixed charges hits the smallest consumers hardest, since bills are based less on usage but increasingly upon a flat-fee structure.

From the perspective of an energy company, such tactics are appealing: They reduce the risk to revenues of either a linear pricing model or even a non-linear model in which both variable and fixed costs are covered. Faced with the threat of lower sales receipts due to the three-fold challenge of greater efficiency, the impact of random output of renewable generation and the under-utilization of dispatchable generation, having a predictable return on assets is desirable. Indeed, in many instances, US public utilities and their regulators are implementing mechanisms to decouple revenues from commodity sales in order to promote efficiency and conservation (Synapse 2016). But apart from the above incentives, can we infer that the technology of renewable electricity generation favors the largest consumers? From the standpoint of Newbery's positivist theory of regulation, one can appreciate how the cost structure of renewable electricity, driven almost entirely by its fixed capital costs, leads to regressive pricing. With a marginal cost of producing each extra kWh of zero, whether a household takes 2000 KWh or 10,000 KWh per annum is immaterial to the producer. It creates no additional cost for the energy company. The technology of renewable electricity as a fixed cost system favors fixed amount pricing, in effect, charging for *access* to power rather than how much one consumes. Faced with the costs and risks of accommodating intermittent random renewable electricity output under dispatch priority, lower prices because of zero marginal cost generation and additional transmission charges inherent to distributed generation, utilities have an incentive to make revenues more predictable. Reducing the risk to cash-flows, it may even lower the cost of capital (Michelfelder et al., 2019)

Correlation, of course, does not imply causation. But the observed relationship across the EU between a regressive pricing structure and renewable electricity underscores the outcomes from generating electricity in this manner. Further, the regressive pricing structure is exacerbated through the effects of taxes and levies. Regressive pricing redistributes income from poorer households to support investment in renewable electricity by wealthier households and investors. In some countries, such as Germany, the phenomenon is made worse by the exemption of large industries from paying taxes and levies. But even without such exemptions, the trend for large industries to go off-the-grid, leaves greater costs to be recovered in the retail sector (Khalilpour and Vassallo, 2015). A utility totally reliant upon purchasing renewable electricity output from households and investors becomes in effect a conduit for transferring income from net buyers to net sellers ("prosumers"). As reported in the German Press, the inequity of the German renewable energy support has become increasingly apparent. Critics have asked how can it be fair for someone living in small flat to subsidize investment by a wealthy home-owner's in roof-mounted solar panels through his electricity bills (Der Spiegel, 2013)?

Looking at alternatives, we observe that from a normative perspective on regulation, there is no reason why the retail price components—energy costs, network charges and taxes levies—should be imposed in a regressive manner. Although the variable costs of renewable energy are zero, to ensure total revenue equals total costs, pricing could still be tied to the volume of consumption. As we have seen, in a few countries at least, a flat pricing structure is applied. One solution may be to charge different fixed fees to different consumers, although even an optimal two-part tariff having a price per unit exceeding marginal cost and a fixed fee, may exclude some customers (see, Brown and Sibley, 1986). Supporting renewable electricity in an equitable manner may require a discriminatory two-part tariff: not based upon willingness to pay but rather upon the ability to pay. Putting a greater burden of fixed costs upon the largest consumers and households, perhaps through a lump-sum tax, might be useful and justified on equality of loss to marginal utility. To minimize the dead weight loss but align costs with revenues, the Ramsey pricing rule may be considered (Ramsey, 1927) though it still may lead to disputes over the allocation of common costs. Almost everyone recognizes the promotion of renewable electricity as a key policy objective. Through exploring alternative pricing structures, its costs may be shared in an equitable manner.

Finally, it is important to reflect on the relationship between market liberalization and regressive pricing. The problem is not that market liberalization favors or necessarily leads to regressive pricing. Rather, renewable electricity came after markets were already liberalized, and once it was is in place, it lent itself to regressive pricing. Competitive markets and the dash for cheap, gas-fired electricity were good for consumers. But when we move to a zero marginal cost system of wind and photovoltaics, the market favors a non-linear pricing system. I return to the analogy at the beginning of the chapter: it is like a motorist with the smallest, least consuming car paying more per liter for petrol than the driver of a large gas-guzzler. If we wish to have a greener future, perhaps we need to consider alternative market designs and structures.

Acknowledgment

The author wishes to thank Dr. Niall Farrell of the Potsdam Institute for his very helpful comments and suggestions and Dr. Laura N. Haar of the University of Manchester for reviewing and checking calculations.

References

Andor, M., Frondel, M., Vance, C., 2015. Installing photovoltaics in Germany: a license to print money? Econ. Anal. Policy 48, 106–116.

Andor, M.A., Frondel, M., Vance, C., 2017. Germany's Energiewende: A Tale of Increasing Costs and Decreasing Willingness to Pay. IAEE Forum, pp. 15–18. Q4.

Ayoub, N., Yuji, N., 2012. Governmental intervention approaches to promote renewable energies—special emphasis on Japanese feed-in tariff. Energy Policy 43, 191–201.

Bertram, G., 2006. Restructuring the New Zealand electricity sector: 1984-2005. In: Sioschansi, F., Pfaffenberer, W. (Eds.), Electricity Reform: An International Perspective. Elsevier, Amsterdam.

Bijker, W., Hughes, T., Pinch, T. (Eds.), 1987. The Social Construction of Technological Systems: New Directions in the Sociology and History of Technology. MIT Press, Cambridge (MASS) and London (UK).

Brazilian, M., Nakhooda, S., Van de Graaf, T., 2014. Energy governance and poverty. Energy Res. Soc. Sci. 1, 217–225.

Brinckerhoff, P., 2012. Solar PV cost update, U.K. Department of Energy & Climate Change. https://assets.publishing.service.gov.uk/government/uploads/system/uploads/attachment_data/file/43083/5381-solar-pv-cost-update.pdf.

Brown, J.S., Sibley, D., 1986. The Theory of Public Utility Pricing. Cambridge University Press, New York.

Chawla, M., Pollitt, M.G., 2013. Energy-efficiency and environmental policies & income supplements in the UK: evolution and distributional impacts on domestic energy bills. Econ. Energy Environ. Policy 2 (1), 21–40. 2013.

Cialani, C., Mortazavi, R., 2018. Household and industrial electricity demand in Europe. Energy Policy 122, 592–600.

Couture, T.D., Cory, K., Kreycik, C., Williams, E., 2010. *A Policymaker's Guide to Feed-in Tariff Policy Design*, National Renewable Energy Lab, https://www.nrel.gov/docs/fy10osti/44849.pdf.

Ecofys, 2014. Subsidies and Costs of EU Energy. Commissioned by the EU Commission. Available from: https://ec.europa.eu/energy/sites/ener/files/documents/ECOFYS%202014%20Subsidies%20and%20costs%20of%20EU%20energy_11_Nov.pdf.

ECOFYS—Fraunhofer ISI, 2016. Prices and the Cost of Renewable Energy: Final Report.

EEA Report, 2016. EEA Report No 4/2016 Renewable Energy in Europe 2016 Recent Growth and Knock-on Effects.

Erbach, G., 2016. Promotion of Renewable energy sources in the EU EPRS | European Parliamentary Research Service, Members' Research Service, —PE 583.810.

European Commission, 2013. Brussels, 5.11.2013 SWD(2013) 439 Final Commission Staff Working Document, European Commission Guidance for the Design of Renewables Support Schemes.

European Commission, 2016. Report From the Commission to the European Parliament, the Council, the EUROPEAN Economic and Social Committee and the Committee of the Regions, Energy Prices and Costs in Europe {SWD(2016) 420 Final}.

Farrell, N., Lyons, S., 2015. Who should pay for renewable energy? Comparing the household impacts of different policy mechanisms in Ireland. Energy Res. Soc. Sci. 7, 31–42.

Frondel, M., Sommer, S., Vance, C., 2015. The burden of Germany's energy transition: an empirical analysis of distributional effects. Econ. Anal. Policy 45, 89–99.

Grosche, P., Schroder, C., 2014. On the redistributive effects of Germany's feed-in-tariff. Empir. Econ. 46, 1339–1383.

Haar, L., Haar, L., 2017. An option analysis of the European Union renewable energy support mechanisms. Econ. Energy Environ. Policy 6 (1), 131–147.

Hughes, T., 1993. Networks of Power: Electrification in Western Society, 1880–1930. Johns Hopkins University Press, Baltimore, MD.

Khalilpour, R., Vassallo, A., 2015. Leaving the grid: an ambition or a real choice. Energy Policy 82, 207–221.

Labandeira, X., Labeaga, J., López-Oteroa, X., 2012. Estimation of elasticity price of electricity with incomplete information. Energy Econ. 34 (3), 627–633.

Michelfelder, R.A., Ahern, P., D'Ascendis, D., 2019. Decoupling impact and public utility conservation investment. Energy Policy 130, 311–319.

Neuhoff, K., Bach, S., Diekmann, J., Beznoska, M., El-Laboudya, T., 2013. Distributional effects of energy transition: impacts of renewable electricity support in Germany. Econ. Energy Environ. Policy 2 (1), 41–54. https://doi.org/10.5547/2160-5890.2.1.3.

Newbery, D.M., 2009. Theories of Regulation, Privatization, Restructuring and Regulation of Network Utilities. MIT Press, London.

Platts PowerVision, 2018. Latest Electric Power Market Insights. Available from: https://www.spglobal.com/platts/en/commodities/electric-power.

Ramsey, F., 1927. Contribution to the theory of taxation. Econ. J. 37, 47–61.

Reuters, 2016. Germany Reaches Deal With EU on Green Energy Levy for Industry Power Stations. Reuters. August, 29.

Der Spiegel, 2013. Germany's Energy Poverty; How Electricity Became a Luxury Good. https://www.spiegel.de/international/germany/high-costs-and-errors-of-german-transi tion-to-renewable-energy-a-920288.html.

Statista, 2011. Annual electricity consumption per square metre in England in 2011 (in kilowatt-hours). Available from: https://www.statista.com/statistics/318402/annual-elec tricity-consumption-england-uk-per-square-metre/.

Thomas, N., Ward, A., 2017. British Gas Warns Energy Policy Weighs Heavily on Bills. Financial Times.

Wagner, G.G., Goebel, J., Krause, P., Pischner, R., Sieber, I., 2008. Das Sozio-oekonomische Panel(SOEP): multidisziplinäres Haushaltspanel und Kohortenstudie für Deutschland— Eine Einführung (für neue Datennutzer) mit einem Ausblick (für erfahrene Anwender). AStA Wirtschafts- und Sozialstatistisches Archiv 2 (4), 301–328.

Yurchenko, Y., Thomas, S., 2015. EU Renewable Energy Policy: Successes, Challenges, and Market Reforms. Public Services International Research Unit (PSIRU).

Websites consulted

https://www.ceer.eu/eer_publications/annual_reports.

https://cundall.com/Cundall/fckeditor/editor/images/UserFilesUpload/file/WCIYB/IP-11%20-%20Comparison%20of%20building%20energy%20benchmark%20to%20total%20UK%20energy.pdf.

https://eur-lex.europa.eu/legal-content/EN/TXT/?uri=COM:2016:769:FIN.

https://ec.europa.eu/energy/sites/ener/files/com_2013_public_intervention_swd04_en.pdf.

https://ec.europa.eu/energy/sites/ener/files/documents/report_ecofys2016.pdf.

http://appsso.eurostat.ec.europa.eu/nui/submitViewTableAction.do.

https://ec.europa.eu/eurostat/web/energy/data/database.

https://www.gov.uk/government/statistics/fuel-poverty-detailed-tables-2018.

https://www.ofgem.gov.uk/system/files/docs/2017/12/feed-in_tariff_FiT_annual_report_
2016-17_0.pdf.

https://www.ons.gov.uk/peoplepopulationandcommunity/housing/datasets/
ratioofhousepricetoresidencebasedearningslowerquartileandmedian.

https://www.ovoenergy.com/guides/energy-guides/how-much-heating-energy-do-you-use.
html.

http://programs.dsireusa.org/system/program/detail/1235.

https://www.reuters.com/article/us-eu-stateaid-germany/eu-probes-green-charge-exemp
tions-for-german-industry-idUSBRE9BH0EZ20131218.

https://in.reuters.com/article/germany-renewables/germany-reaches-deal-with-eu-on-green-
energy-levy-for-industry-power-stations-idINB4N17M01V.

http://www.res-legal.eu/compare-support-schemes/.

http://www.res-legal.eu/compare-support-schemes/.

https://www.wind-energy-the-facts.org/overview-of-the-different-res-e-support-schemes-in-
eu-27-countries.html.

Further reading

Council of European Energy Regulators, 2015. Status Review of Renewable and Energy Efficiency Support Schemes in Europe in 2012 and 2013.

McCormick, M., 2018. Brussels Freezes UK Back-Up Power Subsidy Scheme. Financial
Times.

Speck, S., 1999. Energy and carbon taxes and their distributional implications—policy and
practice in Britain and Germany. Energy Policy 27(11).

CHAPTER

Energy poverty research: A perspective from the poverty side

10

Ray Galvin

University of Cambridge, Cambridge, United Kingdom
RWTH Aachen University, Aachen, Germany

Chapter outline

1 Introduction 221
2 Poverty as discussed in energy poverty literature 223
   2.1 Energy inefficient buildings 224
   2.2 The targeting approach to energy poverty 225
   2.3 The fear of increasing CO_2 emissions 226
3 Method and approach 227
4 Using the 10% indicator 229
   4.1 The logic of the analysis 229
   4.2 The cost of adequate household energy services using the 10% indicator .230
   4.3 Calculating the additional amounts required, under the 10% indicator 231
   4.4 How much would this cost high-income households? 233
5 Would this redistribution be tolerated? 234
6 Implications for CO_2 emissions 237
   6.1 The CO_2 implications of more progressive taxation 238
   6.2 CO_2 emissions and increased incomes among poorer households 240
7 Conclusions 243
References 245
Further reading 248

1 Introduction

In this chapter I explore how energy poverty[a] could be addressed and mitigated, *from the perspective of poverty*. I do this because, as I read more and more of what is now a vast corpus of literature on energy poverty, I find very little, if any, discussion of how to eliminate its most persistent cause, namely economic poverty.

---

[a]I use the terms energy poverty and fuel poverty interchangeably in this chapter.

Inequality and Energy. https://doi.org/10.1016/B978-0-12-817674-0.00010-2

As noted in Chapter 7, it is widely agreed in energy policy studies that the three main determinants of energy poverty are: (a) low household income; (b) energy inefficiency of buildings and appliances; and (c) high fuel prices. A fourth factor is often also seen as a determinant: (d) specific household characteristics.[b] Interestingly, (b) and (d) are by far the most frequently and intensively researched in energy poverty literature.

Despite its emphasis on other factors, energy poverty research overwhelmingly agrees that low household income is a major cause of energy poverty. Of course, not *all* low-income households are in energy poverty. Some live in state-of-the-art, energy-efficient, south-facing council flats that can be heated with hardly energy consumption at all. Many others have responsible private landlords who take pride in the high thermal quality of their properties.[c] But in the UK at least, most low-income households have very little choice as to what quality of dwelling they can live in. They often cannot possibly afford to heat their home to what most of us would regard as a comfortable, healthy standard, and it is even harder for them to save to buy their own house or flat and then invest in insulation, energy efficient windows and an efficient heating system (Institute for Fiscal Studies, 2019).

Increasing the income of these households would not necessarily enable *all* of them to be comfortably warm at home, since many UK dwellings are almost impossible to heat (see Nicola Terry's analysis in Chapter 6!). But it would eliminate the most basic determinant of energy poverty.

While virtually all energy poverty studies mention low household income, they almost all treat it as a given. It is like a closed box, just there, as a fact of life. I know of no energy poverty study that seriously questions why there need to be so many low-income households in the richest societies the world has ever known. Nor is there any attempt to engage with the question of how to fix this problem specifically, as one tool among many to alleviate energy poverty.

In this chapter I will offer a first such attempt. My method is rudimentary and includes a number of approximations. However, the figures I produce are easily within the likely range of what would be obtained in finer grained studies that would take account of greater variation in data. In any case, since no sophisticated research has been done in this area, we have to start somewhere.

I focus here on the UK. I use data on household incomes from the Department of Work and Pensions (DWP, 2019) and the Institute for Fiscal Studies (Crib et al., 2018) to estimate household income bands, and from the Department for Business, Energy and Industrial Strategy (DBEIS, 2019) for energy consumption estimates.

It should be noted that I am considering only the costs of energy services that are used in the home. The energy required for transport, such as commuting to work, is

---

[b]Specific household characteristics can also include ill health; social relations; and tenancy relations (cg Middlemiss and Gillard, 2015).
[c]I myself have been a landlord for some 35 years.

another issue in energy poverty research (Mayer et al., 2014). But because of all the subtle variations in this (drive, cycle or walk to work; do not go to work; take free or paid public transport, etc.) I leave it as another challenge for future research. The existing studies and definitions I draw on here all deal with domestic energy poverty only, relating to energy services used in the home.

My aims in this analysis are, firstly, to see how much money would need to be redistributed to poor households in the UK, to lift them out of energy poverty as it is defined by a specific indicator. I will do this over three scenarios: using a very low, a low and a modest threshold for the level of economic poverty that would be expected to be associated with energy poverty. My second aim is to see what the costs would be to the highest earning households, to enable these redistribution scenarios to happen via taxation. My third aim is to explore how reasonable these redistribution scenarios would be, given the different tax rates there have been in the UK over the past decades. My fourth aim is to see what effects these redistribution scenarios might have on the UK's consumption-based CO_2 emissions.

In Section 2 I look briefly at how energy poverty research to date talks about the issue of economic poverty. In Section 3 I outline my method and approach. In Section 4 I apply this to UK income bands in the 2017–2018 tax year, showing how much money would need to be redistributed in each of the three scenarios. In Section 5 I discuss the tax implications of such redistribution in the light of recent UK tax history. In Section 6 I estimate the effects of these redistributions on consumption-based CO_2 emissions, and in Section 7 I offer conclusions.

Readers who do not find debates among energy poverty researchers interesting may skip Sections 2 and 3, and go directly to Section 4 where the analysis actually starts.

2 Poverty as discussed in energy poverty literature

Research to date on household energy poverty in high-income countries very seldom discusses how one of its three basic causes, economic poverty, could be reduced. Most studies mention economic poverty as a fact of life that is simply there (e.g., Anderson et al., 2012; Boardman, 2012; Brunner et al., 2012). Some, such as Lucie Middlemiss' study in Chapter 5, are concerned to map and explore the complexities and vicious circles that energy poverty can lead to in times of economic austerity. This brief review cannot do justice to the wealth of insightful and compassionately motivated studies on energy poverty. Instead, I focus quite narrowly on how the topic of low income is perceived in these studies.

In the studies that bring low income into the discussion, different issues tend to take priority over it. These include: the energy (in)efficiency of residential buildings; micro-targeting of vulnerable households; and the perceived danger of increasing CO_2 emissions. I discuss here just one or two studies of each of these types to illustrate these approaches.

2.1 Energy inefficient buildings

Among the studies that focus on energy (in)efficiency of dwellings, an interesting example is Santamouris et al.'s (2013) detailed and impressive household-by-household investigation of fuel-poor homes in different regions in Greece. Santamouris and colleagues used cluster analysis to divide these homes into two groups, which may be termed 'disadvantaged' and 'not disadvantaged'. Households in the disadvantaged cluster had lower incomes, had suffered an average 24% fall in income since the start of Greece's financial crisis, lived in buildings an average of 30.5 years old (or last renovated), and had consumed 37% less heating energy than the research team calculated was needed for the (harsh) winter of the study period. In short, their dwellings were outdated in terms of thermal standards and they did not have enough money to pay for decent heating in such dwellings.

The main recommendation the study makes is to target the worst of the dwellings for energy efficiency upgrades. A second recommendation is that 'homes with a propensity to consume more energy should be targeted using behavioural strategies combined with economic penalties and incentives' (op cit.: 485). Regarding money to spend on heating, the study recommends more regulation in the energy market. There is no suggestion of any strategy to increase the incomes of these low-income households.

Of course, in the situation of Greece's severe financial crisis, it might seem very difficult to increase households' incomes. Yet this is one of the two main roots of the problem (the other being inadequate dwellings—which also, of course, can only be put right with money). If good *science* finds low income is a principal cause of a problem, one could argue that good science should then recommend incomes should be raised—which would serve as a challenge to the political forces that are suppressing the incomes of the poorest households. If the solution to inadequate incomes lies on the fiscal, political level, that hardly seems a good reason for energy poverty studies to shy away from it.

The UK is not suffering an economic crisis anything like that of Greece. The UK's GDP continues to increase (at least at the time of writing) and its average household income is among the world's highest, at just over £31,000 per year (DWP, 2019). Yet energy poverty research there also emphasises that increasing energy efficiency is the solution, rather than considering that increasing poorer households' incomes might at least be part of the solution. In Boardman's (2012) words, 'while fuel prices and low incomes are constituent factors, *the real cause of fuel poverty is the energy inefficiency of the home*' (op cit.: 2012: 143, italics added). She adds that 'the cause of fuel poverty' is 'the failure to invest capital to improve energy efficiency' (op cit.: 143). It could be argued, however, that if incomes were improved, households would then have the agency to choose to invest in energy efficiency or, if renting, to look for a better dwelling. As long as tenants are willing to take any dwelling on offer because they cannot afford to choose, landlords have little incentive to improve their properties.

Middlemiss (2017) is one of the few energy poverty researchers to critique the fact that policy and research on energy poverty is tending to focus so single-mindedly on energy efficiency. There seems to have become a fixation on improving energy efficiency as the *only* solution, even though research on energy poverty continues to mention that low income is one of its three main causes. For example, the Joseph Rowntree Foundation is noted for its concern for low-income households in the UK. However, its report on energy poverty in the UK (Stockton and Campbell, 2012) maintains that the priority should be increasing the energy efficiency of hard-to-treat houses occupied by the most financially disadvantaged. There is no suggestion that a parallel policy strategy might be to increase these households' incomes.

A similar approach is evident in the UK government's Warm Front scheme, which ran from 2000 to 2013 (Watson and Bolton, 2013) and directly benefited some 2 million households. The scheme offered free or heavily subsidised loft and cavity wall insulation, draft-proofing, hot water tank insulation and improved heating systems. To qualify for the scheme, a household had to be on a welfare benefit indicative of being vulnerable to extreme poverty. Reviews of the scheme highlight its many positive outcomes, as well as its drawbacks. One of these is the difficulty of accurately targeting the households most vulnerable to energy poverty, namely those with the worst combination of low income and energy inefficient dwellings (National Audit Office, 2009). Another is that the wall insulation offered by the scheme only addressed dwellings with cavity walls, while the dwellings of millions of low-income households would need much more expensive solid wall insulation. A further drawback is the fact that energy poverty continued to increase and possibly even tripled during the period of the scheme, mostly due to large increases in energy prices *while incomes of the poorest sectors did not increase in parallel* (Sovacool, 2015). In other words, these households' lack of money was acting as a persistent driver of increasing energy poverty, despite attempts to address the problem on the level of energy efficiency.

2.2 The targeting approach to energy poverty

A large number of recent studies on energy poverty are concerned to identify specific types, groups, regions, districts or characteristics of households that are, or are likely to be, in energy poverty. For example, Bouzarovski and Tirado Heredo (2017) set out to identify differences in vulnerability to energy poverty between different regions in Hungary, the Czech Republic and Poland. They used household energy expenditure in relation to income and energy needs, as a proxy for estimating the extent of what they termed 'energy deprivation'. They found the incidence of energy deprivation was not always fully captured by the three widely accepted determinants of energy poverty, namely low income, energy inefficient dwellings, and high fuel prices. Rather, they found specific pockets of energy deprivation among specific types of households in specific regions, such as, in the Czech Republic: 'farmers, pensioners, individuals living on their own and low-income households more generally' (op cit.: 47). Studies such as this and Bouzarovski and colleagues' other work (e.g., Bouzarovski and Petrova, 2015; Thompson et al., 2017) effectively add the fourth determinant of energy poverty, specific household needs—which include the types, ages and special needs of people in the household, the geographical region, the rural-urban divide, etc (and see Reinhard Madlener's discussion of this in Chapter 11).

Studies such as this bring the important insight that energy poverty is often very specific to particular households or types of households, which may be missed by too rigid, standardised definitions of energy poverty. It adds a further caution to the approach of this chapter: even if we were to increase the incomes of all economically poor households to a modest standard *and* increase the energy efficiency of their homes, this would not fully eradicate energy poverty, since some households have specific needs that require more energy.

A danger of fully embracing a targeting approach, however, is the socioeconomic implications of targeting. On the one hand, governments like finely targeted welfare interventions because they cost much less than universal welfare and can sometimes be accommodated within the 'competing and contradictory pressures within government' (Sovacool, 2015) that characterise austerity budgeting. In the words of Robinson et al. (2018), there are 'limited alleviation resources' that must be carefully targeted where they are most effective.

But on the other hand, the targeted welfare approach lacks the advantages of the universal welfare approach that lifted high-income countries out of extremes of economic inequality in the 1930s–1970s. As Middlemiss (2017) points out, targeting of households deemed deserving, and only for specific technical assistance, also fails to give these people agency, the means of taking control of their own lives so that they can make their own choices as to how to prioritise their spending to meet their needs as they see these.

Deacon (2005) describes how post-World-War II universal welfare programs that directly increased the income of poorer households, were then systematically eroded and replaced by targeting of 'deserving' cases, as neoliberal economics came to dominate in the 1980s. The consequences of this have not only changed the type of society we live in, but have also influenced popular discourse, to the extent that we often assume that targeting is the correct approach because it appears the most economically efficient (Cox, 1999; Rodgers-Vaughn, 2013). However, there are strong arguments for universal social welfare approaches and already the failure of neoliberal-inspired targeting strategies is strongly evident (Jensen, 2009; Alston, 2018).

2.3 The fear of increasing CO_2 emissions

Some energy poverty studies note the injustice or hardship faced by low-income households suffering energy poverty, but emphasise the danger of increased CO_2 emissions if these households' incomes were to increase. This approach is often connected with a concern for global energy justice: if poor households in rich countries increase their incomes so as to be able to enjoy the energy services others enjoy, they will emit more CO_2, contributing to climate change and disadvantaging poorer countries. For example, Bartiaux et al. (2018, 1232) state:

> *But at the same time, in this era of climate change, and intense awareness thereof, policies toward distributive justice cannot equalise energy consumption between energy-poor households and energy (much) richer ones, whether within or across countries, for this would greatly increase energy consumption and the greenhouse gas emissions associated with climate change.*

Similarly, Walker et al. (2016) argue that the idea of increasing energy services access for the poor in wealthy countries can only be permitted in the context of reducing global greenhouse gas emissions. These authors maintain we must consider 'the scope of UK fuel poverty policy and its interaction with low carbon objectives' together, and not as separate issues (op cit.: 136).

Ürge-Vorsatz and Tirado Herrero (2012) also explore the relationship between alleviating energy poverty and mitigating climate change, framing these as essentially in tension. Their working assumption is that reducing energy poverty by increasing poor households' incomes would increase global CO_2 emissions and compromise climate change goals. They suggest that direct financial help schemes for fuel poor households 'do not provide a long term solution to the energy deprivation challenge – in fact, they may lock-in households in energy poverty if implemented on their own because they remove incentives to invest in energy efficiency at the household level – and do not reduce carbon emissions either.' (op cit.: 89).

While I appreciate the concern to avoid increasing CO_2 emissions, it seems unfair that low-income UK households' immediate and often desperate need for warmth in winter can only be considered in the context of finding global solutions to climate change.

As with other studies mentioned above, the solution offered by Ürge-Vorsatz and Tirado Herrero (2012) is to increase the energy efficiency of sub-standard building stock, so as to bring synergy between the two goals of alleviating fuel poverty and reducing CO_2 emissions. But renovating this huge building stock is neither easy nor cheap. Even if it were pursued vigorously, most energy-poor households would have to wait years before their turn came round.

In any case, I will argue in Section 6 that the assumption behind these studies is simply wrong. A fiscally progressive approach to increasing the income of low-income households would significantly *reduce* global CO_2 emissions.

3 Method and approach

Researchers estimate the level of energy poverty by either 'objective' or 'subjective' indicators (Waddams Price et al., 2012). Objective indicators are measurable criteria such as household income, while subjective indicators are household's own stated perceptions of how easy it is to heat their homes or pay their electricity bills. I use here an objective indicator, rather than the subjective indicator I used in

Chapter 7. I do this because the method I am using focuses on the incomes households earn, rather than their own assessment of whether they earn enough to pay for adequate energy services.

There are four well-known objective indicators of energy poverty. The first, which I use in this study, is known as the '10% indicator' (Heindl and Schuessler, 2015; Roberts et al., 2015; DECC, 2016; Imbert et al., 2016; Papada and Kaliampakos, 2016). This indicator says 'a household is energy poor if it has to spend 10% or more of its income on adequate energy services' (Romero et al., 2018). Income here means income before tax and housing costs. Some studies do not 'equivalise' the income used, i.e. adjust it to household size and composition (Legendre and Ricci, 2015), but I will do so here, as these figures are readily available.

The 10% indicator was the first energy poverty indicator to be developed (see its history in Liddell et al., 2012) and over the years it has been found wanting in several respects. It can give the absurd result that a billionaire is in energy poverty because she must spend more than 10% of her income on heating all her castles, ski lodges, penthouses and country villas. I will avoid this problem by only considering households in modest homes, i.e., those who would have to spend more than 10% of their income to heat and run appliances in a home of a type that lower income households are likely to occupy.

A second problem with the 10% indicator, as we will see below, is that even if all low-income households had *just* enough income to cover energy costs with 10% of this income, this income would still be very low. Some may no longer be in energy poverty, but they would still be in poverty by some definitions. One widely accepted definition of poverty, or 'at-risk-of-poverty', is having an income that is less than 60% of the median, and this measure is used in EU studies of poverty (for EU-wide statistics on this see Eurostat, 2019). Hence I will take the issue one further step and explore the fiscal issues in lifting the income of all UK household to at least 60% of the current national median household income.

A second energy poverty indicator is the 'low income, high cost' (LIHC) indicator, which was developed by Hills (2011) and has, like the 10% indicator, been used in several European countries (Boltz and Pichler, 2014; Imbert et al., 2016). With the LIHC, a household is deemed to be in energy poverty if its income is below a certain threshold relative to the country's median income, *and* its energy costs are higher than a defined threshold level of energy expenditure (Romero et al., 2018). Hills (2011) defined the income threshold as 60% of the median equalized income after housing and modelled energy costs. He defined the energy expenditure threshold as the median energy expenditure of all the houses in the country. My use of the at-risk-of-poverty indicator, mentioned above, goes some way toward this indicator. However, I would argue, along with Middlemiss (2017), that the LIHC indicator underestimates energy poverty because it only includes households with relatively high energy costs. I want to include households in dwellings that do not demand as much energy as the median, but who, nevertheless, still cannot afford to heat them.

Another objective indicator is the 'minimum income standard' (MIS) (Heindl and Schuessler, 2015; Moore, 2012; Romero et al., 2018). The MIS says 'a household would be energy poor if it does not have enough income to pay for its basic energy costs, after covering housing and other needs' (Romero et al., 2018). This is suited to a field study of specific households, since 'other needs' are very specific and can only be known on a household-by-household basis.

A fourth indicator is 'after fuel cost poverty' (Legendre and Ricci, 2015). This looks at the number of households that would fall into poverty if they payed energy bills as high as would be needed for adequate energy services.

While acknowledging its shortcomings, I use the 10% indicator in this analysis because it is straightforward to use in a broad-brush approach such as this. We only have to know the distribution of income bands among the lower earning, and can get averaged results by treating all households as if the energy efficiencies of their dwellings and electrical appliances are about the level of the average for a low to modest income household. However, I make a low, medium and high estimate of the income that would be needed to meet the 10% threshold.

4 Using the 10% indicator

4.1 The logic of the analysis

To achieve these aims using the 10% indicator I use the following logic:

(a) Estimate the cost of providing energy for adequate heating and electrical appliances in an average, *modest-sized* UK home.

(b) Multiply this by 10 to find the minimum income an average household would need, to be able to pay for this energy with 10% of their income.

(c) Looking then at all the households with incomes below this level, calculate how much money would need to be added to each of their incomes, to reach this minimum income level;

(d) Add up all these amounts: this will give the total amount that would need to be redistributed to all low income households to stop them being in energy poverty under the 10% indicator.

(e) Look at the highest earning households—those earning more than £1000 per week—and ask what percentage of their income would cover this amount.

(f) Ask whether the redistribution of this level of income from rich to poor would be unreasonable, given the levels of redistribution that have been tolerated and applauded in British society since the end of the Second World War.

(g) Estimate the increase or decrease in consumption-based CO_2 emissions that might result from this redistribution of money from rich to poor.

The reader will see immediately that there are approximations in this method—though all these could be eliminated in a properly funded, in-depth study where the fine details of household statistics are painstakingly considered. An 'average' UK household consists of 2.4 occupants, and I assume in this study that both the richest and poorest households are of this size. Many low income households have four, five or more members, while others have just one. An actual redistribution of income would have to take this into account in order to ensure that each household did achieve the minimum required income. However, in terms of the *total amount* needing to be redistributed, it is probably fair to assume that the differences will average out across the poorer spectrum of households.

Further, no account is taken here of the different thermal qualities of dwellings. Nicola Terry's analysis in Chapter 6 leads me to conclude that poor people disproportionately occupy thermally bad homes. Most of these people might therefore need to spend more than one-tenth of their income to heat their homes and run their appliances, after such a redistribution of income. On the other hand, the 10% indicator has been developed with this in mind, and there are also some more averages hidden here. Some houses are almost impossible to heat no matter how much fuel one burns, while others need only a little more than at present, and some can be brought up to a good standard with cheap cavity wall and loft insulation. So the redistribution of income envisaged here would not lift all households out of energy poverty and would give others more than they need. But then the second determinant of energy poverty comes in to play, namely the energy efficiency of dwellings and appliances. The approach of this chapter is not intended to negate the need for higher energy efficiency, but to shift it to a different context: if poverty and energy inefficiency are determinants of energy poverty, we need to deal with them both, not just efficiency.

4.2 The cost of adequate household energy services using the 10% indicator

I begin with steps (a) and (b) of the method outlined above. In 2017, average gas consumption in UK households was 12,609 kWh and average electricity consumption was 3760 kWh, (DBEIS, 2019). Following the findings of BRISKEE (2018) and Nicola Terry's observations in Chapter 6, it is probably safe to assume that the homes of lower income households are, on average, of poorer thermal quality than those of higher income households. On the other hand, as Lawrence Haar notes in Chapter 9, they are also, on average, significantly smaller and likely to contain fewer electrical appliances—factors that would put downward pressure on energy consumption. I will assume, then, that the average household of lower income families would need about 70% of the current national average level of domestic energy consumption, to keep it comfortably warm and well-running, i.e., 8826 kWh/year for gas and 2632 kWh/year for electricity.

At a cost of £0.06/kWh for gas and £0.14/kWh for electricity, plus a standing charge of £150, this amounts to an annual energy cost of £1048, or a weekly cost of £20. Hence to escape being in fuel poverty by the 10% indicator, a low income home would need a weekly income, before housing costs, of about £200. This is, of course, assuming no special interventions to upgrade the thermal quality of these people's dwellings.

This figure suggests that only about 6.3% UK households are suffering energy poverty, since about 6.3% of persons live in households with less than £200 per week before housing costs (DWP, 2019). It is very close to the figure for the 'subjective' energy poverty indicator, i.e. the percentage who say they are unable to heat their homes adequately, at 6.1% in 2016 (Eurostat, 2019, and see Chapter 7).

However, a study by DECC (2016) found 11.6% suffering energy poverty by the 10% indicator in 2015. Tables of income bands (DWP, 2019) show that the weekly income of the poorest 11.6% of households is around £260. I therefore use the threshold figure of £260 as a further scenario.

Neither of these scenarios—with £200 or £260 per week—would lift all households out of being 'at-risk-of-poverty' under the EU poverty line indicator of 60% of median household income before housing costs. The 60% line is £304 per week (DWP, 2019). 18.0% of UK households are currently below this line.

I will therefore perform the rest of the analysis using three weekly income threshold figures: £200; £260; and £300. Using the first two will give upper and lower limits of the likely range of the amount of money that would need to be redistributed to energy poor households, to lift them out of energy poverty according to the 10% indicator. Further, although it is convenient for analytical purposes to equate low-income houses one-to-one with energy poverty, the real world is not that tidy. Some low-income households are in council houses of high thermal quality (see Chapter 6), and some households of modest but comfortable means are in thermally disastrous dwellings. This gives a further rationale for doing the analysis one step further up the income scale, to the line that would lift all households out of the at-risk-of-poverty zone.

This explains steps (a) and (b) in our method. I now move to steps (c) and (d).

4.3 Calculating the additional amounts required, under the 10% indicator

The vertical columns in Fig. 1 represent the numbers of persons who are in households that have a net weekly income, before housing costs, of the amounts shown on the horizontal axis. So, for example, the highest bar on the graph indicates that 1.174 million persons live in households that have a weekly income between £320 and £329.99. This means that, in each of these households, adding up the incomes of all the members gives a total that lies between these two figures.

Note that the graph does not include bands for persons in households with a weekly income of £1000 or higher, as this would extend the graph a very, very long way to the right. According to the Sunday Times Tax List, the person with the highest taxable income in 2018 earned £336 million, or about £6.46 million per week. Assuming Fig. 1 is 10 cm long, to fit this person's income in would require me to extend the graph 646 m to the right—about half a mile.[d]

---

[d]This only refers to income declared for tax. According to the Sunday Times Rich List, the person with the most wealth in 2018 had a net worth of just over £21 billion. Piketty's (2014) research indicates that large fortunes typically return an annual income of about 10%, suggesting this person's income could have been around £2.1 billion, or £40.4 million per week. To accommodate this on the graph in Fig. 1 I would have to extend it a further 4 km to the right—about 3 miles.

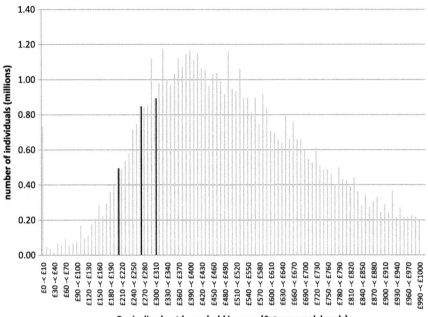

FIG. 1

Number of persons in households with net weekly incomes before housing costs, 2017 (the graph omits the 6.4 million persons in households with weekly income over £1000).

Author's re-drawing of DWP (Department of Work and Pensions), Households Below Average Income, 2017/18: The Income Distribution; 2019. https:/www.gov.uk/government/statistics/households-below-average-income-199495-to-201718.

There are 6.44 million people in this extended band to the right, i.e., with incomes over £1000 per week. On average they live in households with a weekly income of £3876, as I will discuss below.

The dark vertical lines show the incomes calculated above: £200, £260 and £300 per week. To lift people out of energy poverty in their existing homes by the 10% indicator (or above being at-risk-of-poverty) all the bands to the left of the corresponding dark vertical line would have to move right, to at least as far as the corresponding dark line. Hence, for all these people we have to calculate the difference between £200 (or £260 of £300) and their current income.

We now calculate the amounts required to bring all the incomes of all the households that have incomes below £200 per week up to this level. The formula is:

$$T = \sum_{b=1}^{N} (200 - I_b) \cdot P_b / S \qquad (1)$$

where T is the total amount required; b is the income band; P is the number of people within that income band; I is the income of the households within that band; and S is the average number of persons per household, in this case 2.4. For the other two cases, the figure 200 in Eq. (1) is replaced by 260 and 300 respectively. The results for the three cases are given in Table 1.

We see, then, that if 6.3% of UK households are in energy poverty, with incomes lower than £200 per week, it would require an annual redistribution of £6.987 billion to raise their incomes to the £200 threshold. Alternatively, if the figures are 11.6% and the threshold is £260, an annual sum of £14.114 billion would be required. To lift all households to an income of £300 per week would require an annual sum of £22.181 billion.

This takes care of steps (c) and (d). In step (e) we ask how much this would cost per household if it were paid for by the highest earning 10% of UK society, i.e., those earning £1000 per week or more.

4.4 How much would this cost high-income households?

The highest earning 10% of UK households enjoyed incomes above £1000 per week in 2017–2018. According to data from DWP (2019) the average weekly income of this group was £3876, or an average annual income of £201,558. This could be even higher because of hidden income, tax havens, etc. (Zucman, 2015), but it is the amount that income tax is based on, so it is the figure we must work with here.

How much would it cost these high-income households if we redistributed income to low-income bands to eliminate energy poverty by the 10% indicator? For the low-cost case, where we redistribute only to the 6.3% households with incomes below £200, as Table 1 shows, the total cost would be £6.987 billion. This would cost each of these high-earning households an average of £2588 per year, or £50 per week. This is 1.28% of their income.

For the higher-cost case, redistributing to the 11.6% of UK households with incomes below £260 would cost each high-earning household just over twice this

Table 1 Amounts required to increase the incomes of all UK energy-poor households to the theoretical level where they are no longer in energy poverty in 2017–2018, based on two alternative scenarios: that 6.1% and 11.6% of households are in energy poverty

Percentage in energy poverty (or at-risk-of-poverty) estimated by	70% of average consumption	DECC estimate	EU at-risk-of-poverty indicator
Energy poverty (%)	6.3	11.6	18.0
Income required per week	200	260	300
Total money required (millions of £s) per week	134.383	271.418	426.563
Total money required (billions of £s) per year	6.987	14.114	22.181

Author's calculations based on data from DWP (Department of Work and Pensions), Households Below Average Income, 2017/18: The Income Distribution; 2019. https:/www.gov.uk/government/statistics/households-below-average-income-199495-to-201718.

much, i.e. an average of £5258 per year or £101 per week, which is 2.59% of their income.

For the case of lifting all households out of the at-risk-of-poverty zone, i.e., to incomes of at least £300 per week, the cost is £22.181 billion. This would cost each high-earning household £8266 per year, which is £159 per week, or 4.10% of their income.

This takes care of step (e). In step (f) I will ask whether society would tolerate such redistribution of wealth.

Note that none of these amounts of redistribution would necessarily eliminate energy poverty, since the method employed here does not address the wide range of thermal qualities of different dwellings, nor the variation in specific household needs. However, it does address the first of the four determinants of energy poverty, namely low income. If the redistributions estimated here were in fact carried out, poverty itself would no longer be an element in energy poverty. With the second scenario and perhaps even with the first, on average, all households would have enough to consume 70% of average household energy consumption, i.e. enough to heat a typical UK modest sized house and use a reasonable array of electrical appliances. With the third scenario a large number of households would have enough to heat a large or thermally somewhat sub-optimal home—though not the worst of British homes!

This deals with parts (c), (d) and (e) of our method. I now move to part (f) and ask whether society would tolerate a redistribution of money on the scales indicated here.

5 Would this redistribution be tolerated?

To begin with, I am not necessarily advocating a direct, Robin Hood style transfer of monies from high-earning to low-earning households. Instead, transfers such as this can be made via government run welfare schemes, job creation, paid training courses, income supplements, pension increases, or even the increasingly popular idea of a basic income (Lowrey, 2018). The important point, though, is that money is put into the hands of low-income households *so that they have the agency to decide what to do with it.* This is of course anathema to neoliberal doctrine, but it was regarded as much more normal during the post-war egalitarian period, and in fact transfers like this continue to happen today—which is why the Gini coefficient after tax and welfare redistribution is significantly lower than the Gini prior to redistribution (see Chapters 7 and 11).

The three cases discussed above would require 1.28%, 2.59% or 4.10%, respectively, of the income of the highest earning 10% to be redistributed to poor households by some such combination of redistribution methods. Would society tolerate this?

Table 2 gives the current UK income tax bands, excluding Scotland, and Table 3 gives tax bands for Scotland. Note that the Scottish rates are slightly more progressive than the UK rates: low income earners pay 1% less, while high income earners pay 1% more. It is interesting that there are not loud complaints about the highest earning bands paying what amounts to an average of around £2000 per year more each in Scotland than in the rest of the UK.

Table 4 shows how the higher UK tax bands (excluding Scotland) would look, if all the money was raised by increasing the marginal tax rate on the highest tax band, for each of the three cases. The additional amounts are slightly higher than the percentages of 1.28%, 2.59% and 4.10% calculated above, since only the portion of a person's income above £50,000 would be taxed at the higher rate. Regarding the Scottish tax regime, it would not make much difference to the total outcome if the same percentages as in Table 4 were also used for the Scottish higher tax bands.

Table 2 Current personal income tax bands for the UK excluding Scotland

Band	Taxable income	Tax rate (%)
Personal allowance	Up to £12,500	0
Basic rate	£12,501 to £50,000	20
Higher rate	£50,001 to £150,000	40
Additional rate	Over £150,000	45

From HMRC (Her Majesty's Revenue and Customs), Income Tax Rates and Personal Allowances; 2019. https:/www.gov.uk/income-tax-rates.

Table 3 Current personal income tax bands for Scotland

Band	Taxable income	Scottish tax rate (%)
Personal allowance	Up to £12,500	0
Starter rate	£12,501 to £14,549	19
Basic rate	£14,550 to £24,944	20
Intermediate rate	£24,945 to £43,430	21
Higher rate	£43,431 to £150,000	41
Top rate	Over £150,000	46

From HMRC (Her Majesty's Revenue and Customs), Income Tax Rates and Personal Allowances; 2019. https:/www.gov.uk/income-tax-rates.

Table 4 How the higher UK (excluding Scotland) tax bands would look, for each of the three scenarios

Band	Taxable income	Current tax rate (%)	Case 2: EP of 6.1%	Case 2: EP of 11.6%	Case 3: all up to 60% of current median
Higher rate	£50,001 to £150,000	40	41.5	42.7	44.3
Additional rate	Over £150,000	45	46.5	47.7	49.3

How do these higher tax rates compare to historical tax rates in the UK and elsewhere? Fig. 2 gives top marginal tax rates for the UK, the US and Germany from 1900 to 2013, when the UK's rate was reduced from 50% to 45%. It shows that, apart from the years of the Blair government, the UK's top marginal tax rate is lower today than it has been at any time since 1918, when it first rose to 53%. It then rose further, to 60% immediately after the First World War, fell back to 50% in the 1920s prior to the Wall Street share market crash, then rose steadily to a peak of 98% during and immediately after the Second World War, fell again to around 90% and returned to 98% in the early 1970s. There were then three big reductions. The first came in 1979, shortly after the collapse of the Bretton Woods international financial regime (see Chapter 1), when the rate was reduced to 75%. The second fall, to 60%, came in 1984 when Margaret Thatcher's government introduced neoliberal economic policy. Then in 1988 the rate was reduced to 40% as the Blair government deepened neoliberal economic policies. This rate persisted until 2010, when it was increased to 50%. The Conservative government under David Cameron reduced it again to 45% in 2013.

The top marginal tax rate in the US has followed a similar trajectory—the data for Fig. 2 does not extend to the Trump Presidency's recent top marginal income tax reduction to 35%. Fig. 2 also shows that Germany's top marginal tax rate was reduced much earlier than the UK's and US's after the Second World War, but its current level of 45% is lower than any period since 1935 apart from a brief spell at 42% in 2005 and 2006.

Seen in historical perspective since the end of the Second World War, then, the UK's current marginal tax rate of 45% is very much on the low side. It could increase

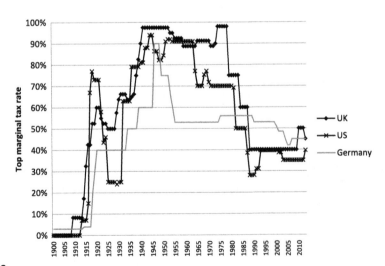

FIG. 2

Top marginal income tax rates for the UK, the US and Germany, 1900–2013.

Data source: http:/piketty.pse.ens.fr/en/capital21c2.

to 60% or even 75% without being in the very high zone associated with the Bretton Woods era. In any case, since that era saw unprecedented levels of economic equality, there may be further arguments for increasing it if a more equitable society is the goal (Piketty, 2014).

A further important point is that the total tax take as a proportion of GDP has been falling since it reached a peak of around 43% in the early 1980s, and is now around 36% (UK Public Revenue, 2019). The government has, year by year, been depriving itself of the revenue required to address the big social challenges of our times. It is not that the country as a whole is poor, since GDP is historically very high, but that the government has been taking a smaller and smaller share for redistribution and pubic services.

At the same time, tax revenues from indirect services such as value added tax (VAT) and sales tax have been steadily increasing, from a low of about 11% of GDP in 1965 to the present level of about 16% of GDP (UK Public Revenue, 2019). Indirect taxes are inherently regressive, i.e. they fall more heavily on the poor than on the rich. For example, with a VAT rate of 20%, a poor household may pay an effective tax rate of 20% on their entire income since all must be spent on basic living costs. So, on the one hand the government has been depriving itself of the income that would give it the means to alleviate poverty, while on the other hand it has pursued fiscal policies that can only deepen poverty.

I am suggesting, then, that increasing the top marginal tax rate by a few percent would give the government the financial means to massively reduce the grinding poverty that is associated with energy poverty. Some might argue that this is arbitrary: why target the money at energy poverty when there are so many other pressing needs?

In fact, however, this money would not only go to alleviating *energy* poverty—only one-tenth of it would, as we are talking about enabling households to meet their domestic energy needs with just 10% of their income. Households would have the other nine-tenths for alleviating many of the other crippling effects of poverty, and they could make their own decisions as to what their priorities are.

But what would the effect be, on CO_2 emissions, of putting all this money into the hands of currently low-income households? We now move to step (g) of the analysis.

6 Implications for CO_2 emissions

The Department for Energy and Climate Change (2011) estimated that the lowest-earning 30% of households in England consume only 70% of what is theoretically needed to heat their homes (Boardman, 2012). As noted in Section 2, some energy poverty researchers maintain the UK's total CO_2 emissions would increase if all the UK's households whose energy consumption is currently limited through poverty were suddenly able to consume at a higher level.

On the other hand, this would be offset by households now being able to spend on energy efficiency measures. As the BRISKEE (2018) survey showed, in 8 European countries, one of the reasons poor households over-consume energy and produce higher CO_2 emission levels than they otherwise might is that they cannot afford basic energy efficiency investments: a new refrigerator, a new boiler, loft or under-floor insulation, or air-tight windows.

But the point missed in this kind of analysis is that *the money redistributed to such households has to come from somewhere*. If it comes from making income tax more progressive—such as slightly higher taxes on the top-earning 10% of society—a new set of effects on CO_2 emissions will come into play.

6.1 The CO_2 implications of more progressive taxation

There is now abundant evidence that CO_2 emission levels in high income countries are positively correlated with economic inequality (Jorgenson et al., 2015, 2016; Knight et al., 2017; Chancel and Piketty, 2015). In other words, as the gap between rich and poor widens in these countries, CO_2 emissions from consumption increase (controlling for differences in GDP). This is partly because wealthy households tend to overconsume (Chancel and Piketty, 2015) and partly due to the inability of poorer households to increase their energy efficiency, as noted above. Hence, if the Gini coefficient reduces, we can expect to see a reduction in consumption-based CO_2 emissions.

Note that a country's consumption-based CO_2 emissions take into account the CO_2 emitted in the production of goods that are made outside that country but used in that country. The CO_2 emissions from goods made in that country but consumed in another country are subtracted from the total. The UK imports a large proportion of its carbon-intensive goods. A study by the House of Commons Energy and Climate Change Select Committee (2012) estimated the UK's consumption-based emissions were 1.1 Giga-tonnes (Gt) per year, or 17t per person, compared with 0.7Gt produced on UK territory, or 10.8t per person.

Regarding *wealth* inequality, Knight et al. (2017) found that each 1% increase (or decrease) in wealth inequality is associated with an increase (or decrease) in consumption-based CO_2 emissions of around 0.8% (more accurately, an 'inequality elasticity of emissions' of 0.8, where income inequality is measured by the Gini coefficient). Jorgenson and colleagues' findings for *income* inequality were more variable, with an elasticity of about 0.8 for post-communist eastern European countries, and an annually increasing elasticity for western countries, reaching 0.3 in 2009 but continuing to rise. I will take a modest approach and ask what the effect on emissions would be if the UK's income inequality of consumption-based CO_2 emissions has not continued to rise, but remains at 0.3.

The UK's Gini coefficient of income inequality (after tax transfers) in 2017 was 33.1% (Eurostat, 2019) and its consumption-based CO_2 emissions were 1.1Gt (UK House of Commons Energy and Climate Change Committee, 2012). For an elasticity of 0.3, the relationship between the Gini coefficient (as a percentage) and emissions is given by:

$$C = 0.385G^{0.3} \qquad (2)$$

where C is CO_2 emissions and G is the Gini coefficient by (readers who want to see the maths may consult the footnote[e]). Using Eq. (2) we can estimate the level of consumption-based CO_2 emissions associated with any value of the Gini coefficient.

To be able to use Eq. (2) we need to find how the Gini coefficient changes when the tax rate on the highest-earning 10% is increased. To do this I draw the reader's attention to Fig. 6 in Chapter 1, which shows an almost perfect correlation between the Gini coefficient and the percentage share of income of the top earning 10%, plotted for 43 OECD countries (data from OECD, 2018). I reproduce this here in Fig. 3, but with the axes swapped and the Gini expressed as a percentage. This shows how the Gini coefficient varies as the income share of the top-earning 10% varies. These are between-country variations, but they are also likely to apply to year-by-year variations within the same country, because the relationship between the Gini and the income share of the top-earning 10% is largely mathematical.

The modelling equation in Fig. 3 is:

$$G = 1.3562P - 1.5316 \tag{5}$$

where G is the Gini coefficient expressed as a percentage, and P is the share of income of the top-earning 10% (after tax transfers). Eq. (5) shows that for every 1% reduction in the share of income of the top-earning 10%, the Gini coefficient decreases by about 1.36 percentage points.

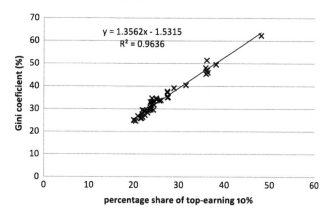

FIG. 3

Gini coefficient as a percentage, plotted against percentage share of total national income of top earning 10%, for 43 OECD and BRIC countries.

From OEDC, Income Distribution and Poverty; 2018. https:/stats.oecd.org/Index.aspx?DataSetCode=IDD.

[e]An income inequality elasticity of CO_2 emissions of 0.4 is expressed mathematically as

$$\frac{dC}{dG} \cdot \frac{G}{C} = 0.3 \tag{3}$$

where C is CO_2 emissions and G is the Gini coefficient. The general solution to this differential equation is:

$$C = K \cdot G^{0.3} \tag{4}$$

Substituting initial values for $C = 1.1$ and $G = 33.1$ gives a value for K of 0.385.

Since the top earners are already being taxed at the rate of 45%, each additional percentage point of tax increase reduces their remaining, after-tax income by almost 2%. Because this group is receiving 42% of all UK household income after tax, a 2% reduction in their income will reduce their after-tax share of national income by 0.84% (since 42% of 2% is 0.84%).We recall from Table 4 that the three scenarios involved tax increases for the top-earning 10% of 1.5%, 2.7% and 4.3% respectively. This would reduce these household's after-tax income by 1.26%, 2.27% and 3.61% respectively. Using the figure of 1.356 from Eq. (5), this would reduce the Gini coefficient by 1.71, 3.08 and 4.90 percentage points, i.e., to 31.4%, 30.0% and 28.2%.

Now substituting these values of the Gini coefficient in Eq. (2), the consumption-based emissions would fall to $1.083\,Gt$, $1.068\,Gt$ and $1.048\,Gt$ of CO_2, i.e. a fall of $17\,Mt$, $32\,Mt$ and $52\,Mt$ respectively.

Using these figures for consumption-based CO_2 emissions, Table 5 summarises the results including estimated reductions in CO_2 emissions from each of the three redistribution scenarios.

We see, then, in Table 5, that the first scenario—where we redistribute sufficient money from the top-earning 10% to the bottom-earning 6.3% to remove the economic poverty from their energy poverty—could (theoretically at least) lead to consumption-based CO_2 reductions of 17 Megatons (Mt), or around 1.55%, mostly because it would make it harder for the rich to over-consume. The second scenario, where we redistribute to the bottom-earning 11.6%, could lead to a reduction of $32\,Mt$ or 2.91%. The third scenario, where we eliminate at-risk-of-poverty, suggests a reduction of $52\,Mt$ of CO_2, or 4.73%.

A critic might claim this is one-sided, since the 6.3%, 11.6% or 18.0% of formerly poor households would now be spending more on heating and operating electrical appliances, thereby increasing their CO_2 emissions. How would this effect CO_2 emissions?

6.2 CO_2 emissions and increased incomes among poorer households

As noted above, the only survey that has tested this issue on a large scale is the EU-funded BRISKEE (2018) household survey of eight EU countries. It found that the poverty of low-income households prevents them from *reducing* their CO_2 emissions. This is because, without money, they lack the agency to take the initiative and begin to increase the energy efficiency of their dwellings and appliances.

However, in case the BRISKEE finding is wrong or tells only half the story, let us assume none of these formerly poor households invest in energy efficiency technology but instead simply consume more energy. Let us assume that one-tenth of all the extra money they get goes on increased home heating, so that, on average for all the scenarios, they are now above the energy poverty line as it is defined in the first two scenarios, and just above the at-risk-of-poverty line for the third scenario.[f]

---

[f]Note that the tax rate on the first £12,000 per year is zero (see Tables 2 and 3), so all the money received by the poorest 6.3% of households is available for spending, as is the case for the other two groups if household income is made up of the incomes of more than one household member. Overall, then, I assume in the calculations on extra energy consumption that the full one-tenth of these households' income, untaxed, is available for spending on energy.

Table 5 Modelled figures for reduction in CO_2 emissions following from two scenarios of tax increases on the UK's highest earning 10%, for energy poverty mitigation scenarios

Percentage in energy poverty estimated by	All households can achieve 70% of average consumption	DECC estimate	EU at-risk-of-poverty threshold
Energy poverty (%)	6.3	11.6	18.0
Income required per week	200	260	300
Total money required (millions of £s) per week	134.383	271.418	426.563
Total money required (billions of £s) per year	6.987	14.114	22.181
Tax increase on top 10% income earners	1.5%	2.7%	4.3%
Reduction in income share of top 10% of income earners	1.26%	2.27%	3.61%
Reduction in Gini coefficient (percentage points)	1.47	2.96	4.69
Estimated reduction in consumption-based CO_2 emissions (%)	1.55%	2.91%	4.73%
Estimated reduction in consumption-based CO_2 emissions (Mt)	**17 Mt**	**32 Mt**	**52 Mt**
Increase in CO_2 emissions if all low-income homes spend one-tenth of redistributed money on increased heating	**0.7 Mt**	**1.4 Mt**	**2.2 Mt**
Increase in CO_2 emissions if all low-income homes have income elasticity of energy consumption of 0.7 and were consuming 70% of national average prior to redistribution	0.68 Mt	1.6 Mt	2.8 Mt

Author's calculations based on income data from DWP (Department of Work and Pensions), Households Below Average Income, 2017/18: The Income Distribution; 2019. https://www.gov.uk/government/statistics/households-below-average-income-199495-to-201718; OEDC, Income Distribution and Poverty; 2018, https://stats.oecd.org/Index.aspx?DataSetCode=IDD; A. Jorgenson, J. Schor, X. Huang, J. Fitzgerald, Income inequality and residential carbon emissions in the United States: a preliminary analysis, Hum. Ecol. Rev. 22 (1) (2015) 93–105; A. Jorgenson, J. Schor, K. Knight, X. Huang, Domestic inequality and carbon emissions in comparative perspective, Sociol. Forum 31 (S1) (2016) 770–86; K. Knight, J. Schor, A. Jorgenson, Wealth Inequality and Carbon Emissions in High income Countries, Soc. Curr. 0 (0) (2017) 1–10.

For the first scenario, where only the lowest-earning 6.3% are the beneficiaries, this means an extra £699 million is spend on heating annually. This would result in an extra 10 billion kWh of heating energy per year, assuming a cost of 7 pence per kWh. Assuming the heat source is all natural gas, which produces 0.070 kg of CO_2 per kWh, this would result in an extra 700,000 t of CO_2 emissions per year. This is an increase of 0.06% on the UK's current annual total of 1.1 Gt of consumption-based CO_2 emissions per year. In other words, *in a worst possible case scenario, the reduction in CO_2 emissions from the rich would be about 24 times as great as the increase in CO_2 caused by poor households finally having enough to spend on adequate energy services.*

For the second scenario, where the lowest-earning 11.6% benefit, the increase is 1.40 Mt, against a decrease among the rich of 32 Mt, a ratio of 1–23. For the third scenario the ratio is also 1–23. *So for all scenarios, the likely reduction in CO_2 emissions brought about by more progressive taxation would far outweigh the increase from low-income households having more to spend on energy services.*

An Irish study in 2017 provides yet another route to estimating the possible increase in CO_2 emissions from low-income households, assuming UK households behave like their Irish counterparts. Harold et al. (2017) investigated income elasticities of household energy consumption (the percentage increase in household energy consumption for each 1% increase in household income). For households in the lowest-earning 10% this elasticity averaged 0.7 for the years 1987–2010, meaning that, for each 1% increase in income, energy consumption increased by 0.7%. What would this mean for our three scenarios?

For the first scenario, the lowest earning 6.3% increase their total income from £1589 million to £1723 million, an average increase per household of 8.46%. Using the elasticity figure of 0.7, their energy consumption would therefore increase by 5.92%. If they had already been consuming the same amount of energy as the average UK household (which severely overestimates their energy consumption, as by definition the fuel-poor cannot do this, on average), their increase in consumption would add up to a national increase of 0.37%—which equates to about 970,000 tonnes of CO_2 emissions. If, however, they were previously consuming 70% of the UK average, their increase equates to 680,000 tonnes of CO_2, just short of the 700,000 tonnes estimated above.

For the third scenario, the lowest earning 18.0% increase their total income from £8720 million to £9147 million, an average increase per household of 4.90%. Their energy consumption would therefore increase by 3.43%. Again, if they had been consuming at the national average level, this would increase national energy consumption by 0.62%—which equates to about 4 Mt of CO_2. This would be 2.8 Mt if they had been consuming 70% of the UK average.

For the second scenario the increase is 1.6 Mt of CO_2. Hence, using the Irish elasticity figure gives a very comparable increase in consumption when low-income households' incomes are increased above the energy poverty threshold (compare results in Table 5, last two rows). So, if UK homes were to behave like Irish homes, the results would be about the same. The reduction in national consumption-based

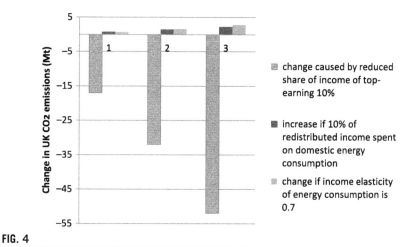

FIG. 4

Estimated changes in UK CO_2 emissions if poorest households' incomes increased above £200, £260 and £300 per week through progressive income tax changes.

Source: Author's calculations in main text.

CO_2 emissions from more progressive taxation would outweigh the increases by a factor of around 20–25.

The estimated changes in CO_2 emissions are shown graphically in Fig. 4.

In fact, for an unimaginable situation in which *all* the money redistributed to the 6.3% is spent on heating, the increase in CO_2 emissions would still be just 7 Mt per year, compared to the reduction by the rich of 17 Mt per year. The argument that redistributing money from the rich to the poor would be risky for climate change mitigation efforts is plainly wrong. The poor are not the problem, where CO_2 emissions are concerned.

7 Conclusions

In this chapter I have broken several unwritten rules of energy poverty research. Most overtly, I have explored the first of the three (or four) widely accepted determinants of energy poverty, namely low income. I have asked what the consequences might be of increasing the incomes of all the UK's lowest-income households to three different levels: £200, £260 and £300 per week before tax and housing costs. I chose the first of these levels because about 6.3% of UK households have incomes below that level, and 6.3% is about the minimum estimate of the number in energy poverty. This would enable these households to spend at least £20 per week on domestic energy consumption without spending more than 10% of their income. This would enable them to spend about 70% of the UK average.

I chose the second threshold of £260 because this would lift the poorest 11.6% of UK households up to this level, and this is the percentage of households the Department of Energy and Climate Change estimated to be in energy poverty in 2015.

However, these increases in income would not eliminate energy poverty because of the uneven distribution of energy efficiency among the dwellings of low-income people, and indeed of the UK generally, plus the uneven distribution of specific household needs. I therefore chose the third threshold, of £300 per week, which represents 60% of UK median household income—a standard EU measure of 'at-risk-of-poverty'. This would benefit 18% of UK households. There would still be a percentage in energy poverty due to severely energy inefficient dwellings and special household needs, but the first of the three (or four) determinants of energy poverty would have been decisively eliminated – there would be no more households at-risk-of-poverty.

A further rule I broke was to suggest a classical, 1950s–1970s-style, universalist solution to poverty, namely to redistribute income to all poor households rather than just specific forms of help for a carefully selected target group. This goes against the deeply ingrained, widely and uncritically accepted neoliberal assumption that we should address poverty by finely targeting just the most needy households in highly technical, restricted ways.

A third rule I broke was to suggest we get the money for this redistribution by increasing the marginal tax rate of the highest earning 10%. This flies in the face of the neoliberal doctrine – again so deeply entrenched that people often assume it is the truth—that taxes on the highest earning should be as low as possible so as not to discourage them from putting their economic genius to work for the greater good.

In fact, of course, the tax increases I suggested would not lift the top marginal tax rate to anything like as high as it was in the years of the Thatcher government, when it was 75% and then 60%.

I have also attempted to show that breaking these rules would bring unexpected benefits. One is that net global CO_2 emissions would most likely fall: *the reduction in consumption-based CO_2 emissions brought about by the reduction in income of the highest earners would be about 23 times as great as any likely increase in emissions due to poor households being able to spend more on domestic energy consumption.*

I admit these figures are somewhat tentative and may be subject to a substantial margin of error. But I cannot see anything in the approximations I have made that would risk them being misleading. It is abundantly clear that, if we fund an increase in the income of the poorest UK households by means of a small increase in the top marginal tax rate, global CO_2 emissions will fall rather than rise.

A further benefit of such an approach is that it would have positive effects way beyond a massive reduction in energy poverty. Imagine living in a country where there is virtually no poverty. Actually this is easy for people over 60 years old who were brought up in the US, Canada, New Zealand, Australia and perhaps the UK, because this was normal at that time in those countries – even though those countries were nowhere near as wealthy per capita as they are now.

I conclude by mentioning the title of Benjamin Sovacool's book, *Visions of Energy Futures: Imagining and innovating for low-carbon transitions* (Sovacool, 2019). A 'vision', he suggests, is 'a description of what could occur in the near-term, mid-term or long-term future.' Citing Berkout (2006) he suggests a vision can 'map a possibility space by identifying a realm of plausible alternatives and the means for reaching them.'

I have attempted in this chapter to identify a realm of plausible alternatives to energy poverty as we know it in the UK today.

References

Alston, P., 2018. Statement on Visit to the United Kingdom, by Professor Philip Alston, United Nations Special Rapporteur on Extreme Poverty and Human Rights. Available for download at https://www.ohchr.org/EN/NewsEvents/Pages/DisplayNews.aspx?NewsID=23881&LangID=E.

Anderson, W., White, V., Finney, A., 2012. Coping with low incomes and cold homes. Energy Policy 49, 40–52.

Bartiaux, F., Vandeschrick, C., Moezzi, M., Frogneux, N., 2018. Energy justice, unequal access to affordable warmth, and capability deprivation: a quantitative analysis for Belgium. Appl. Energy 225, 1219–1233.

Berkout, F., 2006. Normative expectations in systems innovation. Tech. Anal. Strat. Manag. 18 (3–4), 299–311.

Boardman, B., 2012. Fuel poverty synthesis: lessons learnt, actions needed. Energy Policy 49, 143–148.

Boltz, W., Pichler, F., 2014. Getting it right: defining and fighting energy poverty in Austria. ICER Chron., 19–24.

Bouzarovski, S., Petrova, S., 2015. A global perspective on domestic energy deprivation: overcoming the energy poverty–fuel poverty binary. Energy Res. Soc. Sci. 10, 31–40.

Bouzarovski, S., Tirado Heredo, S., 2017. Geographies of injustice: the socio-spatial determinants of energy poverty in Poland, the Czech Republic and Hungary. Post-Communist Econ. 29 (1), 27–50.

BRISKEE (Behavioural Response to Investment Risks in Energy Efficiency), 2018. Research Project of the Fraunhofer Institute. https://www.isi.fraunhofer.de/en/competence-center/energiepolitikenergiemaerkte/projekte/briskee_331589.html.

Brunner, K., Spitzer, M., Christanell, A., 2012. Experiencing fuel poverty. Coping strategies of low-income households in Vienna/Austria. Energy Policy 49, 53–59.

Chancel, L., Piketty, T., 2015. Carbon and Inequality From Kyoto to Paris Trends in the Global Inequality of Carbon Emissions (1998–2013) & Prospects for an Equitable Adaptation Fund. Paris School of Economics. http://www.ledevoir.com/documents/pdf/chancelpiketty2015.pdf.

Cox, H., 1999. The market as god: living in the new dispensation. The Atlantic, March 1999, 1–12.

Crib, J., Norris Keiller, A., Waters, T., 2018. Living Standards, Poverty and Inequality in the UK: 2018. The Institute for Fiscal Studies. https://www.ifs.org.uk/uploads/R145%20for%20web.pdf.

DBEIS (Department for Business, Energy and Industrial Strategy), 2019. Energy Consumption in the UK: 2018 Update. https://www.gov.uk/government/organisations/department-for-business-energy-and-industrial-strategy.

Deacon, B., 2005. From 'safety nets' back to 'universal social provision' global social policy. SAGE Publ. 5 (1), 19–28.

DECC (Department of Energy and Climate Change), 2011. Energy Trends (June 2011 and September 2011 Edition). https://webarchive.nationalarchives.gov.uk/20130106091008/http://www.decc.gov.uk/en/content/cms/statistics/publications/trends/trends.aspx.

DECC (Department of Energy and Climate Change), 2016. Annual Fuel Poverty Statistics Report 2016. https://www.gov.uk/government/statistics/annual-fuel-poverty-statistics-report-2016.

DWP (Department of Work and Pensions), 2019. Households Below Average Income, 2017/18: The Income Distribution. https://www.gov.uk/government/statistics/households-below-average-income-199495-to-201718.

Eurostat, 2019. Relative Median at-Risk-of-Poverty Gap. https://ec.europa.eu/eurostat/web/products-datasets/product?code=sdg_10_30.

Harold, J., Cullinan, J., Lyons, S., 2017. The income elasticity of household energy demand: a quantile regression analysis. Appl. Econ. 49 (54), 5570–5578.

Heindl, P., Schuessler, R., 2015. Dynamic properties of energy affordability measures. Energy Policy 86, 123–132.

Hills, J., 2011. Fuel poverty: the problem and its measurement. CASE report, 69. Department for Energy and Climate Change, London, UK.

House of Commons Energy and Climate Change Select Committee, 2012. Consumption-based emission reporting: Government Response to the Committee's Twelfth Report of Session 2010–12, The Stationary Office Limited, London.

Imbert, I., Nogues, P., Sevenet, M., 2016. Same but different: on the applicability of fuel poverty indicators across countries—insights from France. Energy Res. Soc. Sci. 15, 75–85.

Institute for Fiscal Studies, 2019. Living standards, poverty and inequality in the UK: 2018. Institute for Fiscal Studies, London. http://www.ifs.org.uk.

Jensen, J., 2009. Redesigning citizenship regimes after neoliberalism. Moving towards social investment. In: Morel, N., Palier, B., Palme, J. (Eds.), What Future for Social Investment? Institute for Future Studies, pp. 27–44.

Jorgenson, A., Schor, J., Huang, X., Fitzgerald, J., 2015. Income inequality and residential carbon emissions in the United States: a preliminary analysis. Hum. Ecol. Rev. 22 (1), 93–105.

Jorgenson, A., Schor, J., Knight, K., Huang, X., 2016. Domestic inequality and carbon emissions in comparative perspective. Sociol. Forum 31 (S1), 770–786.

Knight, K., Schor, J., Jorgenson, A., 2017. Wealth inequality and carbon emissions in high income countries. Soc. Curr. 4 (5), 403–412.

Legendre, B., Ricci, O., 2015. Measuring fuel poverty in France: which households are the most fuel vulnerable? Energy Econ. 49, 620–628.

Liddell, C., Morris, C., McKenzie, S., Rae, G., 2012. Measuring and monitoring fuel poverty in the UK: national and regional perspectives. Energy Policy 49, 27–32.

Lowrey, A., 2018. Give People Money: The Simple Idea to Solve Inequality and Revolutionise Our Lives. Penguin, London.

Mayer, I., Nimal, E., Nogue, P., Sevenet, M., 2014. The two faces of energy poverty: a case study of households' energy burden in the residential and mobility sectors at the city level. Transp. Res. Proc. 4, 228–240.

Middlemiss, L., 2017. A critical analysis of the new politics of fuel poverty in England. Crit. Soc. Policy 37 (3), 425–443.

Middlemiss, L., Gillard, R., 2015. Fuel poverty from the bottom-up: characterising household energy vulnerability through the lived experience of the fuel poor. Energy Res. Soc. Sci. 6, 146–154.

Moore, R., 2012. Definitions of fuel poverty: implications for policy. Energy Policy 49, 19–26.

National Audit Office, 2009. The Warm Front Scheme. House of Commons. https://www.nao.org.uk/wp-content/uploads/2009/02/0809126.pdf.

OEDC, 2018. Income Distribution and Poverty. https://stats.oecd.org/Index.aspx?DataSetCode=IDD.

Papada, L., Kaliampakos, D., 2016. Measuring energy poverty in Greece. Energy Policy 94, 157–165.

Piketty, T., 2014. Capital in the Twenty-First Century (Translated from the French by Arthur Goldhammer). Belknapp-Harvard University Press, Cambridge, MA.

Roberts, D., Vera-Toscano, E., Phimister, E., 2015. Fuel poverty in the UK: is there a difference between rural and urban areas? Energy Policy 87, 216–223.

Robinson, C., Bouzarovski, S., Lindley, S., 2018. 'Getting the measure of fuel poverty': the geography of fuel poverty indicators in England. Energy Res. Soc. Sci. 36, 79–93.

Rodgers-Vaughn, B., 2013. Pastoral counseling in the neoliberal age: hello best practices, goodbye theology. Sacred Spaces 5, 5–45.

Romero, J., Linares, P., López, X., 2018. The policy implications of energy policy indicators. Energy Policy 115, 98–108.

Santamouris, M., Paravantis, J., et al., 2013. Financial crisis and energy consumption: a household survey in Greece. Energy Build. 65, 477–487.

Sovacool, B., 2015. Fuel poverty, affordability, and energy justice in England: policy insights from the warm front program. Energy 93, 361–371.

Sovacool, B., 2019. Visions of Energy Futures: Imagining and Innovating for Low-Carbon Transitions. Earthscan-Routlege, London.

Stockton, H., Campbell, R., 2012. Time to Reconsider UK Energy and Fuel Poverty Policies? Joseph Rowntree Foundation.

Thompson, H., Bouzarovski, S., Snell, C., 2017. Rethinking the measurement of energy poverty in Europe: a critical analysis of indicators and data. Indoor Built Environ. 26 (7), 879–901.

UK Public Revenue, 2019. Public Revenue Since 1990. https://www.ukpublicrevenue.co.uk/revenue_history.

Ürge-Vorsatz, D., Tirado Herrero, S., 2012. Building synergies between climate change mitigation and energy poverty alleviation. Energy Policy 49, 83–90.

Waddams Price, C., Brazier, K., Wang, W., 2012. Objective and subjective measures of fuel poverty. Energy Policy 49, 33–39.

Walker, G., Simcock, N., Day, R., 2016. Necessary energy uses and a minimum standard of living in the United Kingdom: energy justice or escalating expectations? Energy Res. Soc. Sci. 18, 129–138.

Watson, C., Bolton, P., 2013. Warm Front Scheme. House of Commons Library, Science and Environment Social and General Statistics.

Zucman, G., 2015. The Hidden Wealth of Nations: The Scourge of Tax Havens. University of Chicago Press, Chicago.

Further reading

HMRC (Her Majesty's Revenue and Customs), 2019. Income Tax Rates and Personal Allowances. https://www.gov.uk/income-tax-rates.

Rayner, J., 2019. Don't talk about 'food poverty'—it's just poverty. The Guardian. 16 May 2019, https://www.theguardian.com/food/2019/may/16/dont-talk-about-food-poverty-jay-rayner.

Reflections

Sustainable energy transition and increasing complexity: Trade-offs, the economics perspective and policy implications

11

Reinhard Madlener

Institute for Future Energy Consumer Needs and Behavior (FCN), School of Business and Economics/E.ON Energy Research Center, RWTH Aachen University, Aachen, Germany

Chapter outline

1 Introduction 252
2 The various energy inequality and justice dimensions to be considered 254
  2.1 Spatial heterogeneity and trends of (primary) energy use: A global perspective 255
  2.2 Spatial heterogeneity and structural economic change: A regional perspective 255
  2.3 Temporal income and energy inequality: Let the "*Gini* out of the bottle" ...257
  2.4 (Social) Life-cycle analysis of energy justice 259
  2.5 Heterogeneity of energy rebound and sufficiency 260
3 Broadening our understanding of energy poverty: The economics perspective 261
  3.1 Some reflections on terminology and the energy/fuel poverty and energy justice debate 261
  3.2 Measurement issues 263
  3.3 Taxation, transfers, and subsidies: How to best re-balance the level-playing field? 267
  3.4 Social/economic welfare considerations 272
4 Smart systems: Efficiency, participation and equity considerations 275
  4.1 Smart grids as enabling technologies 275
  4.2 Smart meters and real-time pricing (RTP) 275
  4.3 Sustainable energy communities 275

Inequality and Energy. https://doi.org/10.1016/B978-0-12-817674-0.00011-4

4.4 Prosumer households and aggregate constructs (microgrids, virtual power plants, energy communities/clouds etc.) 276

4.5 The multi-tenant prosumer concept (MTPC) 277

5 Economic growth, productivity gains, structural change 278

6 Conclusions 279

References 281

Further reading 286

1 Introduction

In recent years, the problem of energy poverty has attracted considerable attention and thus regained momentum since it was first identified in the late 1970s (Boardman, 1991; Boardman, 2012; Koh et al., 2012). Likewise, the literature on energy justice has been growing rapidly. In lockstep with this burgeoning amount of literature, the empirical evidence has grown,[a] and attempts have been made to clarify the terms and concepts used in different disciplines (e.g. Heffron and McCauley, 2017; McCauley et al., 2019). So it is time to pause and reflect on some strands of the literature on energy justice and poverty, to discuss the appropriate use of inequality indicators at different scales, and to point out some thematic areas and developments that seem to require more attention.

In this chapter, on the one hand, I want to bring some economic perspectives into the discussion of energy poverty and energy justice. The economics literature on (energy) justice is widely scattered and not that easy to grasp. On the other hand, I will argue that in the context of the transition to sustainable energy systems, much greater system complexities than before will arise, creating new challenges on how to measure energy inequality and energy poverty at various scales and for different social groups. Furthermore, a rapid transition to low-carbon systems in response to anthropogenic global warming necessitates new policies, ways of doing business, consumer behavior and lifestyles etc. It also requires huge investment in new infrastructure and divestment in existing systems (exnovation[b]), raising issues about the future character of energy supply as a public good. In contrast to other, often more

---

[a]Empirical research shows that energy poverty is a serious problem for many countries (for the EU, see e.g., Phimister et al., 2015 for Spain; Lacroix and Chaton, 2015 and Chaton and Lacroix, 2018 for France; for China, India and selected ASEAN states, see the special issue edited by Phoumin Han et al., forthcoming in *Energy Policy* 2019; for a number of countries see the 2018 *Applied Energy* Special Issue "Low Carbon Energy Systems and Energy Justice" and the 2017 *Energy Policy* Special Issue "Exploring the Energy Justice Nexus").

[b]*Exnovation*—the opposite of innovation—is a term coined by John Kimberly (1981) in the context of organizational and managerial innovation. In recent years, it has also been used in the sustainability literature and on energy justice and the energy transition. It stands for processes in the course of eliminating specific technologies, routines, and techniques to bring about energy transitions based on stakeholder evaluations of technology (see, more recently, David, 2017, 2018a,b, respectively).

recent energy justice literature, Jacobson et al. (2005) repeatedly emphasize the relevance of energy infrastructure for a comprehensive justice analysis, although it is not included in their empirical analysis which focuses on aggregate energy and electricity consumption inequalities only.

Global warming has a lot to do with the excessive combustion of fossil fuels for power generation, heating and cooling, and transportation based on the internal combustion engine. Bloomberg New Energy Finance (2016) suggests that in order to limit global warming to 2 °C, investment needs over the next 25 years amount to more than US$200 billion (cited in Heffron et al., 2018). Others suggest the costs for the US alone could run into tens of trillions of dollars (Nersisyan and Wray, 2019). Both the range and the magnitude of estimates presented in the literature are vast. These enormous investment and related financing needs beg at least two questions: (1) Is inequality regarding energy use rising, both in developing and developed countries? (2) How can a just energy transition be achieved (and incentivized)? Evidence shows that access to affordable and safe energy has been growing also in developing nations (Mainali et al., 2014). This is an important part of the energy justice debate but is not the whole story.

Numerous countries throughout the world are aiming to decarbonize their economies and societies. Given that most greenhouse gas (GHG) emissions arise from energy use (in all sectors of the economy, i.e., residential, services, industry, transport, etc.), this raises questions of how such policy targets can be achieved in an economically efficient but at the same time also fair manner. In light of the fact that the majority of GHG emissions today (not historically) are attributable to the developing countries, despite their typically much lower per capita energy use and emissions, and to cities (where per capita energy use levels are higher and half of the world's population lives today), a number of questions are raised regarding how the energy transition can be guided by justice considerations.

Energy services—based on the four principles "availability", "acceptability", "affordability" and "accessibility" (WEC, 2018; Sovacool, 2013; Sovacool et al., 2017; Cherp and Jewell, 2014)—are fundamental for well-being and social welfare. There seems to be widespread consensus that the sustainable energy transition process will, due to digitalization, also lead to more electricity-based energy services, but that the "electric future" may not necessarily be sustainable (cf. IEA, 2019, p. 412). Again, there are important economic dimensions here that need to be brought to light.

Heterogeneity in energy use within a country can often be as large as across nations, so that aggregate national indicators disguise important disparities—among social groups within a country (including the rich versus the poor), between regions, or between the urban and rural population. Spatial patterns of economic development and activity are reflected in heterogeneities regarding the structure, infrastructure and quantities of energy used in different regions. Some of this variability in energy use can be explained by the disparities in income levels across the population. Partly, differences also arise from differences in climatic conditions (heating and cooling needs), the type/s of energy carriers predominantly used and the level of access to these (Grubler et al., 2012).

Depending on the system boundary chosen, energy systems beyond the conventional heating and electricity systems of individual households tend to be large, complex systems. This makes it very challenging to provide support tools for policymakers that are sufficiently comprehensive, undisputed and easily communicated (if that is possible at all in the field of discussions on concerns related to ethics and morality). In view of the need for a sustainable energy transition, calls have become louder to conceptualize and politicize energy justice not only based on (more or less cumulative) energy consumption data (Jacobson et al., 2005), but also across the social life-cycle (Fortier et al., 2019) or the entire energy life-cycle (Healy and Barry, 2017; Heffron et al., 2018).

Economists argue that policies ought to be cost-efficient not just for saving resources in general (opportunity cost principle), but also because then the burden of the social consequences as well as the distribution effects are smaller (Heindl et al., 2015, p. 249). Put differently, there is no reason to believe that inefficient solutions will be just. In that sense, economic efficiency is less an antagonist of distributional justice than one might have thought (every dollar can only be spent once), but nevertheless trade-offs do exist.

In summary, energy justice and other social-science-based research frameworks have begun to engage with questions of justice and fairness that need to be addressed in the transition to a sustainable energy system. This needs to be further informed by an economic perspective, and one that takes the enormous complexities of such a transition and its implications more fully into account.

The remainder of this chapter is organized as follows. Section 2 explores the spatial, temporal and life-cycle dimensions of energy inequality and justice. Section 3 reflects on some terms and concepts, as well as methodological and measurement issues, and it then discusses social welfare considerations as a key element in the economics literature with relevance to the energy justice debate. Section 4 addresses the role of smart energy technologies and systems, and smart metering in particular, as well as clean energy communities, prosumer households and entities such as microgrids and virtual power plants, for energy justice and fuel/energy poverty. Section 5 discusses economic growth, technological progress and structural change as themes that also have a strong influence on energy poverty and justice, due to the ways that energy, labor and capital as the most important input factors of production are combined over longer time periods. Section 6 concludes and suggests some policy implications.

2 The various energy inequality and justice dimensions to be considered

Energy use can vary greatly along a number of dimensions: over time and space; sectorally; intra−/internationally; and intra−/inter-regionally, including by continent. It can vary in terms of quantity but also structurally, in both relative and absolute terms, depending on income and wealth—but also production and consumption activities and lifestyles –, and at least partly depending on climate (heating, cooling) and resource availability (and who has access to which sources) (e.g., Grubler et al., 2012, p. 134f).

2.1 Spatial heterogeneity and trends of (primary) energy use: A global perspective

Fig. 1A and Fig. 1B provide a global overview of primary energy consumption by world region, both on a per capita basis as of 2014 (Fig. 1A), and in absolute terms from 1970 to 2010, with projections to 2040 (Fig. 1B). The per capita range spans from 0.37 t of oil equivalent (toe) in Africa to 5.81 toe in North America. As can be seen in Fig. 1B, there is a fast, ongoing catch-up process among developing nations and world regions compared to rich but more recently quite stagnant OECD country regions.

2.2 Spatial heterogeneity and structural economic change: A regional perspective

Apart from temporal development and country heterogeneity, the subnational level of analysis can also yield important new insights. Weber and Cabras (2017), using a Delphi method with 18 experts, investigate 117 socio-environmental conflicts across Germany arising from extractive activities, energy production, and infrastructure projects. Most conflicts had arisen from the struggle of societal groups to achieve a low-carbon economy but also a more equitable society. The study reveals contradictions of the green economy strategy adopted by the government. The 100% increase in residential electricity prices, on the one hand, has increased "energy poverty especially in densely populated areas, where inhabitants have little possibility to install subsidized wind power or solar energy installations" (Weber and Cabras, 2017, p. 1227). On the other hand, policymakers tried to push large-scale projects such as high voltage power lines or pumped storage hydro power plants, which were unpopular and triggered resistance of various stakeholder groups at the local level. Overall, the government seems to have focused excessively on the increase of renewable energy production, but has created a number of negative effects affecting energy justice, and has failed to reduce energy-related CO_2 emissions significantly.

Spatial heterogeneity of energy use has also been investigated in terms of economic indicators, such as value added and employment, at the regional scale. For instance, Höwer et al. (2019) analyzed the regional heterogeneity of the macroeconomic effects of the German energy transition (*Energiewende*), using a heuristic approach to measure spatial heterogeneity. The proposed heuristic is particularly useful to model effects with high spatial heterogeneity and small overall net impact. Empirical regional studies of energy market impacts are often unfeasible due to a lack of data with sufficient spatial granularity, particularly regarding energy demand and distribution infrastructure. These authors applied the heuristic to results from an input–output analysis on value creation and employment effects by the energy transition in selected manufacturing sectors in the German federal state of North Rhine-Westphalia. The heuristic is applicable for regionalizing other macroeconomic results, for example, from survey-based and other energy modeling studies.

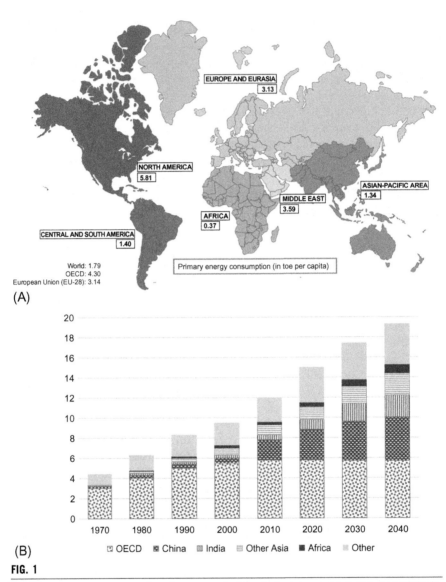

FIG. 1

(A) Primary energy consumption, in tonnes of oil equivalent, by world region, per capita in 2014. (B) Primary energy consumption, in tonnes of oil equivalent, by world region, per capita, trends 1970–2010, with projections until 2040.

From BP, 2015 and 2018. BP Statistical Review of World Energy 2015 and 2018. BP, London. Available at: www.bp.com (retrieved June 28, 2019).

Spatial health inequalities, often caused by energy externalities (e.g. regarding mining or pollutant emission concentrations), are of particular importance in this context. As Bouzarovski and Simcock (2017) argue, "both increase the likelihood of energy poverty emerging in certain communities, and moreover intensify the consequences of the condition. While all energy poverty might be considered as a form of energy justice, the injustice is most *severe* if it is spatially concentrated in localities of relatively poor health." (Bouzarovski and Simcock, 2017, p. 645).

2.3 Temporal income and energy inequality: Let the "*Gini* out of the bottle"

In recent years, a number of inequality measures traditionally used for unequal income distribution have also been applied to energy or electricity consumption, in particular that of private households, and increasingly also to show time dynamics and/or spatial heterogeneity. These include the Lorenz curve and related Gini coefficients (cumulative percentage of income over cumulative percentage of consumption of a certain population), the Theil's H index (decomposition of inequality, taking intra-regional gaps and inter-regional differences into account, enabling the sources of inequality to be investigated), and the Atkinson index (a measure of unfairness of income inequalities based on clearly specified social welfare norms).

In discussions on income inequality within a country or between several countries, one should not only look at the current status, but also the time trend/s. When making an assessment of whether the income inequality level is to be considered as high or low, one can compare the inequality measure before and after taxes and transfers, the difference being a proxy for the regressive or progressive redistribution of the tax system. For the study of energy inequality, one can analogously compare the situation before and after energy taxes and subsidies/levies. The Economist (2019, p. 77) recently published a cross-country comparison of Gini coefficients before and after taxes and transfers, showing not just the large heterogeneity in the Gini coefficients found (based on 2016 or later data), but also that countries vary greatly regarding how much income inequality is being offset. The top five countries in terms of pre-tax inequality are Ireland, Mexico, Chile, Greece and the US; post-tax the top five are Mexico, Chile, Turkey, the US and Lithuania. In terms of redressing the imbalance via taxes and transfers, the Scandinavian countries stand out, which is unsurprising, but likewise do some smaller Western and Central European countries, such as Austria, Belgium, the Czech Republic, France, Ireland and Slovenia (in Ireland, the taxes paid by multinational companies are of high relevance). At the other end of the spectrum are countries like Chile, Mexico and Turkey. But the interpretation of such inequality scores should be made with great caution. Simple, single numbers can of course not do justice to the great complexity of the social and economic systems that lie behind these, and due to the many relevant considerations—including infrastructural ones—it is not straightforward to discuss energy justice on that basis and without a much deeper analysis and interpretation of the results obtained.

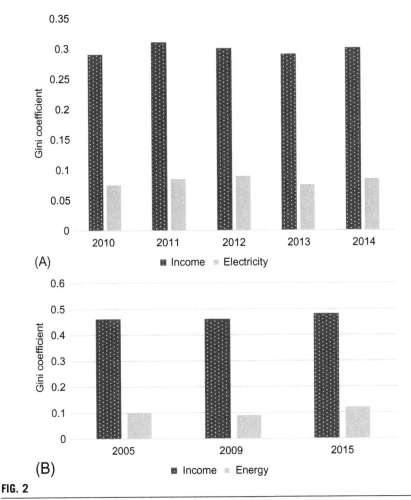

FIG. 2

(A) Development of the income and electricity inequality, measured by Gini coefficients, in Germany, 2010–2014. (B) Development of the income and energy inequality, measured by Gini coefficients, in the US for selected years between 2005 and 2015.

(A): Own calculations, based on SOEP data for Germany; (B): Own calculations, based on SOEP data for Germany and EIA data for the US.

Fig. 2A and Fig. 2B show how the Gini coefficients for income and energy (US), and income and electricity (Germany), respectively, have evolved in recent years in these two major economies. The Gini index ranges from zero to unity; zero implies that income is shared equally, and one implies that one person takes it all. As can be seen, the coefficients were increasing moderately in the US from 0.46 to 0.48 (income) and 0.10 to 0.12 (energy) between 2005 and 2015, whereas for Germany, the income inequality is substantially lower at about 0.3 and stagnant, and the electricity inequality was rising slightly from about 0.07 to 0.08.

Still, identifying income and energy inequalities across populations (also across the two dimensions of time and space) can only be a first step to find out and discuss whether and what kind of policy actions should be adopted to redress the imbalance. Or, as Bouzarovski and Simcock (2017) expressed it in the context of their spatial justice framework, it is "not only about revealing energy-related inequalities, but also *evaluating* them" (p. 645). This raises the issue of responsibility for inequality (how it is produced, and by whom), requiring us to investigate the underlying causes and mechanisms that generate inequality. If allocating responsibilities based on ethical considerations is paired with utility-driven behavior in a complementary manner, then this will yield better results than if they are seen as antagonists. Also, up to now, it is probably fair to say that a clear and systematic responsibility assignment is missing in almost all social-sciences-based energy research, including energy poverty research. This is well in line with economic thinking in general, and utility and welfare optimization in particular, where differences in individuals' choices are desirable whereas factors outside of an individual's control may require policy intervention (a common justification for such intervention being *market failure*—e.g., caused by lack of information, externalities, or abuse of market power).

2.4 (Social) Life-cycle analysis of energy justice

In the course of the ongoing energy transition towards sustainability, with a major goal being decarbonization of the energy system, energy justice evaluations need to be more broadly based than just on (typically national) energy consumption data, but ideally also covering the social life-cycle (Fortier et al., 2019)—i.e., comprehensively addressing energy justice concerns by different stakeholder groups at all life-cycle stages—or covering the entire energy life-cycle ("cradle-to-grave", "well-to-wheel"; cf. Healy and Barry, 2017; Heffron et al., 2018).

Schlör et al. (2015), for instance, investigate the social impact of the German *Energiewende* describing human wellbeing based on the capability approach of Amartya Sen (Sen, 1976, 1985, 1997, 2017; Day et al., 2016). Schlör and colleagues use "social life-cycle analysis" (S-LCA) based on five major criteria (welfare; health and safety; social participation; democracy and freedom; and decent life) plus 24 so-called "impact issues" to analyze the social effects of the production of rare earth elements, i.e., crucial non-renewable (mineral) resources for the *Energiewende*. The use of S-LCA to evaluate energy justice along the life-cycle of products and systems is also suggested by Fortier et al. (2019), who propose a number of indicators for doing so. They point out that the UNEP framework deliberately allows for flexibility in designing the assessment and setting the indicators that determine social impact. However, a comprehensive approach such as this involves many complex and interacting issues, especially if it takes the concerns of different stakeholders into account. This makes it difficult if not impossible to make credible comparisons between different scenarios. It is likely, therefore, that multi-criteria decision analysis would be needed for explicitly evaluating multiple conflicting criteria in decision-making.

The economic dimensions of this must be addressed in tandem with questions of justice, inequality and sustainability. However, such a procedure for studying the social sustainability of energy systems in a more comprehensive way than usual, and one that also takes into account the concerns of different stakeholder groups at all life-cycle stages, will be challenging. Given that justice is inherently complex and contested, and "will inevitably mean different things in different settings, cultures and political arenas" (Jenkins, 2018), energy justice will probably become a struggling and floundering concept despite hopes that is easier to use than the (somewhat broader) notions of "environmental justice" and "climate justice".

2.5 Heterogeneity of energy rebound and sufficiency

Energy rebound[c]—the "Achilles heel of energy efficiency" (the author)—has been discussed more recently also in the context of poverty and sufficiency (Herring, 2009; Sorrell et al., 2018). Madlener and Hauertmann (2011) have shown, based on a representative panel of 11,000 households in Germany, that energy rebound among non-homeowners is higher for low-income households, at 49%, compared to 31% for wealthier households, thus challenging the idea of tackling rebound where it is largest. Note that rebound also tends to be larger in developing countries than in the developed world, as there are more unsaturated needs (see, e.g., van den Bergh, 2011). Chan and Gillingham (2015) have thus argued that energy rebound (and energy efficiency as its counterpart), which enhances social welfare should not necessarily be minimized, i.e., that it is not bad per se (which it is from a resource consumption and environmental degradation perspective, but not necessarily overall, taking all utility gains and externalities from it into account).

Sorrell et al. (2018) investigate rebound effects arising in the context of sufficiency, combining insights gained from economics and social psychology. They point out that energy rebound may raise or diminish social welfare, depending on the situation (see Chan and Gillingham, 2015) and that energy sufficiency actions are themselves associated with indirect rebound effects. They further discuss a more comprehensive approach to energy sufficiency being "voluntary downshifting", i.e., reducing household income deliberately by either moving to a lower-paid job or reducing working hours. Such a voluntary downshifting would not only affect fuel poverty measures in different magnitudes, but also shift expenditure and time-of-use patterns in complex ways, which may lead to a less-than-proportionate reduction in energy use (and related emissions). Likewise, widespread downshifting will reduce energy prices, which in turn would incentivize other groups to increase their energy consumption, and may lead to unintended and undesired effects.

Oberst et al. (2019) analyze for the case of German households whether residential energy prosumers—i.e., producer-consumers with partial self-consumption of

---

[c]Rebound denotes the phenomenon that energy efficiency increases could lead to behavioral responses that reduce, eliminate or even overcompensate the energy savings expected from the efficiency improvement in the absence of such responses.

distributed energy production—differ in a statistically significant way from normal consumer households in terms of housing characteristics and socio-demographics (typically, these prosumers are homeowners and thus can be expected to be wealthier than the average population). They set out to study the effect of residential energy prosuming on households' energy consumption behavior but could not find empirical evidence of a "prosumer rebound effect" on energy consumption.

3 Broadening our understanding of energy poverty: The economics perspective

In this section I shed some light on terminology, measures and unanswered questions, after which I reflect on the concepts of social and economic welfare and how issues of economic efficiency and justness typically tend to be addressed in the economics literature.

3.1 Some reflections on terminology and the energy/fuel poverty and energy justice debate

The two terms "energy poverty" and "fuel poverty" frequently have some overlap, are used interchangeably or are confused in the literature (Li et al., 2014). Energy poverty has often been defined as the lack of access to modern energy services. The notion of fuel poverty was coined in the UK in the 1970s. Lewis (1982) defined it as the inability to afford adequate warmth in the home (see also Walker and Day, 2012), Boardman (1991) defined fuel poverty as having to spend 10% or more of income on total household fuel costs to achieve a satisfactory indoor temperature regime (the so-called "ten percent rule", TPR). Hills (2012) later defined fuel poverty by "low income high costs (LIHC)". Here, a household is considered fuel-poor if it earns <60% of the median income and would have to spend more than median household expenditure on fuel.

Electricity in particular is seen as crucial for education, health, jobs and sustainable development overall (see also Birol, 2007). Household electrification has been shown to significantly reduce energy poverty and, consequently, to raise energy equity (Pereira et al., 2011). Birol (2007) sees three major challenges for the global energy system: (1) a growing risk of disruptions to energy supply; (2) the threat of environmental damage caused by energy production and use; and (3) persistent energy poverty. He argues that the first two of these challenges have attracted considerable attention among energy economists, whereas "energy under-development" has seen much less interest. Modern energy services, he argues, help to meet the basic needs of humans, such as food and shelter, which therefore must be at the heart of any strategy to alleviate poverty.

In recent years it has been claimed that energy justice—with its three dimensions *distribution*, *procedure* and *recognition* (e.g., Williams and Doyon, 2019; Heffron and McCauley, 2017, see Fig. 1)—is indeed relevant for any energy transition that is supposed to be sustainable (an unjust transition is not sustainable, cf. Williams and Doyon, 2019; Heffron and McCauley, 2017, p. 665). Justice in energy transitions has been reviewed lately by Williams and Doyon (2019). They state that energy justice "has emerged as an agenda to include more social science related disciplines within energy research" (p. 147). They also point out that until recently, with the introduction of the notion of "spatial energy justice" (Bouzarovski and Simcock, 2017), spatial aspects have been largely lacking in the discussion on energy justice—an issue I addressed in Section 2.2 and return to below.

Sovacool and Dworkin (2014) define "universal energy justice"—in contrast to particular forms of energy justice—as a "global energy system that fairly disseminates both the benefits and costs of energy services, and one that has representative and impartial energy decision-making" (p. 436; see also LaBelle, 2017). This definition implies universal applicability and burden sharing among a global citizenry. Both concepts are embedded in political and economic systems, and therefore also in institutional routines. Sovacool and Dworkin (2014) list eight elements that determine the pursuit of universal energy justice—quasi a universal checklist for energy justice. It is also important to note the tension that can arise between local and global-universal aspects of energy justice (Galvin, 2019). Energy-related developments on the local scale almost invariably have global, universal repercussions—such as when energy-poor communities in developing countries increase their energy consumption.

Alvial-Palavicino and Ureta (2017) describe an attempt made in Chile to "economize" energy justice by incorporating equity claims into tariffs (so-called "tariff equity law"). They elaborate on how perceived shortcomings in equity and social justice were turned into "market failures" that can be cured by market-based mechanisms. The authors point out that this is too simplistic and show the dangers inherent in uncritically economizing claims to improve energy justice. Ethical concerns regarding the risks and benefits of energy production and supply are reduced to redistributions among consumers, thus limiting structural reforms.

As noted above, a further dimension of energy justice is its spatial component. Bouzarovski and Simcock (2017) argue that spatial justice (in the sense of geographical inequality and inequity) is required in order to provide an explicit spatial focus to research (see also Section 2.2), allowing us to unearth cross-sectoral and entire-energy-chain injustices that lead to elevated energy poverty risks which are not accounted for in analyses focusing only on inequalities between social demographic groups. Based on an extensive literature review covering 126 scientific papers on energy poverty and justice, predominantly focusing on Europe, they distinguish between four distinct mechanisms that operate at multiple scales: (1) *landscapes of material deprivation* (domestic energy deprivation is unevenly distributed in space); (2) *geographic underpinnings of energy affordability* (domestic energy deprivation is strongly influenced by energy prices and household incomes); (3) *vicious cycles of vulnerability* (wider material and economic inequalities also indirectly contribute to spatial inequalities in how energy is demanded, consumed and experienced); and (4) *spaces of misrecognition* (justice as recognition concerning the respect—or lack of respect—given to different identities in social, cultural, and political relations). Note that (1)–(3) are distributive spatial inequalities, whereas

(4) refers to recognition injustice (or "misrecognition"). Bouzarovski and Simcock claim that spatial energy justice research "goes deeper to highlight how domestic energy deprivation is fundamentally intertwined with, and produced through, geographical inequities and flows that are engrained in the economic, infrastructural and cultural make-up of society" (Bouzarovski and Simcock, 2017, p. 640). In that sense, a spatially sensitive energy justice-based policy approach would be capable of recognizing that some geographical areas are more vulnerable than others in terms of energy and fuel poverty (Bouzarovski and Simcock, 2017, p. 641). They also point out that visibly marking out people or places as deprived or vulnerable can be stigmatizing. For example, "area-based" energy efficiency schemes can potentially stigmatize entire neighborhoods and the people living there. Also, energy poverty might be particularly stigmatizing in countries such as the US, where relatively high levels of energy consumption are often considered as "normal", or in Scandinavian countries, where having a warm and cosy home is highly valued (Bouzarovski and Simcock, 2017, p. 644; referring to Sovacool, 2009 and Wilhite et al., 1996; among others).

Sovacool et al. (2017) propose a decision-making framework for fostering energy justice, taking an anthropogenic view of justice excluding other species—in contrast to the environmental justice literature (cf. Williams and Doyon, 2019, p. 147). Most economists—except maybe ecological economists—would probably welcome this narrowing down to humans, simply for the fact that economic/nonmarket evaluations are often carried out using willingness-to-pay or willingness-to-accept concepts, which are hard to apply to animals except maybe those which can participate in experiments (with non-monetary incentives and situations). Nevertheless, the interdependencies of species, including humans, in the planetary ecology cannot be ignored (Thuiller et al., 2011). As the latest World Biodiversity Council report on biodiversity shows, humans are currently reducing the number of animal and plant species at an astonishing rate (IPBES, 2019), and so "Anthropocene" seems to be an appropriate term for the era we have entered.

An *economic dimension* can be integrated with many issues and concepts related to the social and technical dimensions of energy justice. An economic dimension would include issues such as discounting (intertemporal preferences, individual and social time value of money), economic welfare, and intergenerational equity—in terms of cost–benefit considerations and analysis (see also Sovacool et al., 2017, p. 688).

3.2 Measurement issues

Energy justice measures, such as the "Energy Trilemma Index"[d] of the World Energy Council (WEC, 2018), the per capita total primary energy supply (TPES) shares by country (IEA, 2019), the "Energy Justice Metric" and use of Ternary plots (Heffron

---

[d]The three core dimensions of the WEC's definition of energy sustainability are: energy security, energy equity and environmental sustainability.

et al., 2018), or attempts to use Lorenz curves for showing population share/energy (or electricity) versus income inequalities (e.g., Jacobson et al., 2005; Wu et al., 2017) help to point out certain inequalities, but cannot do justice to the enormous complexity of the matter at hand. Wu et al. (2017) show that actual energy consumption-based inequality is systematically different from inequality measurement using household income or expenditure data, and argue that the expansion of energy infrastructure does not go hand in hand with the mitigation of energy inequality, and that energy affordability ought to be improved by income growth and targeted "safety-net programs", instead of energy subsidies. Specifically, measuring energy consumption for specific devices (e.g., space heating and/or cooking devices), they find that expenditure-based inequality measurement reveals greater disparity than the measurement of inequality based on physical quantities. The bottom line is that (energy) policymakers need to tackle deprivation or undesirable levels of income inequality directly, as the optimal outcome will not come automatically when the (energy) infrastructure is updated.

Zheng (1997) summarizes what is relevant for the definition of aggregate poverty measures, and that these, in line with what Sen argued, ought to satisfy clearly specified properties (or axioms), along which the desirability or usefulness of a poverty measure should then be evaluated. Zheng reviews the extensive literature on aggregate poverty measures, also including several useful indicators not covered by previous surveys. He emphasizes that poverty research and discussions, in order to minimize controversies about results, always need to address how poverty is perceived (*identification* of poverty) and then how it is measured (*aggregation* of poverty). As he puts it, "... the answers to many 'seemingly' simple questions may not be readily available and are far from obvious." (p. 124).

Apart from the type of data to be used in measuring energy inequalities, the problem of choosing an appropriate measure also arises. As explained in Chapter 1 of this book, different measures have been employed to quantify inequality, such as the Lorenz curve (or 3D Lorenz surface for following the development over time) and Gini coefficients, Theil's H, or the Atkinson index (see also Section 2.3). These have also been used to measure inequalities in energy consumption. While they may be useful for getting an impression of heterogeneity, they often focus only on sum totals of energy consumption, and in no way can do justice to its underlying complexity. Others have used (ratios of) population percentiles, deciles, quintiles or quartiles as measures of heterogeneity.

Aside from measuring *energy inequality*, yet another complication comes from the right choice of *fuel poverty* measure. As Hills (2012) nicely summarizes, there are at least seven different measures of fuel poverty that have been suggested in the literature, with very different properties. There is the *10% poverty line* (expenditure on energy services $\geq$10% of income), the *2 $\times$ median energy expenditure* (expenditure on energy services $\geq 2 \times$ the median expenditure), the *2 $\times$ median share* (share of energy expenditures relative to income $\geq 2 \times$ the median share of the expenditures in the sample), *2 $\times$ average energy expenditures* (expenditures on energy services $\geq 2 \times$ the average expenditures), *2 × average share* (share of energy expenditures relative to income $\geq 2 \times$ mean share of expenditures in the sample), the *MIS (minimum income standards) based poverty line* (residual income after expenditure on energy services and housing costs $\leq$ the MIS, after housing costs and expenditure on energy services), and the high cost/low income or *HCLI criterion* (households that spend more than the median on all energy services and fall below the poverty line of 60% of median income after expenditures on energy services have been subtracted from income). All of these measures will give different figures with regard to how many people need to be considered as fuel-poor. Nevertheless, such figures (in isolation or in combination) can at least be used to assess whether there is a significant (and across different measures robust) problem of fuel poverty. In this respect, fuel poverty measures can also be seen as an important additional criterion for developing energy supply systems further. In other words, in economic analysis, cost or profit considerations on the one hand, and productivity on the other, are not the only relevant criteria in economic assessments of desirability. Further, there might be important trade-offs to deal with between the two, which could be evaluated by participatory stakeholder analysis (and, say, preference elicitation by means of multi-criteria evaluation and multi-criteria decision analysis, respectively).

For the case of Germany, which has been for years one of the leading countries aiming for a sustainable energy transition, an illustration is provided. Fig. 3 shows the relation between mean income and mean energy expenditures for different types (groups) of private households in 2011, compared to the groups of "fuel poor" based on different poverty measures. Table 1 shows a detailed account of the households potentially subject to fuel poverty (in percentage terms). The share of those households that are potentially subject to fuel poverty amounts to 25.1% when applying the 10% rule (TPR) as the poverty line. In contrast, the "2x median expenditure" and "2x mean expenditure" measures yield only comparatively low shares of 4.6% and 2.9%, respectively (obviously failing to capture poorer households). In comparison, the "2x median share poverty" and "2x mean expenditure share poverty" measures yield values of 11.2% and 4.9%, respectively, whereas the MIS measure points to 8.8% of households being subject to fuel poverty and the HCLI BHC measure 10.5% (the income poverty line is here at €952). Finally, the HCLI AHC measure identifies 12.6% as fuel poor.

Heindl and Schüssler (2015) discuss various problems with measures of affordability and fuel poverty, including dynamic properties for both types of measures. They point out that the literature on affordability measures has little overlap with that on poverty measurement. By means of a simulation, they find odd dynamic behavior of some measures, which policymakers should be aware of in order to avoid erroneous policy recommendations derived from these measures, several of which are used in practice. These authors recommend the adoption of two check-up measures, the position invariant burdening (PIB) and the impoverishment (IMP) measure. PIB requires that the values from a measure of affordability do not decline when the actual expenditure on the respective good/s increases in society in the absence of

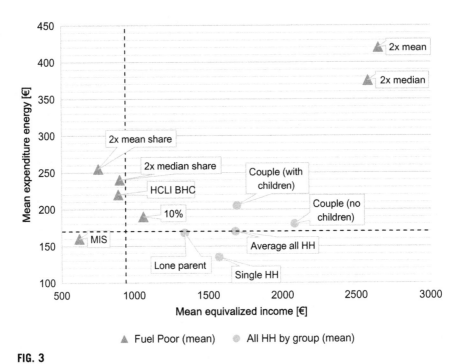

FIG. 3

Mean of equivalized income and all expenditures on energy services, 2011, data sample of German households and several subgroups; within the subgroup "fuel poor", evaluated based on different fuel poverty thresholds (further values given in Table 1): corresponding percentage shares of fuel poor for different poverty measures.

Based on data from Heindl, P., 2013. Measuring Fuel Poverty: General Considerations and Application to German Household Data, ZEW Discussion Paper Series No. 13-046. ZEW Mannheim, p. 19.

a change in income and individual positions. It ensures that any up- or downward shift of an income or expenditure distribution will change the affordability measure "in the right direction", as otherwise the number of people falling below a chosen poverty threshold will be misrepresented (e.g., an increase or decrease in energy price has an effect on the position of the households in an expenditure distribution, but will also shift the distribution itself). IMP requires that it must be possible to push a household below the poverty line by decreasing the household's income (cf. with what has been said above on "energy sufficiency and rebound"). Put differently, shrinking real income can render goods unaffordable, a fact that measures—and energy policies based on such measures—should take care of. Heindl and Schuessler recommend the use of either a (truncated) TPR measure because of its simplicity and ease of use, or a measure they dubbed "LIHCt"—an HCLI measure which includes TPR as a first condition instead of the condition "expenditure above median" and which satisfies both PIB and IMP.

Table 1 Percentages of fuel poor households of different household types, based on different fuel poverty measures, for Germany

(in %)	Single HH	Couple, no children	Lone parent	Couple, with children	Other HH
10% NE BHC	41.3 (42.7)	18.1 (20.3)	38.5 (39.8)	12.1 (14.6)	25.7 (28.2)
2 × median	1.6 (1.6)	4.6 (5.0)	5.4 (4.5)	7.9 (7.8)	7.9 (7.4)
2 × median share of exp.	8.3 (8.9)	8.3 (9.5)	20.5 (22.4)	16.0 (18.5)	17.8 (20.3)
2 × mean	0.8 (0.8)	2.7 (2.9)	3.8 (3.2)	4.9 (4.4)	7.9 (7.4)
2 × mean share of exp.	3.4 (3.7)	3.5 (4.4)	10.8 (12.0)	7.2 (8.5)	5.3 (5.4)
MIS	12.6 (12.3)	4.7 (5.5)	18.4 (20.6)	7.1 (8.1)	16.4 (17.0)
HCLI EI BHC	7.4 (7.8)	8.1 (9.0)	20.7 (24.4)	14.6 (16.6)	16.4 (17.2)
HCLI EI AHC	9.6 (10.7)	8.8 (9.8)	25.2 (29.4)	17.8 (20.3)	19.1 (19.8)

Weighted according to population share in parentheses.
BHC, *before housing costs;* AHC, *after housing costs;* MIS, *minimum income standards;* HCLI, *households with high fuel costs and low income;* EI, *equivalized income;* NE, *non-equivalized.*
Based on data from Heindl, P., 2013. Measuring Fuel Poverty: General Considerations and Application to German Household Data, ZEW Discussion Paper Series No. 13-046. ZEW Mannheim, p. 21; see also Heindl, P., 2015. Measuring fuel poverty: general considerations and application to German household data. FinanzArchiv 71, pp. 178–215.

3.3 Taxation, transfers, and subsidies: How to best re-balance the level-playing field?

The OECD recently published a report on the distributional effects of (excise) energy taxation for 21 OECD countries (Flues and Thomas, 2015). It shows that, overall, taxes on mobility tend to be *progressive* (i.e., in favor of poorer households), whereas taxes on heat are *neutral* (or proportional) and on electricity *regressive* (i.e., in favor of wealthier households) (cf. Fig. 4). Also, larger households were found to spend a higher share of their income and expenditure on energy taxes (especially transport fuel taxes) than smaller ones; rural households spend, on average, higher shares of income and expenditure (on all energy taxes) than households in urban areas; and elderly households (household head >60 years) spend relatively less of their income and expenditure than younger households. However, it should be noted that distributional effects, especially of transport fuel taxes, differ across countries—in that they are in some countries regressive, in others neutral or even progressive—and that taxes on transport fuels are found to be more progressive when measured against an expenditure rather than an income measure. Note further that car usage depends on income and the availability of public transport. So if poorer households are indeed less likely to own a car, taxes on transport fuels tend to be more progressive, while a lack of public transport will lead to more regressive effects of transport fuel taxation (Flues and Thomas, 2015; on this issue see also Sterner, 2012). Apart from taxes, transfers (levies, fees etc.) on energy services also play a role.

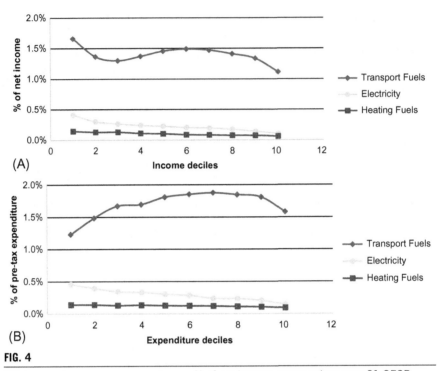

FIG. 4

(A) Average taxes on energy carriers as % of net income, mean values over 21 OECD countries. (B) Average taxes on energy carriers as % pre-tax expenditure (right plot), mean values over 21 OECD countries.

Author's plots from Flues, F., Thomas, A., 2015. The Distributional Effects of Energy Taxes, OECD Taxation Working Papers No. 23. OECD, Paris, p. 8.

In Germany, the EEG-Umlage (see also footnote e)—a levy imposed on non-exempted electricity consumers aimed at financing the feed-in tariffs paid to renewable energy providers—has been investigated regarding its distributional impact (e.g., Grösche and Schröder, 2013; Heindl, 2014; Bach et al., 2018). The levy has risen substantially in recent years, reaching a total annual volume of more than €20 bn overall, or 6.88 €-ct/kWh of electricity consumed in 2017, and slightly declining since (see Fig. 5). Electricity prices for households in Germany are among the highest in Europe and the world. Given that electricity cannot as easily be substituted as other fuels, and given that energy is an essential good for satisfying the basic needs, it is obvious that poorer households spend a relatively higher share of their income on energy than richer households (see also Chapter 9).

Figs. 6 and 7 show how the residential electricity price in Germany (today at around 30 €-ct/kWh) is composed, and how the components' shares have changed over time. As shown in Fig. 6, electricity distribution costs in 2018 accounted for 21.1% of the electricity price charged to a typical German household, network

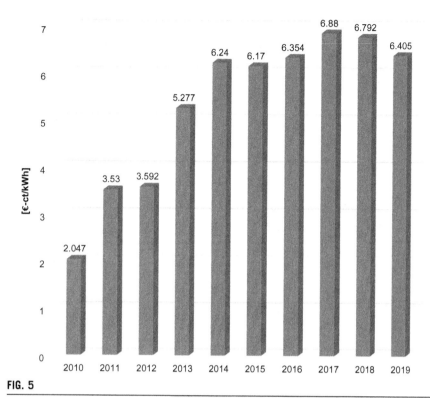

FIG. 5

Development of the German EEG Levy, i.e., the surcharge on electricity use (€-ct per kWh) to finance the cost of feed-in tariffs for producers of renewable energy.

From BMWE, 2018. EEG in Zahlen: Vergütungen, Differenzkosten und EEG-Umlage 2000 bis 2019. Bundesministerium für Wirtschaft und Energie, Berlin. Available at: https://www.erneuerbare-energien.de/EE/ Redaktion/DE/Downloads/eeg-in-zahlen-pdf.pdf%3F__blob%3DpublicationFile (retrieved June 23, 2019).

charges about 24.7%, and taxes, duties and transfers 54.2%. Fig. 7 shows that over the last 12 years the wholesale electricity price increased by 26%, grid charges by 5%, and taxes, duties and transfers by an astonishing 110%, leading to a 51% price increase overall.

Jenkins et al. (2016) build on this example, stating that "…[t]he resultant financial burden on lower income communities, who have to pay a relatively higher share of their total income for their energy costs and the additional EEG-Umlage, raises concerns of distributional justice." (Jenkins et al., 2016, p. 176). They state further that "an adapted electricity grid will be necessary to support the transport of decentralized and naturally volatile electricity supplies. Much of the electricity produced from renewable sources comes from wind turbines in the northern regions. Yet typically, the energy-intensive industries are located in the south. Such an imbalance in supply and demand across the grid requires a complex and expensive process of rebalancing […], including extensive networks of new transmission lines. In this regard,

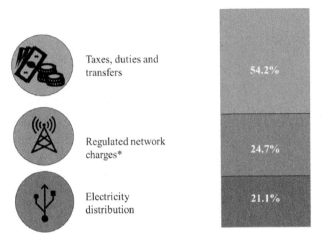

*Average net network charge, including charges for measurement and measuring point operation, may vary significantly from region to region

FIG. 6

Average composition of the electricity price for a household in Germany, 2018.

From BDEW, 2018. Bundesverband der Energie- und Wasserwirtschaft. https://www.bdew.de/.

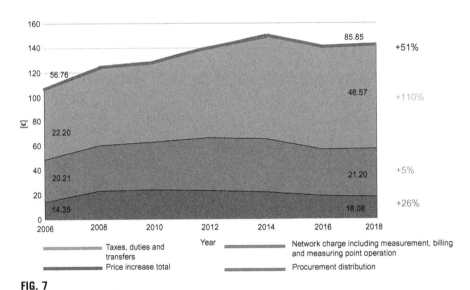

FIG. 7

Change in composition of monthly average household electricity costs in Germany from 2006 to 2018.

From BDEW, 2018. Bundesverband der Energie- und Wasserwirtschaft. https://www.bdew.de/.

distributional justice also manifests as the physical siting of energy infrastructure." (Jenkins et al., 2016). This is another argument that energy considerations alone are clearly insufficient to cover the range of energy justice issues, and that infrastructure also needs to be taken into account, both in terms of physical and financial impacts. In recent years, for instance, the German federal state of Bavaria has blocked both the development of wind farms close to settlements (BGBL, 2014; see also Galvin, 2018) as well as new power lines (both due to public opposition). Jenkins et al. (2016) use Bavaria's resistance against new power lines as an example, and argue that southern regions may have to pay a premium for their electricity (due to the resulting grid congestion), which could further aggravate the disproportional financial burden of the *Energiewende* on lower income clusters. In other words, this introduces a regional dimension of distributional injustice in the southern states of the country, and reminds us nicely about the philosopher David Hume's thought that justice has to do with the distribution of benefits as well as burdens (cited in Jenkins et al., 2016, p. 176). Note that the argument brought forward by Jenkins et al. seems somewhat flawed in light of the fact that the southern states are wealthier than most others (Statista, 2019) and that Bavarians enjoy the benefit of a landscape less spoilt by power lines as well as lower adverse impacts in terms of noise and flickering from wind turbines on nearby settlements. Also, federal states such as North Rhine-Westphalia pay more EEG levies than they receive from the scheme in terms of feed-in tariffs, whereas states like Bavaria receive more than they pay due to their many EEG-subsidized installations, especially solar photovoltaic (PV) systems. Finally, on the federal level, it has been questioned whether it is just to exclude >2000 enterprises in Germany from paying the levy (on the grounds of maintaining international competitiveness), as the burden then has to be carried on fewer shoulders, mainly by private households and small- and medium-sized enterprises.

The funding of low-carbon infrastructure through less regressive means than carbon taxes or flat levies on energy bills, as well as reconfiguring the energy transmission infrastructure and regulation and shifting away from increasingly expensive and centralized fossil fuel plants onto more localized and distributed forms of renewables-based micro generation, have been suggested as having the potential to reduce energy poverty (Hiteva, 2013; Sovacool et al., 2014; cited in Bouzarovski and Simcock, 2017, p. 646). Here, empirical research is desperately needed.

Fig. 8 shows that the lower income groups in Germany are hit harder by an electricity price increase. Household electricity prices in Germany, as noted above, are among the highest in Europe and the world. A substantial part of this serves the funding of the *Energiewende* in the form of guaranteed feed-in tariffs for electricity from renewable energies but also combined heat-and-power generation.[e]

---

[e]The German Act on Granting Priority to Renewable Energy Sources (*Erneuerbare Energien Gesetz, EEG*) stipulates a levy that has risen to about 7 €-ct/kWh, and is expected to go down only slowly in the coming years. Currently, a debate has started regarding the introduction of a CO_2 tax, and on the expected distributional consequences of this.

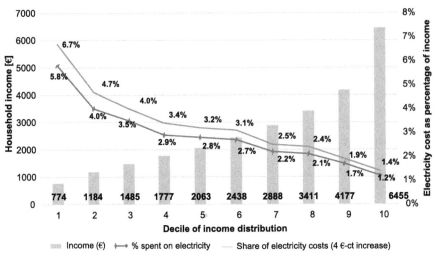

FIG. 8

Income distribution in Germany in 2012 and share of electricity costs in real disposable income, per decile (before/after an assumed electricity price increase of 4 €-ct/kWh).

Based on data from Heindl, P., 2014. Ökonomische Aspekte der Lastenverteilung in der Umweltpolitik am Beispiel der Energiewende: Ein Beitrag zum inderdisziplinären Dialog, ZEW Discussion Paper No. 14-061, ZEW Mannheim, p. 15.

3.4 Social/economic welfare considerations

Social welfare or economic welfare analysis[f]—typically based on the principle of Pareto optimality—is an important pillar of research in economics. The Swedish economists Knut Wicksell (1851–1926) and Eric Lindahl (1891–1960) pioneered new ways of thinking in the social sciences by merging politics and economics. They viewed social welfare as the utilitarian sum, with decreasing marginal utility of income, and "advocate some forms of redistribution aimed at the social optimum" (Silvestre, 2003, p. 531). Wicksell refused (real or perceived) claims that, in the case of private goods, competitive markets will lead to a social optimum. Wicksell and Lindahl envisaged a "stratified society" in which parties are naturally associated with social groups, and for each party the elected representatives are faithful agents of their constituents. In contrast, they saw the *median-voter paradigm* critically, because under this paradigm society is a unidimensional continuum of diverse people

---

[f]See also Hicks (1939). The *economic welfare* of a nation or business can often be assessed by reviewing the level of employment and the average financial compensation that is received by workers employed within the system (Business Dictionary, 2019a). *Social welfare* is not the same as standard of living but is more concerned with the quality of life that includes factors such as the quality of the environment (air, soil, water), level of crime, extent of drug abuse, availability of essential social services, as well as religious and spiritual aspects of life—the overall level of financial satisfaction and prosperity experienced by participants in an economic system (Business Dictionary, 2019b).

voting for the most attractive candidate in a winner-takes-all contest. Candidates only care about getting elected, and all end up representing the same voter, i.e., the median one (ibid.).

Based on ideas of Mill (1848), Wicksell criticized the argument that a government should take up every measure that has a positive social net benefit, referring especially to the law of diminishing marginal benefit. In his view, a redistribution from the rich to the poor—and thus egalitarian distribution of income and wealth—might not be desirable ("communism at its worst"). The British economist Arthur Cecil Pigou (1877–1959) used a similar argument, pointing out that an equal (ized) distribution, in light of heterogeneous income-utility functions), cannot be the optimal outcome. Lindahl (1967) also emphasized that there is the problem of non-homogenous distribution of political power, which might lead to undesirable distributions in the provision of public goods and their financing. This may happen by blurring the costs and benefits of providing public goods, or by openly using the political and societal dominance for achieving higher (relative) burdens of the less privileged society (see Heindl, 2014).

Silvestre (2003) summarizes the seminal ideas of Wicksell (1896 [1958]) and Lindahl (1958 [1919]) on public goods. Both proposed normatively ideal solutions for public action that are based on respect for individual interests and elimination of power abuses. They distinguished between *social justice* (guided by utilitarian principles applied to the distribution of wealth) and *economic justice* (guided by the *benefit principle*, applied to public decisions on what services to supply, and how to finance them) (Silvestre, 2003, p. 528). For both men, economic justice required unanimous agreement between political representatives, leading to what has become known as *economic efficiency*. Both viewed power asymmetries as the main dangers for economic justice, whereas they were much less concerned about principal-agent problems, free riding, and the size of the government. In their view, the policy space is inherently multi-dimensional, because the supply of public services must be simultaneously determined along with the distribution of its costs. In this respect, it would be useful to discuss the financial burden sharing of public investments or subsidies for propelling the sustainable energy transition (e.g., charging infrastructure for electric or hydrogen-fueled vehicles, extension of the power grid) beforehand and in a transparent manner.

The final state depends on a reference state and public decisions, and is characterized by resulting individual and social welfare levels. But, instead of adopting some sort of "welfarist" approach and to normatively evaluate the final state, Wicksell and Lindahl propose attaching different normative notions to the reference state and public decisions. Specifically, the reference state is to be evaluated based on *social justice* terms and the *utilitarian principle*, whereas the public decisions need to be normatively evaluated by a new criterion based on the *benefit principle* (political and economic justice, called *Gerechtigkeit*).

Fortier et al. (2019) introduce a concept for evaluating energy justice across the life cycle that is based on four stakeholder groups: workers, electricity consumers, local communities, and society as a whole. They claim that a holistic assessment across multiple interconnected activities involved in an energy system is needed, and that a set of methods can be applied for the planning of future energy systems maximizing justice across the life-cycles of these systems. "[D]ifferences in impacts between individual categories of stakeholders can be identified, and variations in equity and justice at multiple points in the life cycle of the same energy system can be determined." (p. 212). For several examples they illustrate that when applying this concept, potential wealth (re-)distribution problems might arise, creating winners and losers. For instance, coal miners might lose their jobs in the course of the sustainable energy transition, but new jobs might be created locally and elsewhere, and the local community might enjoy health gains from lower pollutant emissions, and society overall benefit from greenhouse gas mitigation. The net welfare gains of energy policy measures will thus depend strongly on the system boundary chosen. Of course, by the same logic, poor work conditions and negative health implications of cobalt and lithium mines extracting material inputs for renewable energy technologies would also have to be taken into consideration (see also Section 2.4).

Heffron and McCauley (2017)—in their reflections on the use of the energy justice concept across the disciplines—have made an attempt to examine the relationship between energy justice and economics (p. 664), and also how this can be transferred directly into practice by assisting policy-makers in balancing competing aims of the energy trilemma. As a conceptual framework, they suggest a three-phase approach, comprising (1) the core three tenets of energy justice (i.e., distributional, procedural, and recognition justice), (2) an energy life-cycle system analysis, and (3) the application of principles for the practice of energy justice (e.g., based on affordability, availability, sustainability, transparency, responsibility, among others). They criticize further the way, and the history of how, the World Energy Council (WEC)—as "an economics-led institution"—has filled the "energy trilemma" with a much too narrow view of economics when dealing with the notions of energy justice and social justice, putting "affordability" and "accessibility" at the forefront. These authors argue that over-emphasizing cost efficiency has led to a retention of the status quo in the energy system, and a continued reliance on fossil fuels in the short term, at the expense of a focus on the long-term development of low-carbon energy infrastructure and a low-carbon economy, respectively. Much of their argument, however, seems to be based on the insufficient internalization of externalities of energy use (which, however, most economists would also see as an important prerequisite for efficient market outcomes).

Rehner and McCauley (2016) argue that in Germany, energy security of supply (with the four components "availability", "accessibility", "affordability" and "acceptability" of energy) was over-prioritized at the expense of environmental justice (with the two components "social acceptance" and "environmental compatibility"). Instead of the classical energy policy triangle (the "trilemma") they see a dichotomy between prioritizing either energy security or environmental justice, and that policy actors, inspite of their best intentions, often follow either a security or a justice direction but not both.

4 Smart systems: Efficiency, participation and equity considerations

4.1 Smart grids as enabling technologies

Upgrading the distribution grids for electricity is an important factor for creating the flexibility needed to absorb higher shares of variable renewable energy sources (VRES), to better control the power flows, and to better link the pool of flexible resources available (IEA, 2019 p. 307f). Investments in smart grid technologies are substantial (2017: US$33 bn worldwide; cf. IEA, 2019, p. 308), but can help reducing outages and enable the introduction of new business models based on distributed energy resources, so that also poor or deprived households might benefit. In recent years, literature has grown exponentially on the adoption of smart grid technology and how it shapes society (e.g., Luthra et al., 2014; Broman Toft et al., 2014; de A. Dantas et al., 2018; Cardenas et al., 2014; Buchmann, 2017; Faruqui et al., 2010; Tuballa and Abundo, 2016; Colak et al., 2016; Zafar et al., 2018; Ellaban and Abu-Rub, 2016). Up to now, however, few of these seem to be explicitly engaging with fuel/energy poverty or justice issues, so the research needs on the energy poverty and justice impacts of smart grid technologies are still substantial.

4.2 Smart meters and real-time pricing (RTP)

In energy economics, dynamic pricing has often been proposed to incentivize economic efficiency by providing more detailed and more frequent information on energy consumption patterns, and a better signaling of the temporal scarcity of energy supply. However, the better allocation of scarce resources enabled by smart technology and tariffs that better reflect scarcity may or may not correlate positively with the goal of mitigating fuel poverty and energy injustice. Poorer households are often found to be less flexible in reacting to price signals, and need to spend a relatively larger share of their income to satisfy their basic energy needs. Hence smoothing the load curve by dynamic pricing might hit them harder than wealthier households. Further, whether such tariff schemes lead to average energy price increases or not, and whether they are regressive or progressive, depends on their specific design.

4.3 Sustainable energy communities

The development of "clean energy communities" or "citizen energy communities" as important elements in the transformation of the hitherto hierarchical and largely centralized energy system towards a more distributed and decentralized one is an interesting perspective in itself, and also for considering its expected impact on energy poverty and energy justice. "Citizen energy communities" come in a variety of forms—such as energy cooperatives, micro-grids, peer-to-peer trading—and so, depending on the evolution of how energy supply is being organized, energy transition pathways in different countries are expected to vary as well. In particular, they will differ depending on the co-evolution of energy systems, and how economies and societies along the low-carbon transition pathways are driven by social, technological, organizational/institutional and economic contexts involving both (existing/incumbent) actors and regimes (Gui and MacGill, 2018).

Community-scale energy projects have been explored and pursued for very different types of projects in many countries by many community groups. While motives may vary a lot, ranging from concerns about climate change, economic and financial considerations etc., some are also aiming at addressing poverty and social equity problems. Sometimes, such projects are not-for-profit and revenues from such undertakings may be directed to dedicated community funds to support further community endeavors (see also Gui and MacGill, 2018, p. 96).

4.4 Prosumer households and aggregate constructs (microgrids, virtual power plants, energy communities/clouds etc.)

At the micro-level it is prosumer households that will play a particularly important role as new active players in the balancing of energy supply and demand, potentially at all levels of analysis—i.e., the home, neighborhood, community, and larger system area. It is an active field of research whether prosumer households systematically consume more energy than others, and whether their rebound effects are systematically larger (Oberst et al., 2019; Oberst and Madlener, 2014), questioning the magnitude of their contribution to a more sustainable energy future. However, prosumer energy rebounds seem to be small relative to the magnitude of the potential contribution of prosumers (using renewables-based distributed energy resources, DER) in decarbonizing the grid. Still, our Institute's current research is finding that rebound effects tend to become significantly larger when prosumers are denied the option of feeding their electricity into the grid. The development of the grid needs to keep pace with the contribution prosumers can make, or much of it will be lost, also through rebounds.

Digitalization also enables the aggregation of distributed energy resources (DER) in general, and prosumer households in particular, to larger, "virtual" constructs, such as peer-to-peer (P2P) energy communities (cf. Section 3.2) and local energy markets, P2P or virtual energy clouds, virtual power plants (VPPs), or constructs requiring closer physical proximity of the elements, such as microgrids. Whereas this research field is often driven (and still dominated) by electrical engineering and information science research interests, it raises many new research questions also in the social sciences domain, and both for concepts mainly used in developing countries (e.g., offgrid solutions) and in the developed world (autonomy/autarky/self-supply as an alternative to supply via the public transmission and distribution grid). Such systems often enable a more cost-effective and reliable supply of electricity (provision of lighting for education and refrigeration for health services) than do traditional, hierarchical grid systems.

Micro-grids, sometimes also referred to as "offgrid solutions" in the context of developing countries (Castán Broto et al., 2018), enable energy services to be brought to those without access to modern energy sources (still an estimated 700 million people despite significant progress, cf. IEA, 2019, Summary, p. 6), which is often the case in the most remote areas on the planet. Solar photovoltaics play an important role in such new forms of offgrid electrification in rural areas. Such solutions often compete with established energy supply systems based on fuelwood, kerosene, LPG or charcoal, rather than with the central grid (Castán Broto et al., 2018, p. 649). Overall, offgrid solutions allow moving away from the idea of energy supply systems as a collective, typically national public service, and provide an alternative to LPG, which itself is a fossil fuel alternative to traditional cooking fuels such as fuelwood and charcoal. Whether increasing access to LPG helps to raise energy justice or, quite the contrary, reduces households' autonomy and increases the dependence on large multinational enterprises, has been subject to debate (Munro et al., 2017).

Finally, I would raise the questions as to whether smart grids enable higher resilience of energy supply, and whether low-income groups of society will benefit from smart grids as much as wealthier groups (or whether the smart grids and enabled smart new products and services would lead to some upwards distribution of wealth).

4.5 The multi-tenant prosumer concept (MTPC)

Decentralized renewable energy technologies are often installed in owner-occupied one- and two-dwelling buildings, whereas multi-family buildings often have an under-proportional share of these technologies. For instance, only about 1.4% of the total output of PV installed in Germany is produced in the 10 largest cities in Germany, with 11.2 million inhabitants (14% of the entire population) (Will, 2016; cited in Zimmermann and Madlener, 2018). Hence this is another dimension of inequality of an energy issue. While it can be argued that there are different business models for participating in the renewable energy transition (e.g., buying stocks or shares in energy cooperatives) poorer households can be expected to have less access to this type of participation in the *Energiewende*. In Germany, an amendment to the Renewable Energies Act (EEG), known as the Tenant Electricity Act (*Mieterstrom-Gesetz*), was passed by the German federal parliament in 2017 with the aim of allowing tenants an enhanced participation in distributed renewable energy projects. Zimmermann and Madlener (2018) have modeled the potential for what they call the multi-tenant prosumer concept (MTPC). They find that a total of 370,000 multi-family buildings are in principle suitable for the MTPC, and that solar PV turns out to be profitable in all cases, and can profitably be combined with micro-cogeneration for increased system sizes, whereas the combination PV, combined heat-and-power (CHP) and heat pumps is found to be uneconomical.

5 Economic growth, productivity gains, structural change

The long-term evolution of economies along the lines of input factor[g] productivity gains due to technical change helps us to better understand the role of energy. Three studies recently conducted by Frieling and Madlener (2016, 2017a,b) on Germany, the US and the UK can serve as an illustration of the energy transition during the fossil fuel age and before the era of large-scale decarbonization and high shares of renewables that is envisaged for the decades to come. These authors investigated the relationship between a factor-augmenting (i.e., non-neutral) technical change and factor substitution between capital, labor and energy inputs to the economy, also for the long term, trying to understand the role of the different production input factors and technical change on economic growth and welfare. Comparing the US and the UK energy situation in the 19th century, the authors write "Energy consumption in the US is often portrayed to be more than a simple economic decision, it is a part of the American identity [...] firewood was so plentiful in the 19th century US that a roaring hearth was seen as a fundamental right. The average American used five times as much firewood compared to the average Englishman, as deforestation in Europe necessitated a substitution of wood with coal, which spurred the development of new ovens to avoid the noxious coal smoke." For the 20th century, they continue by arguing that "The paradigm of cheap and abundant energy was challenged by the oil crises of the 1970s, which put an end to the gas-guzzling V8 as a standard engine and led to energy saving efforts to curb the ballooning energy expenditures." and then: "Still, the US has enjoyed ready access to cheap energy for much of the 20th century, especially compared to other industrialized countries. The reality of climate change and the economical and ecological imperatives of depleting fossil fuel deposits raise the question of how an economy can adjust to these changing circumstances, and what this means for growth and the use of input factors like labor, capital, and energy." (Frieling and Madlener, 2017a). This shows that such long-term studies on the evolution of economies, measured by the use and interplay of energy, labor and capital can also yield important new insights regarding perceptions and definitions of energy justice—a field that needs yet to be plowed.

Another example from the literature is Fouquet and Pearson (2006) investigating 700 years of lighting in the UK. They show that efficiency increases have drastically reduced the cost of lighting, so that many more people can afford lighting, that efficiency and productivity measures give very different results (raising the question which measure is the right one to be used?; see also Madlener and Alcott, 2009), and that the technologies and fuels used for lighting have changed greatly as well (with comfort, ethical and other implications). They find that "the economic history of light shows how focusing on developments in energy service provision rather than

---

[g]For readers not familiar with economics studies, the term "production factors" or "input factors" refers to the different types of resources that serve as inputs to industrial production: such a labor, capital and materials. More recently, in aggregate economic models, energy is often incorporated as a factor of production as well.

simply on energy use and prices can reveal the "true" declines in costs, enhanced levels of consumption and welfare gains that have been achieved." (p. 139). At the same time, they also warn "against the dangers of over-reliance on past trends for the long-run forecasting of energy consumption given the potential for the introduction of new technologies and fuels, and for rebound and saturation effects" (Fouquet and Pearson, 2006). This warning could also straightforwardly be used in the context of fuel and energy poverty, and energy justice, all the more so since digitalization, decentralization and decarbonization will have tremendous implications on how society will make use of energy sources and services.

6 **Conclusions**

In recent years, both the theoretical and empirical literature on energy inequality, justice and poverty has grown at an astonishing pace. Despite this development, many questions and pitfalls remain, and new questions pop up in light of the ongoing sustainable energy transition. In this chapter, I have argued that the sustainable energy transition can be expected to have significant impacts on economic and energy inequality. One important reason lies in the tremendous need to change the energy mix and to adapt and enhance the energy infrastructure, another in the decentralization, decarbonization and digitalization (the "3 Ds") that the energy transition implies. The fact that energy is an essential good justifies the separate discussion of energy justice and energy/fuel poverty in particular, versus justice and poverty in general.

I argued that apart from the enormous complexities of studying holistically the co-evolution of technology, society and the economy, as well as the numerous regimes and institutions involved, there are measurement issues and ample research needs. In recent years, the interest in using Lorenz curves and Gini coefficients has increased, as has interest in fuel/energy poverty and energy justice issues. These are a useful metric for analyzing energy equity, provided approximately constant marginal utility over a broad section of the population is given, i.e., energy consumption is assumed to have declining, i.e., not constant, marginal returns (Jacobson et al., 2005, p. 1826f). Economists are only about to start contributing to the debate, which seems to have been dominated so far mainly by sociologists, anthropologists and political scientists.

Energy efficiency and energy rebound have also been discussed in the context of energy/fuel poverty. Poor households often live in less energy-efficient, rented homes (on the dilemma of matching low-income people with energy-efficient, green homes, in the context of energy justice, see Agyeman and Evans, 2003), where the landlord-tenant dilemma is an important barrier to investments in fuel-saving technology and where higher transaction costs often hinder the use of distributed energy resources. In many cities throughout the world, affordable housing has become a virulent topic again, simultaneously raising questions concerning justice and energy justice.

Smart energy technologies as enablers of a digitalized, decentralized and decarbonized energy system can also enable better monitoring and hence improved transparency regarding increases or decreases in energy justice and fuel/energy poverty along the sustainable energy transition pathways adopted in various countries. At the same time, however, the technical and socio-economic systems can be expected to also become more complex and less tractable, and thus harder to model and to monitor in terms of energy justice dynamics.

Quantitative (model-based) empirical analysis of the history of long-term evolution of economies helps to guide governments to maneuver toward more sustainable energy futures than futures with economies powered by fossil and nuclear fuels. Interestingly, despite the enormous impacts of a globalized economy, the energy systems of the future will (again) be much more decentralized and distributed than the centralized and hierarchical systems that have dominated over the last 50–100 years. In light of rapid technological innovation, this will likely bring more heterogeneity, and to the consumers more autonomy than before also in the physical sense (and not just in terms of contractual customer choices in liberalized markets). In this respect, citizen energy communities of various forms will have a more important role but also increasing responsibility for "grid-solidaric" energy supply systems. This requires a comprehensive, multi-scale strategic approach that enables research to address both spatially narrow (energy communities, individual households) and system-wide resources and system requirements (grids, centralized renewable energy and reserve capacities) in their mutual ability to strengthen each other, and to allow policy not only to rectify existing (energy) injustices but also dynamically to tackle the causes of these (energy) injustices. Along with adequate institutional setups for such an evolution towards more cellular, modular, interrelated and overall multi-directional energy systems, appropriate social norms and economic incentives are needed in order to foster economic efficiency, reduce energy/fuel poverty, and achieve and maintain energy justice. Such policy-making needs to carefully take into account the potentially very different interests of stakeholders, and to maneuver using taxes and subsidies that per se might have regressive, neutral or progressive effects.

Sustainable energy transitions are "generators of geographically-uneven social, political and environmental displacements which may increase the vulnerability of particular social groups or places: a finding that is of special relevance to the global movement towards a low carbon future." (Bouzarovski and Simcock, 2017, p. 645). Notice that this brings together the temporal and spatial dimensions of energy justice, requiring us to take into account energy prices, infrastructure provision and economic inequality, whereas much of the energy/fuel poverty and justice literature up to now has mainly put the focus on discrepancies among social groups. The (long-term) analysis of past macroeconomic and technological developments can only guide energy policy-makers to a limited extent, and probably it is still much too early to fully comprehend the complex implications of digitalization, but also of decentralization and decarbonization, on energy justice and fuel poverty.

Policy-makers are challenged to find out *ex-ante* what the likely impacts of an induced energy transition are. Energy system and life-cycle analysis that includes cost aspects can be helpful tools in this endeavor. What can be considered as reasonably or practically "just" needs to be found out in each society, but an increase in inequality ought to be avoided in order not to jeopardize social cohesion beyond the *status quo*. In this respect, measures to mitigate energy poverty should cover the entire energy system, rather than concentrating solely on the end-use side of the energy supply chain. We see the beginnings of more holistic (system-wide) thinking and analysis in the energy justice literature, such as Sovacool and Dworkin (2014).

Overall, more comprehensive frameworks are needed for analyzing energy inequality, energy/fuel poverty, and energy justice more holistically along the life-cycle of systems and for the various stakeholder groups involved along the life-cycles. This is a very challenging undertaking in light of increasingly complex systems. It might help to avoid parochial policy that ignores wider implications and repercussions of sectoral or local policies.

Energy economics and political economy can contribute to the scientific and political discourse, as well as to the empirical evidence, and thus should be made an elementary constituent of any energy poverty and justice debate, which has been hitherto largely dominated by other social sciences. It remains to be seen to what extent energy supply systems will remain in the public goods sphere, and what role privately owned distributed energy resources will eventually have in terms of affecting energy inequality and energy/fuel poverty (at all scales).

References

Agyeman, J., Evans, T., 2003. Toward just sustainability in urban communities: building equity rights with sustainable solutions. Ann. Am. Acad. Pol. Soc. Sci. 590, 35–53.

Alvial-Palavicino, C., Ureta, S., 2017. Economizing justice: turning equity claims into lower energy tariffs in Chile. Energy Policy 105, 642–647.

Bach, S., Harnisch, M., Isaac, N., 2018. Verteilungswirkungen der Energiepolitik. Endbericht, DIW Berlin, Berlin. November.

BGBL (Bundesgesetzblatt), 2014. Gesetz v. 15. 7. 2014, BGBl. I 934. Schneider A (2014) Die Windkraft-Länderöffnungsklausel im neuen § 249 Abs. 3 BauGB. NuR36, Bundesanzeiger Verlag, pp. 673–678. 673. www.bgbl.de.

Birol, F., 2007. Energy economics: a place for energy poverty in the agenda? Energy J. 28 (3), 1–6.

Bloomberg New Energy Finance, 2016. Mapping the Gap: The Road from Paris (Finance Paths to a Two-Degree Future). Available at: http://about.bnef.com/whitepapers/mapping-the-gap-the-road-from-paris/.

Boardman, B., 1991. Fuel Poverty: From Cold Homes to Affordable Warmth. Belhaven Press, London.

Boardman, B., 2012. Fuel poverty synthesis: lessons learnt, actions needed. Energy Policy 49, 143–148.

Bouzarovski, S., Simcock, N., 2017. Spatializing energy justice. Energy Policy 107, 640–648.

Broman Toft, M., Schuitema, G., Thogersen, J., 2014. The importance of framing for consumer acceptance of the smart grid: a comparative study of Denmark, Norway and Switzerland. Energy Res. Soc. Sci. 3, 113–123.

Buchmann, M., 2017. The need for competition between decentralized governance approaches for data exchange in smart electricity grids—fiscal federalism vs. polycentric governance. J. Econ. Behav. Organ. 139, 106–117.

Business Dictionary, 2019a. Economic Welfare. Available at: http://www.businessdictionary.com/definition/economic-welfare.html. (Accessed 3 June 2019).

Business Dictionary, 2019b. Social Welfare. Available at: http://www.businessdictionary.com/definition/social-welfare.html. (Accessed 3 June 2019).

Cardenas, J.A., Gemoets, L., Ablanedo Rosas, J.H., Sarfi, R., 2014. A literature survey on smart grid distribution: an analytical approach. J. Clean. Prod. 65, 202–216.

Castán Broto, V., Baptista, I., Kirshner, J., Smith, S., Neves Alves, S., 2018. Energy justice and sustainability transitions in Mozambique. Appl. Energy 228, 645–655.

Chan, N.W., Gillingham, K., 2015. The microeconomic theory of the rebound effect and its welfare implications. J. Assoc. Environ. Resour. Econ. 2 (1), 133–159.

Chaton, C., Lacroix, E., 2018. Does France have a fuel poverty trap? Energy Policy 113, 258–268.

Cherp, A., Jewell, J., 2014. The concept of energy security: beyond the four As. Energy Policy 75, 415–421.

Colak, I., Sagiroglu, S., Fulli, G., Yesilbudak, M., 2016. A survey on the critical issues in smart grid technologies. Renew. Sust. Energ. Rev. 54, 396–405.

David, M., 2017. Moving beyond the heuristic of creative destruction: targeting exnovation with policy mixes for energy transitions. Energy Res. Soc. Sci. 33, 138–146.

David, M., 2018a. Exnovation as a necessary factor in successful energy transitions. In: Davidson, D.J., Gross, M. (Eds.), Oxford Handbook of Energy and Society. Oxford University Press, Oxford, UK. ISBN 9780190633851.

David, M., 2018b. The role of organized publics in articulating the exnovation of fossil-fuel technologies for intra- and intergenerational energy justice in energy transitions. Appl. Energy 228, 339–350.

Day, R., Walker, G., Simcock, N., 2016. Conceptualizing energy use and energy poverty using a capabilities approach. Energy Policy 93, 255–264.

de A. Dantas, G., de Castro, N.J., Dias, L., Henggeler Antunes, C., Vardiero, P., Brandão, R., Rosental, R., Zamboni, L., 2018. Public policies for smart grids in Brazil. Renew. Sust. Energ. Rev. 92, 501–512.

Economist, The, 2019. Net benefits: America's high inequality reflects gross incomes as much as its tax system. The Economist. 77, April 13, 2019.

Ellaban, O., Abu-Rub, H., 2016. Smart grid customers' acceptance and engagement: an overview. Renew. Sust. Energ. Rev. 65, 1285–1298.

Faruqui, A., Harris, D., Hledik, R., 2010. Unlocking the €53 billion savings from smart meters in the EU: how increasing the adoption of dynamic tariffs could make or break the EU's smart grid investment. Energy Policy 38, 6222–6231.

Flues, F., Thomas, A., 2015. The Distributional Effects of Energy Taxes, OECD Taxation Working Papers No. 23. OECD, Paris.

Fortier, M.-O.P., Teron, L., Reames, T.G., Trishana Munardy, D., 2019. Introduction to evaluating energy justice across the life cycle: a social life cycle assessment approach. Appl. Energy 236, 211–219.

Fouquet, R., Pearson, P.J.G., 2006. Seven centuries of energy services: the price and use of light in the United Kingdom (1300−2000). Energy J. 27 (1), 139–177.

Frieling, J., Madlener, R., 2016. Elasticity Estimation for Nested Production Functions with Generalized Productivity, FCN Working Paper Series No. 1/2016. Institute for Future Energy Consumer Needs and Behavior, RWTH Aachen University. March (revised September 2016).

Frieling, J., Madlener, R., 2017a. Fueling the US Economy: Energy as a Production Factor from the Great Depression Until Today, FCN Working Paper Series No. 2/2017. Institute for Future Energy Consumer Needs and Behavior, RWTH Aachen University. February.

Frieling, J., Madlener, R., 2017b. The Turning Tide: How Energy Has Driven the Transformation of the British Economy Since the Industrial Revolution, FCN Working Paper Series No. 7/2017. Institute for Future Energy Consumer Needs and Behavior, RWTH Aachen University. June.

Galvin, R., 2018. 'Them and us': regional-national power-plays in the German energy transformation: a case study in lower Franconia. Energy Policy 113, 269–277.

Galvin, R., 2019. What does it mean to make a moral claim? A Wittgensteinian approach to energy justice. Energy Res. Soc. Sci. 54, 176–184.

Grösche, P., Schröder, C., 2013. On the redistributive effects of Germany's feed-in tariff. Empir. Econ. 46 (4), 1339–1383.

Grubler, A., Johansson, T.B., Mundaca, L., Nakicenovic, N., Pachauri, S., Riahi, K., Rogner, H.-H., Strupeit, L., 2012. Energy primer. In: Team, GEA Writing, (Ed.), Global Energy Assessment: Toward a Sustainable Future. Cambridge University Press and IIASA, pp. 99–150. October 2012.

Gui, E.M., MacGill, I., 2018. Typology of future clean energy communities: an exploratory structure, opportunities, and challenges. Energy Res. Soc. Sci. 35, 94–107.

Healy, N., Barry, J., 2017. Politicizing energy justice and energy system transitions: Fossil fuel divestment and a "just transition" Energy Policy 108, 451–459.

Heffron, R.J., McCauley, D., 2017. The concept of energy justice across the disciplines. Energy Policy 105, 658–667.

Heffron, R.J., McCauley, D., Zarazua de Rubens, G., 2018. Balancing the energy trilemma through the energy justice metric. Appl. Energy 229, 1191–1201.

Heindl, P., 2014. Ökonomische Aspekte der Lastenverteilung in der Umweltpolitik am Beispiel der Energiewende: Ein Beitrag zum inderdisziplinären Dialog, ZEW Discussion Paper No. 14-061. ZEW Mannheim.

Heindl, P., Schüssler, R., 2015. Dynamic properties of energy affordability measures. Energy Policy 86, 123–132.

Heindl, P., Kanschik, P., Schüssler, R., 2015. Anforderungen an Energiearmutsmaße: Ein Beitrag zur normativen und empirischen Definition. In: Großmann, K., Schaffrin, A., Smigiel, C. (Eds.), Energie und soziale Ungleichheit: Zur gesellschaftlichen Dimension der Energiewende in Deutschland und Europa. Springer, Berlin/Heidelberg, pp. 241–262 Kap. 9, S. 241–262.

Herring, H., 2009. Sufficiency and the rebound effect. In: Herring, H., Sorrell, S. (Eds.), Energy Efficiency and Sustainable Consumption. Energy, Climate and the Environment Series. Palgrave Macmillan, London.

Hicks, J.R., 1939. The foundations of welfare economics. Econ. J. 49, 696–712.

Hills, J., 2012. Getting the Measure of Fuel Poverty. Final Report on the Fuel Poverty Review. London.

Hiteva, R.P., 2013. Fuel poverty and vulnerability in the EU low-carbon transition: the case of renewable electricity. Local Environ. 18, 487–505.

Höwer, D., Oberst, C.A., Madlener, R., 2019. General regionalization heuristic to map spatial heterogeneity of macroeconomic impacts: the case of the green energy transition in NRW. Util. Policy 58, 166–174.

IEA, 2019. World Energy Outlook 2018. OECD/IEA, Paris.

IPBES, 2019. Intergovernmental Science-Policy Platform on Biodiversity and Ecosystem Services. https://www.ipbes.net/ retrieved June 24, 2019.

Jacobson, A., Milman, A.D., Kammen, D.M., 2005. Letting the (energy) Gini out of the bottle: Lorenz curves of cumulative electricity consumption and Gini coefficients as metrics of energy distribution and equity. Energy Policy 33, 1825–1832.

Jenkins, K., 2018. Setting energy justice apart from the crowd: Lessons from environmental and climate justice. Energy Res. Soc. Sci. 39, 117–121.

Jenkins, K., McCauley, D., Heffron, R., Stephan, H., Rehner, R., 2016. Energy justice: a conceptual review. Energy Res. Soc. Sci. 11, 174–182.

Kimberly, J.R., 1981. Managerial innovation. In: Nystorm, P.C., Starbuck, W.H. (Eds.), Handbook of Organizational Design. Elsevier, Amsterdam, pp. 84–104.

Koh, S.C.L., Marchand, R., Genovese, A., Brennan, A., 2012. Fuel Poverty: Perspectives From the Front Line, Centre for Energy, Environment and Sustainability: Fuel Poverty Series. University of Sheffield, Sheffield, UK.

LaBelle, M.C., 2017. In pursuit of energy justice. Energy Policy 107, 615–620.

Lacroix, E., Chaton, C., 2015. Fuel poverty as a major determinant of perceived health: the case of France. Public Health 129, 517–524.

Lewis, P., 1982. Fuel Poverty Can Be Stopped. National Right to Fuel Campaign, Bradford.

Li, K., Lloyd, B., Liang, X.-J., Wei, Y.-M., 2014. Energy poor or fuel poor: what are the differences? Energy Policy 68, 476–481.

Lindahl, E. (Ed.), 1958. Knut Wicksell: Selected Papers on Economic Theory. Harvard University Press, Cambridge, MA.

Lindahl, E., 1967. Just taxation—a positive solution. In: Musgrave, R.A., Peacock, A.T. (Eds.), Classics in the Theory of Public Finance. Macmillan, London.

Luthra, S., Kumar, S., Kharb, R., Ansari Md, F., Shimmi, S., 2014. Adoption of smart grid technologies: an analysis of interactions. Renew. Sust. Energ. Rev. **33**, 554–565.

Madlener, R., Alcott, B., 2009. Energy rebound and economic growth: a review of the main issues and research needs. Energy 34 (3), 370–376.

Madlener, R., Hauertmann, M., 2011. Rebound Effects in German Residential Heating: Do Ownership and Income Matter? FCN Working Paper No. 2/2011, Institute for Future Energy Consumer Needs and Behavior. RWTH Aachen University February.

Mainali, B., Pachauri, S., Rao, N., Silveira, S., 2014. Assessing rural energy sustainability in developing countries. Energy Sustain. Dev. 19, 15–28.

McCauley, D., Ramasar, V., Heffron, R.J., Sovacool, B.K., Mebratu, D., Mundaca, L., 2019. Energy justice and the transition to low carbon energy systems: exploring key themes in interdisciplinary research. Appl. Energy 233-234, 916–921.

Mill, J.S., 1848. Principles of Political Economy. John W. Parker, London.

Munro, P., van der Horst, G., Healy, S., 2017. Energy justice for all? Rethinking sustainable development goal 7 through struggles over traditional energy practices in Sierra Leone. Energy Policy 105 (Supplement), 635–641.

Nersisyan, Y., Wray, L., 2019. How to Pay for the Green New Deal. Levy Economics Institute of Bard College (Working Paper No. 931).

Oberst, C.A., Madlener, R., 2014. Prosumer Preferences Regarding the Adoption of Micro-Generation Technologies: Empirical Evidence for German Homeowners, FCN Working Paper No. 22/2014. Institute for Future Energy Consumer Needs and Behavior, RWTH Aachen University. December (revised September 2015).

Oberst, C.A., Schmitz, H., Madlener, R., 2019. Are prosumer households that much different? Evidence from stated residential energy consumption in Germany. Ecol. Econ. 158, 101–115.

Pereira, M.G., Freitas, M.A.V., Silva, N.F., 2011. The challenge of energy poverty: Brazilian case study. Energy Policy 39, 167–175.

Phimister, E., Vera-Toscano, E., Roberts, D., 2015. The dynamics of energy poverty: evidence from Spain. Econ. Energy Environ. Policy 4 (1), 153–166.

Rehner, R., McCauley, D., 2016. Security, justice and the energy crossroads: assessing the implications of the nuclear phase-out in Germany. Energy Policy 88, 289–298.

Schlör, H., Zapp, P., Marx, J., Schreiber, A., Hake, J.-F., 2015. Non-renewable resources for the Energiewende—a social life cycle analysis. Energy Procedia 75, 2878–2883.

Sen, A., 1976. Poverty: an ordinal approach to measurement. Econometrica 44 (2), 219–231.

Sen, A., 1985. Commodities and Capabilities. North-Holland, Amsterdam.

Sen, A., 1997. On Economic Inequality. Clarendon Press, Oxford.

Sen, A., 2017. Collective Choice and Social Welfare, Expanded Edition Penguin, UK (first published 1970 by Harvard University Press, revised version published 1979 by North-Holland).

Silvestre, J., 2003. Wicksell, Lindahl and the theory of public goods. Scand. J. Econ. 105 (4), 527–553.

Sorrell, S., Gatersleben, B., Druckman, A., 2018. Energy Sufficiency and Rebound Effects, Concept Paper. ECEEE Energy Sufficiency. https://www.energysufficiency.org/static/media/uploads/site-8/library/papers/sufficiency-rebound-final_formatted_181118.pdf retrieved June 21, 2019.

Sovacool, B.K., 2009. The cultural barriers to renewable energy and energy efficiency in the United States. Technol. Soc. 31, 365–373.

Sovacool, B.K., 2013. Energy & Ethics: Justice and the Global Energy Challenge. Palgrave MacMillan, London.

Sovacool, B.K., Dworkin, M., 2014. Global Energy Justice: Problems, Principles, and Practices. Cambridge University Press, Cambridge.

Sovacool, B.K., Sidortsov, R.V., Jones, B.R., 2014. Energy Security, Equality and Justice. Routledge, London.

Sovacool, B.K., Burke, M., Baker, L., Kotikalapudi, C.K., Wlokas, H., 2017. New frontiers and conceptual frameworks for energy justice. Energy Policy 105, 677–691.

Statista, 2019. Bruttoinlandsprodukt (BIP) je Einwohner nach Bundesländern im Jahr 2017. https://de.statista.com/statistik/daten/studie/73061/umfrage/bundeslaender-im-vergleich—bruttoinlandsprodukt/ retrieved June 24, 2019.

Sterner, T., 2012. Distributional effects of taxing transport fuels. Energy Policy 41, 75–83.

Thuiller, W., Lavergne, S., Roquet, C., Boulangeat, I., Lafourcade, B., Araujo, M.B., 2011. Consequences of climate change on the tree of life in Europe. Nature 470 (531), 534.

Tuballa, M.L., Abundo, M.L., 2016. A review of the development of smart grid technologies. Renew. Sust. Energ. Rev. 59, 710–725.

van den Bergh, J., 2011. Energy conservation more effective with rebound policy. Environ. Resour. Econ. 48, 43–58.

Walker, G., Day, R., 2012. Fuel poverty as injustice: integrating distribution, recognition and procedure in the struggle for affordable warmth. Energy Policy 49, 69–75.

Weber, G., Cabras, I., 2017. The transition of Germany's energy production, green economy, low-carbon economy, socio-environmental conflicts, and equitable society. J. Clean. Prod. 167, 1222–1231.

WEC, 2018. World Energy Trilemma Index 2018. World Energy Council, London, UK.

Wicksell, K., 1896. Finanztheoretische Untersuchungen nebst Darstellung und Kritik des Steuerwesens Schwedens (Part 1 translated by Chamberlain as "On the Theory of Tax Incidence", Chapter 5 in Sandelin (1997); part 2 translated by J.M. Buchanan as "A New Principle of Just Taxation", in R.A. Musgrave and A.T. Peacock (eds.) (1958), Classics in the Theory of Public Finance). Macmillan, London, pp. 72–118.

Wilhite, H., Nakagami, H., Masuda, T., Yamaga, Y., Haneda, H., 1996. A cross-cultural analysis of household energy use behaviour in Japan and Norway. Energy Policy 24, 795–803.

Will, H., 2016. Geschäftsmodelle mit PV-Mieterstrom, Berlin. Available at:https://www.pv-mieterstrom.de/wp-content/uploads/2016/11/PV_Financing_Mieterstrom.pdf   retrieved April 14, 2018.

Williams, S., Doyon, A., 2019. Justice in energy transitions. Environ. Innov. Soc. Trans. 31, 144–153.

Wu, S., Zheng, X., Wei, C., 2017. Measurement of inequality using household energy consumption data in rural China. Nat. Energy. . https://doi.org/10.1038/s41560-017-0003-1.

Zafar, R., Mahmood, A., Razzaq, S., Ali, W., Naeem, U., 2018. Prosumer based energy management and sharing in smart grid. Renew. Sust. Energ. Rev. 82, 1675–1684.

Zheng, B., 1997. Aggregate poverty measures. J. Econ. Surv. 11 (2), 123–162.

Zimmermann, G., Madlener, R., 2018. Techno-Economic Evaluation of Combined Micro Power and Heat Generation Assets: Implications for the Multi-Tenant Building Market in Germany, FCN Working Paper Series No. 18/2018. Institute for Future Energy Consumer Needs and Behavior, RWTH Aachen University. December.

Further reading

BDEW, 2018. Bundesverband der Energie- und Wasserwirtschaft. https://www.bdew.de/ retrieved June 28, 2019.

BMWE, 2018. EEG in Zahlen: Vergütungen, Differenzkosten und EEG-Umlage 2000 bis 2019. Bundesministerium für Wirtschaft und Energie, Berlin. Available at: https://www.erneuerbare-energien.de/EE/Redaktion/DE/Downloads/eeg-in-zahlen-pdf.pdf%3F__blob%3DpublicationFile retrieved June 28, 2019.

BP, 2015. BP Statistical Review of World Energy 2015. BP, London. Available at: www.bp.com retrieved June 28, 2019.

BP, 2018. BP Statistical Review of World Energy 2018. BP, London. Available at: https://www.bp.com/en/global/corporate/energy-economics/statistical-review-of-world-energy/primary-energy.html retrieved June 28, 2019.

Heindl, P., 2013. Measuring Fuel Poverty: General Considerations and Application to German Household Data, ZEW Discussion Paper Series No. 13-046. ZEW Mannheim.

Heindl, P., 2015. Measuring fuel poverty: general considerations and application to German household data. FinanzArchiv 71, 178–215.

Can economic inequality be reduced? Challenges and signs of hope in 2019

12

Danny Dorling

School of Geography and the Environment, University of Oxford, Oxford, United Kingdom

Chapter outline

1 The roaring 20s 287
2 Modern times 291
3 Hard times 292
4 The crash and the rise of the far right 294
5 The human geography of Brexit 295
6 Inequality, pollution and stupidity 300
7 Inequality extremes in Europe 300
8 Hope 306
References 308
Further reading 310

1 The roaring 20s

In some ways we have returned to the past, to the jazz age of the 1920s, the time of The Great Gatsby when—in 1925 F. Scott Fitzgerald wrote about the Long Island goings on in the long hot summer of 1922. As Ray Galvin explains in Chapter 10 of this volume, today just as in the 1920s, most people in the richest of countries do not question why so many households are poor while a few are so rich. But in between now and then we did, and we redistributed both wealth and income so that by the 1960s and 1970s most people had never been so equal. The rich could no longer be unbelievably profligate and the poor no longer had to shiver in winter for lack of money to heat their homes. But we forgot what we had gained and (in the UK and USA especially) we have returned to the past excesses of economic inequality, although in neither case are things as bad today as they were then.

After 1922 in the United States, and a little earlier across the rest of the rich world, economic inequalities fell, through to the late 1970s, but then they rose and rose again so that just one lifetime later, in October 2006, Knight Kiplinger, Editor in

Inequality and Energy. https://doi.org/10.1016/B978-0-12-817674-0.00012-6

Chief of 'Kiplinger Magazine' could lecture his readers. He claimed that: 'The biggest barrier to becoming rich is living like you're rich before you are'. And then he went on to explain what he meant by this. He wrote that that: 'I often hear complaints from young adults, twentysomethings to those in their early thirties, that they'll never be able to buy a home because they can't afford the down payment. But when I probe them about their budgets, I find that they earn enough to make a down payment in just three or four years – if they cut back on their spending, and if their starter-home expectations are reasonable.' (Kiplinger, 2006). Within just a few months of him writing that, but long enough for a few to have taken his advice, the US housing market crashed and those young people foolish enough to have believed him were financially burnt. Many may still be holding the negative equity if they were able to scrimp and save for those down payments.

Thirteen years later and the Magazine's writers were at it again. This is what they said most recently: 'Becoming wealthy and staying that way takes a certain level of discipline. Sure, an occasional splurge won't put you in the poor house, but frequent frivolous spending can quickly erode your net worth. The frugal habits necessary to achieve financial success and maintain it can be surprisingly simple' (Browne-Taylor, 2019). However, as Sam Pizzigati explained in commenting on all this foolishness, a simplistic view of why the rich are rich which pervades modern America, '….we simply do not want to believe that our rich may have gained their riches through exploiting others or rigging our economy or just finding themselves in the right place at the right time. So we ascribe to our awesomely affluent rich noble qualities that make them ever so deserving of their wealth'. (Pizzigati, 2019)

The key assumption being made here is the same as was presumed in the 1920s, it is that economic inequalities are justified. The rich are rich because they have worked hard, saved hard and invested well. The poor are poor because they have made a series of bad choices and, if only they had been frugal and well-disciplined, they too could be rich. Everyone can't be rich, or so this particular story goes, because not everyone can be well-disciplined, but try hard enough—and it could be you.

Travel back in time to 1922 again, not to the US Jazz age—but to the only very recently created UK which took its current shape that year, the year when The Irish Free State became independent of the rest of the British Union. Writing about that time a few decades later Evelyn Waugh created Lady Marchmain, the matriarch of a fictitious aristocratic pile. In his book 'Brideshead Revisited' based on real aristocratic families, the fictional character Lady Marchmain (Sebastian Flytes' mother) prior to launching into her explanation of how camels and rich men (apparently) can fit through the eye of a needle in England in the early 1920s said this:

> It used to worry me, and I thought it wrong to have so many beautiful things when others had nothing. Now I realise that it is possible for the rich to sin by coveting the privileges of the poor. The poor have always been the favourites of God and his saints, but I believe that it is one of the special achievements of Grace to sanctify the whole of life, riches included. Wealth in pagan Rome was necessarily something cruel. It's not any more. (Waugh, 1945)

Waugh used her words to try to explain how the wealthy attempted (and still attempt) to console themselves in many different ways. One route is religious, suggesting God's favorites are the poor and that balances out the injustice of the wealth of the rich—even, in her words, to the extent of god sanctifying her riches with his 'Grace'. Others claim that their riches are merited by their imagined immense talents, and/or their frugality. Another tactic is to suggest that it has always been like this. They might well say that: "Although lamentable, many having so little, while a few have so much, is inevitable." Charity will have to suffice, they presume. None of these excuses are true, and charity is never enough.

The fictional Lady Marchmain was musing in the early 1920s at the start of an era of immense social change and during the last period when economic inequalities in British society (and just a little later in the USA) began to fall. It was that fall in inequality that most changed her life and her family's. That fall was more important to her than everything else; more important than the First World War or the loss of the British colonies that began in earnest with Ireland. Try to imagine the shock of the rich in the 1920s and 1930s as they watched their riches melt away before their eyes. Try to imagine the New Deal USA after the 1929 stock market crash and when unemployment at first exceeded 14 million people, and just how different that was to the Jazz age only a few summers before. Imagine that and you might possibly be able to imagine a little of what may well be about to come today.

Lady Marchmain was musing just after the First World War, when the best-off tenth in British society were taking almost 50% of all the income in the country, leaving only half for the other 90% of all British people to live on; but that unfair share was beginning to be corrected and, in the late 1920s, it would fall more rapidly than it had ever done before or since. In the early 1920s, within that top tenth of the population, the top 1% were taking almost a quarter of national income every year – half of what the entire top tenth took. Within that top 1% the best-off tenth-of-a-percent, in 1923, were taking 9.29% of all national income, almost 100 times the average income. And within that tiny portion was the group that Lady Marchmain belonged to, the highest 'earning' fraction, the 0.01 (just 1 in 10,000 people), who were taking 3.34%, or 334 times as much as the average person.

Lady Marchmain may have been fictional, but she was based on the few very extremely wealthy families of her day. She and they had never worked, or even imagined ever working. Her income was derived from 'investments' most of which would have been held overseas in the British Empire. She spent her days worrying about not saying the wrong thing in front of the servants, worrying about her errant son Sebastian, for whom all the riches of the world were no salvation, and worrying about whether she would fit through the eye of the needle when her time came.

And then change came (see Fig. 1). When change truly happens it at first strikes seasoned commentators as frankly impossible—a pipe-dream; then undesirable and full of negative consequences; then 'just about possible' once the clamor for change becomes overwhelming. Finally change happens and the memories of all of those effected change with it. Many will say that they believed in the change as desirable all along; they somehow saw it coming and so, too, were on the right side of history.

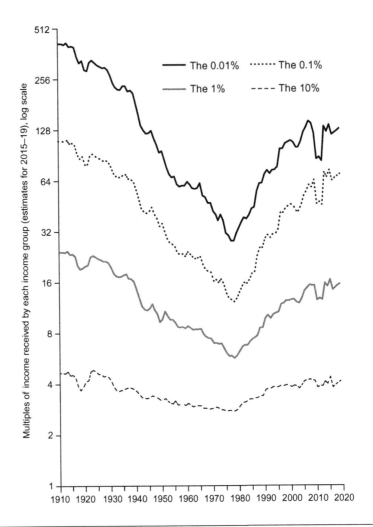

FIG. 1

Income inequality 1910–2019, UK (multiples of income received by each income group and estimates for 2015–19).

From Pre-tax national income share including pension income, individuals over age 20, from: Dorling, D., 2013
Fairness and the changing fortunes of people in Britain. J. R. Stat. Soc. A 176 (1), 97–128; Atkinson, A.B.,
Hasell, J., Morelli, S., Roser, M., 2017. The Chartbook of Economic Inequality; Brewer, M., 2019. What Do We
Know and What Should We Do About Inequality? Sage, London; and 2015–2019 extrapolated from household
data given in Shine, M., Webber, D., 2019. Using tax data to better capture top earners in household income
inequality statistics. ONS, London, February 26.

Then we can all forget that just a few years ago so many, especially those with power and a voice, had so vehemently opposed the change, had justified the status quo, were so very scornful, and ultimately wrong. That matters little. It is just history. What matters is ensuring that we are now at the peak and that we are now starting on our way down. It's a long way down because the peak of inequality we are currently at is so very high. This chapter concentrates on the UK, but what is says can be applied to many of the most unequal of affluent countries in the world today. They are not so unequal due to some mistake, and they will not become more equal without hard work (Dorling, 2018).

2 Modern times

Average income on the graph in Fig. 1 equates with the value 1. Any group taking above average income results in a larger number of other people getting less than 1. By 2014 the best-off tenth of adults aged 20 and over in the UK were taking 4 times average income. As a result, the remaining 90% of people were getting by on only 0.67 times average income (Atkinson, 2007, Dorling, 2013, Atkinson et al., 2017).

In 2014 the best-off 1% took 15 times average incomes, a share almost identical to what they took before the Second World War. The best-off 0.1% took 67 times, a share so high that they had last achieved such excesses only in 1936—other than in the 1 year, 2013, when their take was 7.2% or 72 times average earnings—and that year the best-off 0.01% took 130 times average earnings. It is sobering to realize that 130 times is three times less than in Lady Marchmain's day, although 130 times the average was what the share of families like hers had been 'reduced to' by 1943. The fictional Lady died (appropriately) in 1926, the year of the general strike. She died trying to tell herself that she had god's grace.

We will probably see the fall in inequality begin at the top of the income and wealth scale, not at the bottom, and not until after life for the poorest in the UK has first worsened to be even harder than it was at the peak. This is what happened before in the 1920s and 1930s and it may be what is just starting to happen now. It happens partly because rising wealth inequalities can never continue indefinitely.

In the year to October 2016, sales of individual properties within London 'worth' more than £10m fell by 86%; and outside of London, they fell from 10 properties in 2015 to none being sold for such huge prices in 2016. In May 2017 it became clear that values across London for all properties had fallen slightly. The fall in UK wealth inequality could have already begun. If you measure your wealth in dollars and you live in the UK and own property there, you are already much worse off than in June 2016.

A fall in inequality can begin without policy and political changes, but they help sustain it. In December 2016 the City of Portland, Oregon, announced that it would surcharge companies that paid their CEOs more than 100 times their median workers' pay. In 2016 in the USA the average pay of the top 500 CEOs had fallen to 335 times the income of the average worker. That is incredibly high ('Brideshead Revisited' high) but in 2014 it was 373 times, although note that this compares to 42 times in 1980 (Dorling, 2017a).

Today it is when inequality falls that whole countries actually progress. There is now mounting evidence that since 2008 income inequalities in China have begun to fall and the benefits of prosperity have started to spread (Zhuang and Shi, 2016). In 2017 I was able to write that the World Wealth and Income Database reported the 1% taking a little less than they took a few years ago in the USA, South Africa, UK, Canada, China, Germany, Ireland, Switzerland, Australia, Italy, Japan, France, Spain, Norway, Finland and the Netherlands. Among all the countries enumerated in that database, only in Denmark, Sweden and New Zealand did I find that income inequalities higher at the most recent year of recording than they were in 2007. Today I can now say that Denmark still has not updated its data. Sweden and New Zealand had yet to produce the equivalent statistics, but it looks likely that of the 21 countries for which these data exist only Korea was still seeing income inequalities rise in 2016 (see Table 1). However, a political scandal was then brewing in Korea that brought down president Geun-hye (who is now serving a 25 year prison sentence). Her successor, president Moon, was then also mired in scandals also involving money. High and rising income inequality tends to lead to political trouble and encourages corruption.

3 Hard times

When inequality is high people lose face, they lose confidence, they suffer from comparisons in which it is implied that the vast majority warrant little or no respect. Improvements in life expectancy stall or even reverse, you fear for your children and their future. Life feels like a game of chance with most of the odds heavily stacked against you. Fear divides one person, one family, one social class, from another; loneliness increases, even as we become more crowded in cities.

Our greatest fear is other people, and inequality becomes the enemy between us. As we see increasing inequality, whenever and wherever people experience this, the pressures to move toward more equality increase. However, in the most unequal country of all in Table 1, the USA, the arguments for inequality being justified are still being made most strongly, and corruption is rife as embodied by the man who is currently president (who even set up a 'Trump University' which was later shut down by the authorities with a $25 million fine being imposed). In contrast, Finland is the most equitable country in that table, and has no such stories to tell, although again its data could do with updating!

We too often too easily forget the past, but we have to look a long way back to see a situation as bad as that which we are facing today. If we are very optimistic, then it is possible to believe that income inequality recently peaked in the UK (Dorling, 2018). By several measures the *quality* of life in Britain peaked much earlier, in 1976, which far from coincidently was when income *equality* peaked (Monbiot, 2004; Jackson, 2004). In contrast, by 2019, university research had found that almost a third '…of UK adults with children under the age of 16 were food insecure. The risk was greatest among the unemployed, those with long term conditions or disabilities, and those on the lowest household income.' (Loopstra et al., 2019). This compares to 1 in 5 of all adults being so insecure, 3% of whom often go without food. Between 2004 and 2016 food insecurity among the least well-off almost doubled. Energy insecurity will have done the same, but it is less well measured.

Table 1 Fiscal (taxed) income top 1% share of individual adults (% of national income)

Country	2007	2008	2009	2010	2011	2012	2013	2014	2015	2016	2017
Australia	**9.11**	8.08	8.37	8.55	8.22	8.52	9.04	8.98	9.10	9.07	
Canada	**15.63**	14.38	13.30	13.62	14.60	13.80	13.80	13.70	13.90		
China	15.30	15.20	**15.40**	15.10							
Colombia	**20.49**	20.25	20.17	20.45							
Czech Republic	9.04	**10.32**	9.55	9.07	9.16	9.84	9.33	9.40	9.61		
Denmark	6.12	6.05	5.44	**6.41**							
Finland	8.26	**8.50**	7.46								
Greece	6.13	6.14	6.04	6.61	6.60	6.75	7.01	8.65	**8.76**	8.41	7.79
Hungary	**10.47**	9.64									
Italy	**9.86**	9.66	9.38	10.44							
Japan		10.90	10.42	9.43		9.11					
Korea	11.29	11.38	11.31	11.74	12.02	11.74	11.63	11.78	12.07	**12.16**	
Malaysia			**9.41**		8.95	9.11					
New Zealand	7.76	8.07	7.80	7.36	8.01	**8.94**	7.79	8.06	8.35	8.24	
Norway	**8.54**	7.70	7.11	7.74	7.80						
Singapore	14.06	**15.15**	13.66	13.39	13.85	13.57	13.57	14.02			
South Africa	**20.06**	19.46	18.28	18.54	18.46	19.21					
Spain	**11.24**	9.83	9.30	8.69	9.03	8.58					
United Kingdom	**15.44**	15.40	15.42	12.55	12.93	12.70	14.53	13.88			
Uruguay			14.20	**14.60**	14.40	14.00					
United States	19.87	19.52	18.54	19.80	19.60	**20.78**	19.59	20.20			

Year of highest share is in bold, 21 countries shown for which full data was available. See: https:/wid.world. Note: The USA and China data were released in November 2017 and have not been updated, the data for all the other countries were retrieved in June 2019. The proportions are before tax and welfare redistributions.
From WID, 2019. World Inequality Database. https:/wid.world.

As more and more people go hungry in Britain, as homelessness rises and overall life expectancy begins to fall, we see wealth inequality still rising. Wealth inequality tends to lag the trend of income inequality. On April 12, 2019 Mike Brewer of the University of Essex revealed the most recent income inequality data. All eyes quickly turn to the very last data point, and the up-tick from 2015 to then. Income inequality briefly fell during and immediately after the financial crash of 2008. Now it appears to be rising again. However, there are many reasons to believe that we may now be at a peak, and not at yet another false summit.

As Mike Brewer explained in April 2019 'There is a big difference in what tax data says about the very rich and what household surveys say. Tax data says that the very rich continued to drift upwards until 2008. The financial crisis took a chunk out of top incomes, but by 2015-16 they were clearly on a rising trend'. (Brewer, 2019). Household survey data is of little use in monitoring just how large a take the top 1% extract. This is because those in the top 1% are loathe to fill in voluntary surveys, and if they do, then they often tend to be somewhat modest about their actual total income. Taxation data, of course, due to tax avoidance (legal) and evasion (illegal) also underestimates top incomes as those at the top have the greatest incentive to avoid declaring all their sources of income and have the resources to pay accountants to help them do that. Furthermore, the current Conservative UK government is not encouraging HMRC to release taxation data in a timely or useful fashion (Corlett, 2017).

4   The crash and the rise of the far right

To understand what has occurred in a country such as the UK in very recent years requires understanding the gravity of the economic crash of 2008 and its incredibly long legacy. No one explains the situation better than economist Danny Blanchflower:

> The UK's recovery was the third-slowest peacetime recovery in six hundred years. The South Sea Bubble, which was a speculation mania that ruined many British investors in 1720 when the South Sea Company collapsed, was the next slowest. The Black Death, from 1347 to 1351, which resulted in the deaths of more than 75 million, was the slowest. It was good for productivity, though. Recovery from the Great Recession was even slower than it was from the year without a summer, which occurred in 1816. Severe climate abnormalities caused global temperatures to drop, resulting in major food shortages. These weather conditions appear to have been caused by a huge volcanic eruption in 1815 by Mount Tambora, in the Dutch East Indies, which was the largest eruption in 1,300 years. Policymakers didn't learn the lessons of history and down we all went. It was as bad as that. (Blanchflower, 2019)

The 2008 economic crash was especially bad for the UK, and in particular for England, because successive governments, most recently and significantly New

Labor, had been happy to accept living with high income inequality, hoping that the trickle down they expected from it would appease the poor. It was only toward the very end of the New Labor period that Gordon Brown grudgingly introduced the smallest of tax rises for the very rich. It did not work and New Labor lost the 2010 general election. The Conservative-Liberal coalition that was formed after that election was a union between a Tory party that had moved far to the right and a Liberal party who might have fitted quite well within the old Conservative party.

To try to hold his rapidly splitting and right-ward moving swivel-eyed troops together, David Cameron took his party out of the center-right EPP block in the European parliament in 2014. In the European elections of that year parties to the right of mainstream European Conservative national parties secured over 52% of the popular vote in the UK. These were principally the Conservatives and the extreme right UKIP. From 2014 through to 2016 the British Conservatives were allied in the European parliament with many extreme parties including the German AfD neoNazis, who they ditched a little later as the British public could recognize them for who they were (Rankin, 2016). The British public tend to have less good knowledge of all the other minor far-right European parties now all allied with the UK Conservatives; but they do have a folk memory of what far-right in Germany means.

The rise in far-right voting in Britain has been far greater than in any other country in the EU when far right is defined as voting for a political party allied to a political group to the right of the mainstream EPP European Conservatives. In the European elections of 1979, 1984 and 1989 almost no far-right or extreme-right candidates stood. In 1994 those that stood won 1% of the vote, in 1999 they secured 7.5%; in 2004 a fifth; in 2009 around a third, and in 2014 over half. Thankfully in 2019 the far-right Conservative+UKIP+Brexit Party vote share fell as compared to 2014. Along with the loss of one Unionist seat in Northern Ireland to an Alliance MEP (Member of the European Parliament) this meant that 11 fewer far-right MEPs were sent to the European Parliament in 2019 as compared to 2014.

After the 2019 European parliamentary election the rump of the two UKIP and one Democratic Unionist MEP joined a small group of "Non-Inscrits" (none attached members) now standing alongside the fascist Golden Dawn MEPs of Greece and the extreme-right Jobbik MEPs of Hungary. The Conservative MEPs, now reduced to just four, joined a group called ECR that was mainly made up of the Polish extreme right "Law and Justice Party". Meanwhile the new UK "Brexit Party" was finding it hard to form any alliances at all with any other group of other MEPs and, at the time of writing (June 16, 2019) were not party of any bloc.

5 The human geography of Brexit

When writing in 2019 on whether economic inequality can be reduced, we have to begin with some of the most inequitable countries in the affluent world, of which the UK and USA are among the most inequitable and ask what hope there might be given what is currently occurring in those countries. We could look at Trump in the USA, but his agenda is fairly clear. Looking below the surface of the Brexit saga in the UK gives us some clues in relation to the UK where the right-wing does a slightly better job of hiding its true intentions. This is useful when considering other unequal countries. Often a false narrative is spun very strongly in places where inequality is most high. There is generally greater honesty and trust, both in life in general and in politics, when economic inequalities are low.

The European elections of 2014 were, in hindsight, a pretty good guide to what the outcome of the 2016 Brexit referendum might be. Brexit was a project led by a very small number of people with personal household/family incomes well up into the 1% who simply won a majority too early. They did succeed for a time in successfully selling the story that it was the voice of downtrodden northern working class that they had fired up; by cherry picking a few places and ignoring the low turnout there. In general, in areas where people did not usually vote, such as London and the North of England, those who might have voted Remain in greater numbers were very unlikely to vote (given Remain were assumed to be winning and general lack of interest). Turnout was lowest among young working-class men living in the North of England. But that, and the general performance of the Remain campaign, was not for lack of effort by the Labour party.

Over the course of the referendum campaign, the UK Labour Party spent £4,852,423 pledging to Remain on campaigning, on campaign materials and on trying the oppose the wealth and might of the newspapers owned by a few media moguls and a small number of their very rich friends who were implicitly or explicitly funding the Brexit campaign. Labour's £4.85 million was the large majority of all the Remain spending during the entire referendum campaign. The Conservative "In" campaign only spent £595,475—despite having access to far more wealthy potential donors (Electoral Commission, 2018). The Labour Party was ghosted in TV debates and much of the media to the extent that the vast majority of people had no idea that it was campaigning so hard to Remain and spending so much money to do so. Commentators on the BBC TV, radio and internet news sites still regularly say that both the Labour and the Conservative parties were and are 'equally split' and that Jeremy Corbyn 'really supported Leave'. If he did, why spend £4.85 million pounds doing the opposite? These reports that Labour did not care were almost never properly challenged. They were spending the hard earned money of many tens of thousands of trade unionists and other modestly off people to try to combat the large donations of just a few Brexiteer multi-millionaires.

In the run-up to the referendum Corbyn made 123 public appearances, including 60 in just 22 days, traveling the length and breadth of the country. Angela Eagle, a Labour shadow minister, who later left the front bench, praised him at the time for, 'pursuing an itinerary that would make a 25-year-old tired' (Barker, 2016). You would never have known he had done this if you watched the news or read almost all of the later 'analysis'.

The June 2016 Brexit referendum result was a shock when it came. We might now say that it should not have been, but we must not forget that it was. It was especially a shock to those who campaigned for Leave, many of whom clearly had not expected to win, weren't prepared for winning and had much to lose by winning. That could be seen not just from the faces of Boris Johnson and Michael Gove on the day of the result, but from the fact that the victors did not cover their tracks in terms of the activity they took part in which was later deemed to have been illegal. And, as a consequence, there has been political chaos from the date of the vote, 23 June 2016, right through to the day Britain did not leave, March 29th 2019, and beyond.

Brexit is part of an agenda for enabling the richest to turn Britain into a country that enables the wealthiest to take even higher proportions of the wealth generated in the country. That only became clear to people in Britain in the summer of 2019, after Mrs. May resigned in May, and as Conservative party leadership candidates who had only ever looked after the interests of the extremely wealthy competed with each other over who could leave the EU the quickest.

With a colleague I wrote a book about Brexit during the febrile months of late 2016, 2017 and almost all of 2018 (Dorling and Tomlinson, 2019). However, one thing that was not highlighted strongly enough in that book was that that the Leave vote was a majority middle class vote, 59% of Leave voters were middle class. The exit-poll sample that this vital statistic was estimated from was so large that the 95% confidence limits on that figure are 58%–60%.

In the spring of 2019 local elections were held in which the Conservatives did spectacularly badly, but a new party, the Brexit party had emerged and appeared to be about to sweep the board (see Fig. 2). Fig. 2 is deliberately drawn to look like one of those patronizing graphics which are frequently now used in the TV news in place of just giving the three numbers to be illustrated. Needless to say the BBC did not actually make the comparison being shown below in that figure. The BBC did say that the two main parties had both received a 'string rebuke' from voters.

In the event, as explained above in more detail the Brexit party did badly—but this is only seen when its results were combined with the Tory and UKIP result in the EU 2019 election that were held in the UK. No major news outlet in Britain has explained that the total number of seats in the European Parliament held by these three parties fell by 10 in 2019 as compared to when the European elections were last fought in 2014. Furthermore, none have highlighted the importance of a far-right Unionist MEP in Northern Ireland having lost his seat to a pro-Remain Alliance candidate increasing the overall pro-Brexit loss to 11. Labour, a party which was at the time ambiguous on Brexit lost 10 seats. We have to go back to the 2016 referendum to see what was really transpiring.

Most of the Leave voters of the UK lived in the south of England (including London). This is surprising as it was where only a minority of the UK electorate lived. But it was where the Tory and UKIP voters who most wanted to leave lived in greatest numbers, where the old are concentrated, and where electoral turnout is almost always highest, and was highest then. Following on from the south east region of England where referendum turnout was highest (and which does not include London), the largest turn-outs were—in descending order—the south west of England, Eastern England, and the East Midlands. Nowhere in the North of England had turnout as high as in the South. Nowhere in the North of England could a set of areas be found that was home to as many

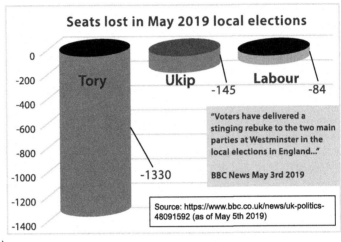

(A)

The Brexit party also continues to gain a lot of traction in the lead-up to the European elections

European election voting intention, change since 23 April

The Brexit party **34** (+6)

Labour **21** (-7)

Liberal Democrat **12** (+5)

Conservative **11** (-3)

Green **8** (+2)

Ukip **4** (+1)

SNP **4** (-1)

Change UK – The Independent Group **3** (-4)

Note: The three dark-shaded political parties are those that are aligned with political groups in the European Parliament to the right of the Conservative mainstream European People's Party (EPP).

(B)

FIG. 2

Local elections and national polls, UK, May 2019. (A) May 2019 local election results. (B) May 2019 European election results.

From Savage, M., 2019. Brexit party may get more EU election votes than Tories and Labour combined—poll. The Guardian, 11 May, https://www.theguardian.com/world/2019/may/11/brexit-party-may-get-more-eu-election-votes-than-tories-and-labour-combined-poll

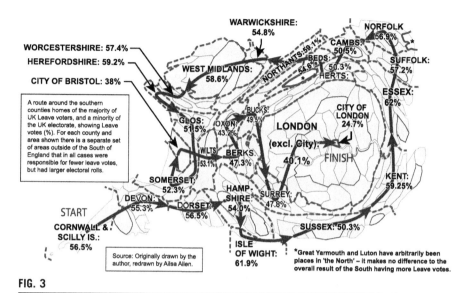

FIG. 3

The Brexit Way (a route past the homes of the majority of UK Leave voters and a minority of the UK electorate). Proportion of the vote is shown. For each county and area shown there is a separate set of areas outside of the South of England that in all cases were responsible for viewer leave votes but had more voters on the electoral roll.

actual Leave voters as in comparable areas of the South of England (an area with a similar electorate). Every single county in the Map below (Fig. 3) has a corresponding set of areas that are not in the South of England, that it can be compared to, which had fewer Leave voters despite a larger electorate in every single case.

The map below shows the 'Brexit Way', a new national walk you can take to understand the Leave voters as it goes past the homes of the majority of Leavers. So many people in the South and East of England voted Leave, that the walk would take you a very long time. It is better to cycle the route to pass along the site of the Brexit victory of Southern England. The Brexit Way could be a permanent reminder to people in future of where the most and the strongest support for Leave came from.

Although the Liberal party and the Greens may have been most vocal in expressing their love of all things European, they raised hardly any funds to support the Remain campaign and carried out almost no campaigning in what turns out to be Leave areas. The 2016 Remain campaign was almost entirely supported financially by the contributions of millions of trade unionist, by Labour party members and by small left-wing donors, mostly then funneled through the Labour Party's own coffers. In contrast, the Leave campaigns were almost entirely funded by members of the 1%. Almost all the key advocates of Leave are in the 1%, from 'European Research Group' chairman Jacob Rees Mogg to Weatherspoon's owner and vocal Brexiteer Tim Martin. The Leave supporters among the 1% own newspapers, write columns in the Daily Telegraph, and have been largely successful in their attempts to portray themselves as champions of the common man, occasionally also acknowledging the existence of women. They have been as successful as Donald Trump has been in controlling the narrative (and he has been successful).

6 Inequality, pollution and stupidity

At first the links between the politics described above and outcomes such pollution and poor educational performance are not obvious, but as we step back and look it becomes more and more clear how everything may be more closely related than you might at first think. The most economically unequal rich countries generate the most pollution as their citizens use the most energy. They are also, not coincidently, worse at maths. Fig. 4 shows the general rise in pollution per head as inequality rises. Better quality data, and taking into account whether a country was tightly densely packed, like Singapore, or relied on flying to move people around like Australia, would probably tighten up the distribution shown in the top half of Fig. 4.

The graph in the bottom half of Fig. 4 shows a remarkably tight correlation between academic performance at mathematics after leaving school and income inequality. The UK and USA appear to do so very badly because children in those countries are "taught to the test" rather than taught actual mathematics. One theory is that this is because, in a very economically divided society, outcome matters so much more. Actual ability is taken far less seriously. Fig. 5 adds to the evidence and is from one of many now widely circulated reports showing that it is, in general, the most affluent tiny slice of the population (the 10%) who pollute the most when they are living and consuming in the most unequal of affluent countries.

7 Inequality extremes in Europe

In February 2019 the *Telegraph* reported that income inequality in the UK was still growing, 10 years after the crash because 'the richest grew their earnings but the poorest were faced with falling benefits.' (Chandler-Wilde, 2019). The newspaper went on to explain to its (on average) more well-healed readers that median household income after tax was now stagnant and still standing at just £28,400 at the latest point in time for which we have data (the financial year ending in 2018). That is just £77.80 a day, to live on, to find the rent, food, clothing, bus tickets, to pay gas and electricity bills, to pay an Internet access charge, to replace a mobile phone, to buy the children clothes, maybe a book or a birthday present occasionally. Fifty percent of the population had to manage on £77.80 a day, or less. If you happen to think that is possible then it is worth knowing that the EU wide indicator for being 'at-risk-of-poverty" is 60% of median income, meaning £46.70 a day in the UK. Imagine living on that!

To calculate the national median household income the Office for National Statistics (ONS) had tallied up all income from wages, pensions, investments and benefits.

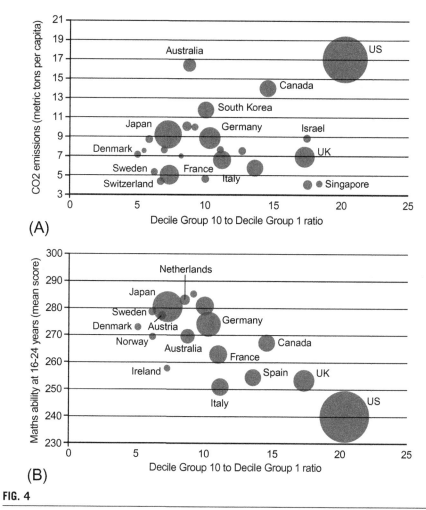

FIG. 4

Economic inequality, CO_2 emission and numeracy. (A) Economic inequality and carbon dioxide emissions, 2011. (B) Economic inequality and the mathematics ability of young adults up to age 24, 2012.

Original figures from Dorling, D., 2017. The Equality Effect. New Internationalist, Oxford.

Between April 2012 and April 2017 median household incomes had been rising by 2.2% a year, slowly gaining back some of the ground lost with the huge falls that came with the 2008 crash. However, in the latest financial year the poorest fifth of households, those families surviving (or not) on £36.50 on average a day, have seen their incomes fall. In contrast, the richest fifth of households most recently saw their incomes rise as their earnings rose and as they also benefited from very low rates of taxation (UK taxes are lower than in most other European countries). The ONS report that all the newspaper articles were relying on explained that this was the second year in a row that income inequality as measured by household surveys had risen.

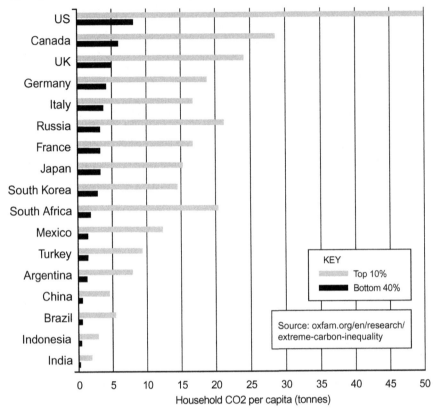

Emission of pollution by income group in selected rich nations, 2015

Household CO2 per capita (tonnes)

KEY

Top 10%

Bottom 40%

Source: oxfam.org/en/research/extreme-carbon-inequality

FIG. 5

Emission of pollution by income group in 2015.

Original figures from Dorling, D., 2017. The Equality Effect. New Internationalist, Oxford relying in turn on Oxfam.

Furthermore, we know that household surveys substantially under-estimate the income of the top 1%.

In March 2019 the Institute for Fiscal Studies (IFS) explained that there were 31,000 people in the UK with incomes within the top 0.1%, each receiving a million pounds a year or more (£3000, £5000 or even £10,000 a day to live on)! The IFS explained that this group accounted for 8% of all PAYE (pay as you earn) income tax and that that, combined with the national insurance receipts of this group for the financial year 2017–18 means that the average tax paid by each member of the 0.1% was £730,000 that year—which demonstrates just how much higher than a million pounds a year most of them were earning.

Echoing the analysis of Danny Blanchflower (quoted above), the Director of the IFS, Paul Johnson explained that 'the Bank of England had looked back through history to find a worse period for relative workers' *pay*. "It's reached the early 1800s. I think it might be heading to the black death soon…" (Partington and Inman, 2019).

In that same month (March 2019), the European Banking Authority released shocking statistics showing that the pay of bankers in the UK had begun to rise again so that now a record 3567 'top UK bankers' were receiving more than a million euros a year, with the average pay of bankers in this group being £1,700,000 a year (£4660 a day). The UK was still home to nine times as many high paid bankers as in the next most 'banker heavy' country in the EU (Germany). A decade after the 2008 crash, three-quarters of all Europe's highest paid bankers were still working and living in London; but the writing was on the wall. At the very same time as these figures were revealed, it was announced that 7000 finance jobs could be lost from London in the very near future due to the Brexit mess (Neate, 2019). The bankers were simply taking the money while they could. They knew the good times were likely to end.

For significant names at the very top, the good times were already over. Martin Sorrel, the then boss of WPP, resigned in April 2018 having been until that moment the highest paid of any FTSE 100 chief executive. His pay was already falling before he quit, and he quit in unusual circumstances (Goodley and Davies, 2018). He was reported to have been paid £70m in 2015, £48m in 2016, and £14m in 2017. A few months later Jeff Fairburn, chief executive of housebuilder Persimmon, had to resign after his £85m pay for 2 years of 'executive work' was labeled by one shareholder as 'grossly excessive'. It would have reached £110m for those 2 years' 'work' had he stayed and carried on being rewarded under the arrangements he had initially negotiated (Evans, 2019).

The mean income of the very highest paid may now be falling as the tallest poppies are cut down, but median FTSE 100 CEO median pay still rose by 11% between 2016 and 2017, and now stands at £3.93 million per year, an increase on £3.53 million in 2016. The mean 2017 pay ratio between FTSE 100 CEOs and the mean pay package of their employees is 145:1, which was higher than the year before (it was 128:1 in 2016), but lower than its peak in 2015, which was 146:1 (High Pay Centre, 2018). However, in 2018, just like the bankers taking their possible last chance to grab the cash before inequality has to fall, median CEO pay rocketed up to £3.9m. (Rutter Pooley, 2019).

When reporting on all these events, the *Financial Times* was reproached by one of the paper's letter writer's Mary Acland-Hood (of London) who suggested that '*A news headline is not the place for value judgments*'. (Acland-Hood, 2019) This was despite the headline being factual: 'Top UK CEOs earn annual wage of average worker in 2½ days'.

Recently, lower ranking bankers and CEOs started taking more when they saw that those above them were being made to take less and that they had little time left to cash in and get out. In the summer of 2018 the Archbishop Justin Welby, a former banker, mentioned that he had been in a meeting in which a group of senior bankers were reflecting upon their diminished compensation. One was reported to have said that 'back in 2007 many of us were on eight-figure salaries — i.e. over £10m [a year…] If you look around this room, there's not one of us who's getting paid more than £5m a year' (Davies, 2018).

By late spring 2019 there was no sympathy for the top CEOs and top bankers who had seen their take drop, and rising anger over any at the top who had taken more than they had received in the year before. In the February of 2019, taking into account inflation in the financial year to April 2018, ONS reported that the average income of the poorest fifth of households had fallen by 1.6%, while the average incomes of the best-off fifth had risen by 4.7%. (Partington, 2019).

In April 2019 the news was released that two million of the country's poorest families were to lose £1000 a year due to further benefit cuts (Giordano, 2019). This was on top of earlier reports that had revealed that: "…the impact of welfare reforms between 2010 and 2018, shows households with lone parents and children are set to lose an average of £5,250 – almost one-fifth of their total net income, compared to a loss of £3,000 for couples with children. This will see the child poverty rate for those in lone parent households increase from 37 per cent to over 62 per cent" (Bulman, 2018), Table 2 shows that by 2017 the UK was already the most unequal OECD

Table 2 The most unequal of OECD countries (by Income Inequality, 2017)

Rank	Gini coefficient	State
1	0.241	Slovak Republic
2	0.244	Slovenia
3	0.253	Czech Republic
4	0.255	Iceland
5	0.262	Norway
6	0.263	Denmark
7	0.266	Belgium
8	0.266	Finland
9	0.282	Sweden
10	0.284	Austria
11	0.284	Poland
12	0.285	Netherlands
13	0.288	Hungary
14	0.291	France
15	0.294	Germany
16	0.296	Switzerland
17	0.297	Ireland
18	0.304	Luxembourg
19	0.307	Canada
20	0.314	Estonia
21	0.328	Italy
22	0.330	Australia
23	0.331	Portugal

Table 2 The most unequal of OECD countries (by Income Inequality, 2017) *Continued*

Rank	Gini coefficient	State
24	0.333	Greece
25	0.339	Japan
26	0.341	Spain
27	0.344	Israel
28	0.346	Latvia
29	0.349	New Zealand
30	*0.351*	*United Kingdom*
31	0.355	Korea
32	0.376	Russia
33	0.378	Lithuania
34	*0.391*	*United States*
35	0.404	Turkey
36	0.454	Chile
37	0.458	Mexico
38	0.470	Brazil
39	0.480	Costa Rica
40	0.495	India
41	0.514	China
42	0.620	South Africa

Gini coefficient after tax and welfare redistribution, 2017 or nearest year.
Source: OECD (2019 data release).

country in Europe by income inequality other than Lithuania (and the statistics for Lithuania fluctuation greatly year on year).

The reports kept flooding in: 'The Social mobility Commission says Government policies have harmed efforts to improve social mobility by axing children's centres, cutting school budgets and limiting access to free childcare' (Hymas, 2019); Academic research has recently revealed that UK 'newspapers deploy deeply embedded Malthusian explanations for poverty' *(*McArthur and Reeves, 2019*)*. There are now many signs of a rise in disgust at the conspicuous excess behavior of a few very rich people in the world. Two years ago a party planner based in Monte Carlo explained: 'We probably got through 200 bottles of £300 a bottle Perrier-Jouët champagne. We'll go through Whispering Angel, and Garrus – their finest rose. We'll go through three kilos of French organic caviar. We also have a live cooking station with caviar and truffles' (Halls, 2017). Would such a report be written today?

In the United States, lavish tax cuts under Trump will probably cause income inequalities to rise soon. However, 2019 changes to the tax rules that prevent the very rich from deducting much of their state taxes from their federal tax liabilities have resulted in the suggestion that the rich of New York (which has higher local taxes) will now move to Florida or Puerto Rico as the changes mean they can now only deduct a maximum of $10,000 a year from their federal tax payment due to what they have to pay locally. But an exodus does not appear to be happening. The price of luxury property in tax 'haven' Puerto Rico is falling. Instead, the New York rich are gritting their teeth and paying their new taxes. In reporting on all this, the *Financial Times* asked, 'After all, if the wealthy leave New York, who is going to live in all those glass skyscrapers in Hudson Yards?' (Financial Times Reporter, 2019). It is worth thinking about who could sleep in all those luxury apartments now, often many stand empty. Many luxury apartments were also recently built in London, and also still stand empty. At least they have been built, they are available if we just have the will to find the way. The way to occupy those apartments is not to strive to be rich as was suggested by Knight Kiplinger at the very start of this chapter. Instead it is to ensure that empty and under-used apartments are taxed so highly that their owners work very hard to ensure they are occupied – or just sell them.

8 Hope

There is anger now. To see whether this is well directed enough turn the clock back just over a century to 1917. That year was the height of the last peak of income inequality, the year in which Richard Tawney wrote an essay that was published in the then educational supplement of *The Times*. He concluded:

> The educational system of today was created in the image of our plutocratic, class-conscious selves, and still faithfully reflects them. Worshipping money and social positions, we have established for the children of the well-to-do an education lavish even to excess, and have provided for those of the four-fifths of the nation the beggarly rudiments thought suitable for helots,[a] who would be unserviceable without a minimum of instruction, and undocile helots if spoilt by more. The result has been a system of public education neither venerable, like a college, nor popular, like a public house, but merely indispensable, like a pillar-box. (Tawney, 1917)

In our Internet age the pillar boxes are no longer indispensable. Public Houses are far less popular. Oxford and Cambridge colleges are no longer seen as venerable, but as snobbish and anachronistic. However, even they are changing, a little more doubtful about their pomp and circumstance, slightly more self-aware. The English educational system has not yet fully returned to its plutocratic past, although England's private schools have never been so highly funded as they are today. Their money comes through escalating fees that rose as the take of the 1% rose and as they began to take more and more of the children of the global 1%—as boarders.

Until the crash of 2008, worshipping money and social position had again begun to be seen as normal. We had sleepwalked into believing it was acceptable to provide beggarly rudiments for the seven-eighths of the nation able neither to pay for private education, nor to afford a home within the catchment area of an 'excelling' state school. We watched as our housing became unaffordable and our streets again became places to sleep. We complained as our health service worsened, but did

---

[a]The underdogs in ancient Sparta. There were seven helots to every citizen.

too little to save it. We lamented the growth of inequality, but refused to even consider outlawing gross inequities.

We can choose now to be at the peak of income inequality. We can choose now to demand it falls each and every year for many years to come. But it will take luck as well as determination if this is to be the time the turn comes. The UK provides just one example, but people are fighting in the USA, in Australia, in Israel and all around the most unequal parts of the rich world for change. As Sam Cooke sang in 1964 'Its been a long time going, but I know, a change gonna come'.

To end, here is a prayer, not a prayer for luck, but for those whose luck has already run out:

A prayer in a time of austerity
We remember all who have died while their income was sanctioned,
who were overcome by any feelings of humiliation or shame,
by fear or distrust, insecurity or loneliness;
or by a sense of being trapped and powerless
under the abuses of power by the State
in a time of austerity.
In a time of austerity we pray,
in solidarity with the 1000s of UK citizens currently suffering sanctions,
which are imposed with the maximum use of the media
to blame decent people for their own unemployment and poverty;
for the millions of UK citizens who are suffering under unmanageable debts due
to high rents, the council tax, the caps and cuts in social security
imposed by Parliament, made worse by sanctions.
We pray too for those in power, and seeking power, that they may find;
The courage to work for and implement social and economic justice,
The will to build a well-being state on the ashes of the welfare state
in which rich and poor and Parliament are in solidarity with each other
The policies to ensure that no one will have to choose between
heating or eating,
the rent or the streets,
life or death
due to the unjust enforcement of debts
against inadequate incomes,
or no incomes at all,
due to a sanction.
In the name of Jesus Christ,
Amen.[b]

---

[b]This prayer was first published in a book edited by Neil Paynter of the Iona Community, of which its author, the Reverend Paul Nicolson, is an associate. It is reproduced here with Paul and Neil's permission. N. Paynter, *In the gift of this new day: Praying with the Iona Community* (Glasgow: Wild Goose Publications, 2015). It was first said outside the DWP Headquarters, Westminster on March 19, 2015, at a demonstration called by UNITE Community, with the relatives of people who had died after their income was stopped by a job centre sanction.

References

Acland-Hood, M., 2019. A news headline is not the place for value judgments. The Financial Times. (8 January). https://www.ft.com/content/67bcf17e-1282-11e9-a581-4ff78404524e.

Atkinson, A.B., 2007. The distribution of top incomes in the United Kingdom 1908–2000. In: Atkinson, A.B., Piketty, T. (Eds.), Top Incomes Over the Twentieth Century. A Contrast Between Continental European and English-Speaking Countries. Oxford University Press, Oxford (Chapter 4. Series updated by the same author).

Atkinson, A.B., Hasell, J., Morelli, S., Roser, M., 2017. The Chartbook of Economic Inequality. https://www.chartbookofeconomicinequality.com/inequality-by-country/united-kingdom/. (See note in figure 1 for more recent sources).

Barker, D., 2016. It's neither in labour nor the UK's interests to blame Jeremy Corbyn for Brexit. New Statesman. (11 July 2016).

Blanchflower, D.G., 2019. Not Working: Where Have All the Good Jobs Gone?. Princeton University Press, Princeton and Oxford.

Brewer, M., 2019. What Do We Know and What Should We Do About Inequality? Sage, London. (graph tweeted on 12 April 2019). https://twitter.com/MikeBrewerEssex/status/1116696685068070912.

Browne-Taylor, A., 2019. 14 Frugal habits of the super rich and famous. Kiplinger Magazine (21 May). https://www.kiplinger.com/slideshow/saving/T037-S001-14-frugal-habits-of-the-super-rich-and-famous/index.html.

Bulman, M., 2018. One and a half million children will fall into poverty due to conservative welfare reforms, finds report. The Independent (14 March). https://www.independent.co.uk/news/uk/home-news/children-poverty-conservative-welfare-reforms-tory-universal-credit-benefits-poverty-a8254171.html.

Chandler-Wilde, J., 2019. Income inequality grew last year as poorest hit by benefits cut. The Telegraph (26 February). https://www.telegraph.co.uk/business/2019/02/26/income-inequality-grew-last-year-poorest-hit-benefits-cut/.

Corlett, A., 2017. Unequal Results: Improving and Reconciling the UK's Household Income Statistics. Resolution Foundation, London (December 2017). https://www.resolutionfoundation.org/app/uploads/2017/12/Unequal-results.pdf.

Davies, S., 2018. Morning Coffee: This Is What Really Happened to Banking Pay since the Financial Crisis. The Court Case Over Whether Your Boss Can Snoop Your Laptop. EFC News, efinancialcareers (EFC) (10 September). https://news.efinancialcareers.com/ca-en/323145/banking-pay-since-the-financial-crisis.

Dorling, D., 2013. Fairness and the changing fortunes of people in Britain. J. R. Stat. Soc. A 176 (1), 97–128.

Dorling, D., 2017a. The Equality Effect. New Internationalist, Oxford.

Dorling, D., 2018. Peak Inequality. Policy PressBristol (Chapter 7.11).

Dorling, D., Tomlinson, S., 2019. Rule Britannia; From Brexit to the End of Empire. Biteback, London.

Electoral Commission, 2018. Campaign Spending at the EU Referendum, Official Report. Electoral Commission, London. https://www.electoralcommission.org.uk/find-information-by-subject/political-parties-campaigning-and-donations/campaign-spending-and-donations-at-referendums/campaign-spending-at-the-eu-referendum.

Evans, J., 2019. Former Persimmon boss was paid £85m in two years. The Financial Times (18 March). https://www.ft.com/content/4c23d282-498e-11e9-8b7f-d49067e0f50d.

Financial Times Reporter, 2019. Is the home of wall street at risk of losing its wealthiest residents? The Financial Times (17 April). https://www.ft.com/content/db6acf6c-60b9-11e9-b285-3acd5d43599e.

Giordano, D., 2019. Universal credit to see 1.9 million people lose more than £1,000 per year, IFS finds. The Independent (April 24). https://www.independent.co.uk/news/uk/home-news/universal-credit-benefits-impact-income-welfare-ifs-report-a8882806.html.

Goodley, S., Davies, R., 2018. Martin Sorrell's WPP exit came amid bullying and sex worker allegations. The Guardian (11 June). https://www.theguardian.com/media/2018/jun/11/martin-sorrells-wpp-exit-came-amid-bullying-and-sex-worker-allegations.

Halls, E., 2017. How to behave (and get invited) aboard the best superyacht party at the Monaco Grand Prix. GQ Magazine (10 May). https://www.gq-magazine.co.uk/article/superyacht-parties-rules-etiquette.

High Pay Centre, 2018. High Pay Centre/CIPD Executive Pay Survey 2018 (15 August). http://highpaycentre.org/pubs/high-pay-centre-cipd-executive-pay-survey-2018.

Hymas, C., 2019. Social mobility has stagnated and is in danger of going into reverse, says Commission. The Telegraph. (30 April 2019). https://www.telegraph.co.uk/politics/2019/04/30/social-mobility-has-stagnated-danger/.

Jackson, T., 2004. Chasing Progress: Beyond Measuring Economic Growth (16 March)The New Economics Foundation, London.

Kiplinger, K., 2006. The invisible rich. Kiplinger Magazine (1 October). https://www.kiplinger.com/article/saving/T047-C014-S002-the-invisible-rich.html.

Loopstra, R., Reeves, A., Tarasuk, V., 2019. The rise of hunger among low-income households: an analysis of the risks of food insecurity between 2004 and 2016 in a population-based study of UK adults. J. Epidemiol. Community Health. https://doi.org/10.1136/jech-2018-211194 (on-line first).

McArthur, D., Reeves, A., 2019. The rhetoric of recessions: how British newspapers talk about the poor when unemployment rises, 1896–2000. Sociology. First Published on-Line, 9 April. https://journals.sagepub.com/doi/full/10.1177/0038038519838752.

Monbiot, G., 2004. Goodbye, Kind World', Guardian, 9 August 2004, referring in turn to Tim Jackson, 'Chasing Progress: Beyond Measuring Economic Growth'. New Economics Foundation (16 March).

Neate, R., 2019. More than 3,500 UK bankers paid €1m a year, says EU report: European Banking Authority says Britain is home to 73% of Europe's millionaire bankers. The Guardian. (11 March). https://www.theguardian.com/business/2019/mar/11/bankers-pay-uk-european-banking-authority-report.

Partington, R., 2019. UK income inequality increasing as benefits cuts hit poorest. The Guardian (26 February). https://www.theguardian.com/inequality/2019/feb/26/uk-income-inequality-benefits-income-ons.

Partington, R., Inman, P., 2019. Big pay rises for top earners deepening inequality, says IFS. The Guardian (14 March). https://www.theguardian.com/business/2019/mar/14/hammond-could-have-ended-austerity-earlier-without-brexit-ifs.

Pizzigati, S., 2019. The Fake Frugality of the Fabulously Fortunate: Scratch a grand fortune, one common media trope likes to suggest, and you'll find a frugal lifestyle, 11 April, "Inequality.org: Blogging Our Great Divide". https://inequality.org/great-divide/fake-frugality-fabulously-fortunate/.

Rankin, J., 2016. Tory MEPs under pressure to ditch Alternative für Deutschland: Rightwing German Party, allied with David Cameron's MEPs, caused outrage by calling for police to use firearms to stop migrants. The Guardian (8 February). https://www.theguardian.com/world/2016/feb/08/tory-meps-pressure-ditch-alternative-fur-deutschland-migrants.

Rutter Pooley, C., 2019. Top UK CEOs earn annual wage of average worker in 2½ days. The Financial Times (4 January). https://www.ft.com/content/6d79d110-0f51-11e9-acdc-4d9976f1533b.

Tawney, R.H., 1917. A national college of all souls. Times Educational Supplement 34 22 February 1917, published in The Attack and Other Papers, (London: George Allen and Unwin, 1953).

Waugh, E., 1945. Brideshead Revisited, The Sacred & Profane Memories of Captain Charles Ryder, 2011 ed. Penguin Random HouseLondon, p. 161. originally published in 1945.

Zhuang, H., Shi, L., 2016. Understanding Recent Trends in Income Inequality in the People's Republic of China, ADB Economics Working Paper Series, No. 489, July. Asian Development Bank, Philippines. https://www.adb.org/sites/default/files/publication/186143/ewp-489.pdf.

Further reading

Dorling, D., 2017b. Do We Need Economic Inequality. Polity, Cambridge.

WID, 2019. World Inequality Database. https://wid.world.

Concluding remarks

Ray Galvin

University of Cambridge, Cambridge, United Kingdom
RWTH Aachen University, Aachen, Germany

A point that comes through repeatedly in these chapters is that energy and wealth do not exist in separate spheres. How energy gets produced, supplied and consumed in our society is strongly determined by how wealthy our society is and how wealth and income are distributed. If the distribution of wealth and income are lopsided, energy production, supply and consumption will also be lopsided. It seems to follow that, if we do our energy research with our back turned toward economic inequalities, we will miss half the story, or more.

In doing research for this book, one of the most unsettling issues became clear to me through reading the works of Jeffrey Winters, Geoffrey Ingham, David Graeber and Thomas Piketty: that all human civilizations tend to drift in the direction of higher and higher inequality, unless something tumultuous or determined intervenes. The drift toward inequality almost has the character of a law of nature, as James Boyce once neatly modeled in the sphere of environmental economics (Boyce, 1994). Imagine a civilization where everybody starts with exactly the same amount of wealth. As soon as one person or group gets a little more than the others, that gives them a little more power to shape the rules and play the markets, which makes them just that little bit richer still, at the expense of the others. That gives them yet more power in the markets and the rule-making, and so on and so on, until the civilization eventually reaches the hugely unequal form that almost every civilization in the history of the planet has had. Sadly, most civilizations have been characterized by extreme hyper-inequality, with just a small percentage owning almost everything and the vast majority existing in abject poverty or slavery, or the practical equivalent of slavery. Sometimes the level of inequality reaches a ceiling, where it cannot go any further because if the poor are too deeply beggared they have no energy or means to continue producing goods for the wealthy to live off.

It has also become clear that the relative egalitarianism in high-income countries in the latter half of the 20th century was an almost unique exception to this general rule.[a] Most of us doing energy research in high-income countries today were either brought up in that period, or are still benefiting from its aftermath, to a greater or lesser extent. That probably makes it hard for us to see how different it was from the norm. It can also obscure from us the fact that our society is heading back toward the norm. Just during the 13 months while this book was being written, the fortunes of

---

[a] The other exception was communist countries, but the totalitarian nature of most of these is not something I would argue for.

most of the world's billionaires increased by billions, but prospects for the vast majority of our young people continued to stagnate or shrink. Even though wage inequality may have stabilized or slightly reduced in some countries, the inequalities in accumulated wealth are now so great that they have become self-reinforcing. As Piketty (2014) shows, for example, the wealthiest US universities increase their assets by around 10% each year, simply because they can afford the cleverest financial managers, while a typical run-of-the-mill university can barely chalk up 2%. Individuals have it even worse. New studies in the UK show that if young people move to big cities where the wages are higher than in their home towns, the higher rents they have to pay often make them worse off than before.

Another point that became clear in the research for this book is that high-income countries like the UK, Germany, New Zealand, Australia, the US and France are extraordinarily rich, in both income and accumulated wealth. Their governments may claim to be short of funds, but the sum of the wealth of their people is huge, as is the sum of their incomes. The argument that there is not enough money in the UK or New Zealand to fix the deep, chronic problem of energy poverty, for example, is patently false. There is plenty of money, billions and billions of untapped wealth. It is simply a matter of shifting it from where it is being hoarded, hidden in tax havens, wasted on CO_2-intensive luxury items like super-yachts, private jets, holiday estates, penthouses and the like, and directing it toward socially useful projects and redistribution.

And this, of course, raises a key question for energy research and social science. Should we expand our focus to include the concern for better redistribution of wealth and income? Or should we bracket this out, treat it as a black box, a fact of life that we all just have to live with, and limit our work to examining just the energy component of society, given that society is very unequal?

This issue comes sharply into focus in energy poverty research. There are more and more studies aiming to help governments identify more and more precisely which households or categories of households are most likely to be suffering energy policy, so that 'scarce' resources can be more effectively 'targeted' to support those who are most affected. Targeting may sound like an efficient use of resources, but there are many complexities and quandaries in trying to find out precisely who is in energy poverty and who is not, and in bringing just the right type and quantity of resources to lift each of them into energy sufficiency. One of the strengths of the UK's winter fuel payment—though it is very small—is that every pensioner gets it and can decide whether to use it how they like. They might decide to buy warmer clothes, turn up the heating, draft-proof the windows, or spend a couple of hours sipping a hot drink in a warm café 5 days a week for the coldest 4 months of the year (it would just pay for that). Some argue that only the poorest should be given the payment, but imagine the bureaucracy and expense of trying to establish who is poor and who is not. In any case, since the Department of Work and Pensions electronically records the payments for Her Majesty's Revenue and Customs, the rich pay 45% of it back in tax.

Trying to identify precisely who is in fuel poverty can also border on means-testing, where government agencies carefully distinguish between the deserving poor and the undeserving poor. This approach tends to resonate with the 'poor laws' of late Tudor England. The government in those days was caught between wanting to do its Christian duty of giving relief to the poor, while at the same time feeling it would be immoral to give a free ride to those who did not deserve it. Lowrey (2018) argues that this conflicting set of attitudes has been deeply embedded in poverty relief programs for centuries, except perhaps in the Nordic countries.

Energy poverty research can inadvertently continue and deepen this tradition. I have argued in this book for a different approach, where we put the targeting aside, expose the myth of scarce resources, and deal directly with the poverty that lies behind energy poverty. This would require the focus of energy poverty research to expand into territory many energy researchers are not familiar with. But as academics, would we not enjoy a new challenge?

Shifting the focus to another area of energy research: social practice theory—often known simply has practice theory—has now been used in energy research for some 15 years. This began as a refreshing, welcome attempt to understand energy consumer behavior in terms of society's routinized, everyday practices, rather than individuals' inner psychological motivations. One of the philosophical pillars of practice theory is Anthony Giddens' (1984) well-argued tenets that societal practices are shaped by social structures, that social structures are shaped by the actions of people, and that well-resourced, powerful actors have far more influence on what these social structures become, than average folk. So the question arises: Who are these well-resourced people, these powerful actors, who are heavily influencing the way society gets structured around energy? In what ways, precisely, are they influencing how energy gets produced, supplied and allocated in society? What actions, machinations and maneuvers are they engaging in, to keep the energy scene the way it is?

Unfortunately, this side of Giddens' notion of practice theory hardly ever came to fruition, if at all, in the versions of practice theory used in energy studies. Instead, we see an almost slavish focus on simply identifying the routinized practices that individual consumers are hooked into. But to be faithful to the Giddensian roots of practice theory, we would have to start researching upward, from the consumers' routinized practices, to the social structures that form the practices, to the powerful actors who influence the rules that make the social structures what they are. In other words, we would need to start naming names.

Historians are always happy to name names. History would be dull if it was just a description of practices. History is more often than not about powerful actors and what they did to influence the shifts in society and world events. Energy studies may need to write a history of the recent past, or even a history of the present, daring to name the names that keep the CO_2 juggernaut in its trundle and deprive millions of households of basic energy services. Practice theory would be a good vehicle to start with, if we would take it right back to its roots.

Another ever-popular energy research approach is the theory of planned behavior (TPB). This focuses entirely on consumers' inner mental processes. Its followers boast that over 50% of the energy behavior of consumers can be explained by the three variables: attitudes; social norms; and perceived behavioral control. But as

Hargreaves (2008) over a decade ago pointed out, the third of these variables, 'perceived behavioral control' is a kind of escape hatch that lumps together everything that is not a mental process, yet without exploring its components. So, for example, if a person cannot heat their home because they do not have enough money to run its inefficient heating system or to upgrade it and insulate the house, this is neither an attitude problem nor a problem of social norms. It is therefore left unexplored, coming under the black box of (lack of) perceived behavioral control.

As such, then, the TPB as such is no use for energy studies that want to take account of economic inequality, social structures, or the powerful actors who have most influence on these. It could, however, be a useful stepping stone, if its researchers would go beyond the psychology of the consumer and her social circle to look critically at all the factors that contribute to lack of 'perceived behavioral control'. These would, of course, include the consumer's economic situation, the social structures that make her poor, and the powerful actors that hold these in place.

I suggest, then, that whatever research framework we use for energy research, we would do well to push it further, beyond its traditional boundaries, toward the actions and machinations at the top of the power structures in which our subject matter is embedded.

There is a further advantage in doing this. One of the points that several of us have tried to make in this book is that the energy injustices we see about us today are not inevitable. Lawrence Haar has shown us that poorer households pay disproportionately for the energy transition in most EU countries; Nicola Terry has alerted us to the sad condition of much of the UK's non-social housing and the gradual reduction in the number of social housing units; Lucie Middlemiss has taken us into the home of a fuel-poor householder who could have been any one of us; Minna Sunikka-Blank has documented how women, especially mothers, often bear the brunt of both energy poverty and the energy transition. I do not think these authors are implying that this is the way it has to be.

These injustices are not inevitable. As I noted above, it is almost as if there is a law of nature that the rich get richer and the poor get poorer and that this escalates toward hyper-extremes, *unless something tumultuous or determined intervenes*. This "unless" is vitally important. Early sociologists like Durkheim and Weber tried to establish sociology as a science with the same kind of credentials as physics and chemistry, with predicable regularities in a rationally understandable universe. They were right to claim sociology can produce robust knowledge of the world, but it was a mistake to assume society works like physical or chemical substances, with predictable regularities that apply at all times and places. Unfortunately, today's social science can also make this mistake—such as when practice theory, at its worst, simply looks for regularities among various social practices and habits. If we take this approach, we can end up assuming the poverty of increasing numbers of people is inevitable and the course of history has to be left to play out its cycles.

What other writers like Giddens (1984), Harré (2009) and Wittgenstein (1953) have impressed upon us is that these regularized social structures do not hold together of their own accord, by some rarefied law of nature, but only because powerful human beings are acting, deliberately, to keep them that way. And if other human beings have the right kind of resources, they can intervene decisively and change these toxic social structures into something new.

Acemoglu and Robinson (2012) have provided a refreshing account of some of the biggest changes in recent social and political history, from the point of view of 'contingencies': unexpected events that cannot be explained simply in terms of chains of cause and effect unfolding within social structure. For example, feudalism in England did not die of its own accord due to its own internal faults, but largely because the great plague intervened and upset the numerical balance of laborers to masters. Another example: unskilled workers in the 1950s earned good salaries and could buy houses and raise families on one salary, not because some inevitable law of nature had tipped the balance in their favor, but due to a number of unique factors that came together at the right time. Chief among them was that millions of committed people had organized, lobbied, campaigned and made sacrifices over decades of struggle, to pressure governments to adopt radically progressive fiscal and social policies. Meanwhile, their allies among economists and political scientists worked hard to develop and win the intellectual arguments to show how these policies would not beggar their economies but enrich them.

This lends itself to an approach to energy studies where we research the chains of causality all the way up to the powerful actors who have disproportionate influence on government and industry, including energy production and supply. In order to have a hope of curbing climate change and establishing energy justice, these specific human actions need to be made known. This can help inform those who want to press for better policies, better redistribution of resources, and management of energy that protects the life of the planet.

References

Acemoglu, D., Robinson, J., 2012. Why Nations Fail: The Origins of Power, Prosperity, and Poverty. Crown Business, New York.

Boyce, J., 1994. Inequality as a cause of environmental degradation. Ecol. Econ. 11, 169–178.

Giddens, A., 1984. The Constitution of Society. University of Los Angeles Press, Berkeley and Los Angeles.

Hargreaves, T., 2008. Making Pro-Environmental Behaviour Work: An Ethnographic Case Study of Practice, Process and Power in the Workplace (Thesis for the Degree of Doctor of Philosophy). University of East Anglia, School of Environmental Sciences.

Harré, R., 2009. Saving critical realism. J. Theory Soc. Behav. 39 (2), 129–143.

Lowrey, A., 2018. Give People Money: The Simple Idea to Solve Inequality and Revolutionise Our Lives. Penguin, London.

Piketty, T., 2014. Capital in the Twenty-First Century (Translated from the French by Arthur Goldhammer). Belknapp-Harvard University Press, Cambridge, MA.

Wittgenstein, L., 1953. Philosophical Investigations. (Translated by G.E.M. Anscombe), Blackwell, Oxford reprint 1967.

Author index

Note: Page numbers followed by *b* indicate boxes and *np* indicate footnotes.

A

Aalbers, M., 25, 67, 146–147
Abbott, J., 106
Ablanedo Rosas, J.H., 275
Abraham, C., 55
Abundo, M.L., 275
Abu-Rub, H., 275
Acemoglu, D., 21–22
Achtziger, A., 63, 152
Acland-Hood, M., 303
Adams, A., 122–123
Aglietta, M., 41, 44
Agyeman, J., 279
Ahern, P., 216
Akyelken, N., 101–102
Albala, P.A., 100–103, 107–108
Aldrich, C., 176
Ali, W., 275
Alia, A., 64
Allen, D., 78–79
Allen, J., 120
Aloise-Young, P., 126
Alstadsaeter, A., 19–20
Alston, P., 65–66, 90, 226
Alvaredo, F., 3–4, 15, 25, 31
Alvial-Palavicino, C., 79–80, 262
Ambrey, C., 80
Ambrose, A., 101–102, 133–137
Anable, J., 79–80
Anderson, W., 106, 223
Andersson, G., 179, 182
Andor, M.A., 191, 195–196
Antoniades, A., 31–35, 41, 64
Araujo, M.B., 263
Aravena, C., 136
Archer, M., 4
Aron, J., 63, 152–153
Ástmarsson, B., 126, 133–134
Atkinson, A.B., 15, 291
Austin, J., 43*b*
Ayoub, N., 193

B

Bach, S., 210, 214–215, 268
Baffert, C., 149

Baker, K.J., 100, 106, 109–110
Baker, L., 80–81, 253, 263
Baker, N., 59
Balfour, R., 120
Baptista, I., 277
Bardhan, R., 174–175, 177
Barker, D., 296
Barker, N., 119
Barnard, H., 178
Barnes, J., 79–80
Barry, J., 80, 254, 259
Bartiaux, F., 76, 80, 226–227
Barton, B., 53–55
Barton, C., 119, 134
Beaumont, A., 129
Beder, S., 61
Beggs, M., 32, 44, 60
Bennett, F., 179
Berkout, F., 245
Berry, A., 80
Berta, P., 182–184
Bertram, G., 191
Bevan, M., 80, 100–102, 106
Beznoska, M., 210, 214–215
Bianchi, S.M., 176
Billig, M., 57
Birol, F., 261
Bjerg, O., 39*np*
Blanchflower, D.G., 294
Bloome, D., 182–184
Boardman, B., 99–101, 150, 179, 223–224, 237, 252, 261
Boccanfuso, V., 64
Bolton, P., 225
Boltz, W., 228
Booth, A., 136–137
Booth, N., 174–175
Botti, F., 175
Boulangeat, I., 263
Bourdieu, P., 4, 40–41, 55
Bouzarovski, S., 76–77, 80, 100–104, 109–110, 146, 149–150, 225–226, 257, 259, 261–263, 271, 280
Boyer-Xambeu, M., 39
Brajković, J., 149

Brandão, R., 275
Braun, B., 32, 34, 37–38, 38*np*
Braun, P., 45–46
Brazier, K., 149–150, 179, 227–228
Brazilian, M., 193–194
Brennan, A., 252
Brenner, N., 45–46
Brewer, M., 179, 294
Bridges, S., 63
Brigg, M., 45–46
Briggs, R., 153
Broad, R., 65
Broman Toft, M., 275
Brooke-Peat, M., 126
Brown, J.S., 217
Brown, K., 103
Browne-Taylor, A., 288
Brunner, K.-M., 106, 223
Buchmann, M., 275
Buckley, P., 175
Bullion, S.D., 62–63
Bulman, M., 304–305
Burgess, J., 54–55
Burke, M., 80–81, 253, 263
Butler, C., 110
Butler, D., 106
Butler, P., 90–91

C

Cabras, I., 255
Campbell, R., 224–225
Cardenas, J.A., 275
Carlsson-Kanyama, A., 176
Carrasco, J.-A., 101–102
Carrette, J., 65
Carrington, G., 53–55
Carroll, J., 136
Castán Broto, V., 277
Cauvain, J., 109–110
Cavanagh, J., 65
Chan, N.W., 260
Chancel, L., 3–4, 15, 25, 31, 68, 238
Chandler-Wilde, J., 300
Chappells, H., 150
Chaton, C., 252*np*
Chatterton, T., 79–80
Chawla, M., 193–194
Cherp, A., 253
Cheyne, C., 80
Chick, V., 32
Christanell, A., 106, 223

Chua, P.P., 176
Cialani, C., 214–215
Cisternas Solsona, D., 101–102
Clegg, S., 58–59
Clower, R., 40
Colak, I., 275
Collins, L., 78–79
Cooke, J.M., 106
Corlett, A., 294
Correia, A., 54–55
Correljé, A., 80
Corsi, M., 175
Cory, K., 194
Costa Dias, M., 179
Coulombel, N., 80
Couture, T., 194
Cox, H., 65, 226
Crawford-Brown, D., 129–130
Crib, J., 222
Cullinan, J., 242
Cuppen, E., 80
Cupples, J., 150
Curl, A., 103–104

D

Dankelman, I., 174–175
Dantas, G.A., 275
D'Arcy, C., 178
D'Ascendis, D., 216
David, M., 252*np*
Davidson, E., 107–108
Davidson, P., 32
Davies, M., 129
Davies, R., 303
Davies, S., 303
Davies, W., 23
Davis, N., 90–91
Day, R., 53–54, 76–77, 79–80, 87,
    100–104, 101*np*, 106, 146, 149, 226–227,
    259, 261
de Castro, N.J., 275
de Chavez, A.C., 101–102, 106
De Miglio, R., 149
Deacon, B., 226
Deane, P., 149
Debnath, R., 177, 185
Delbeke, B., 100–101, 109–110
Deleplace, G., 39
Denny, E., 136
Dequech, D., 32
Devalière, I., 109–110

Dewey, J., 81
Di Muzio, T., 9–10, 25, 32–36, 44–45, 56, 62–64
Diakoulaki, D., 149–150
Diamond, C., 82*np*, 83–84
Dias, L., 275
Diekmann, J., 210, 214–215
D'Ippoliti, C., 175
Disney, R., 63
Dobbins, A., 149
Dodd, N., 35, 39
Dorling, D., 3–4, 287–307
Dow, S., 32
Doyon, A., 261–263
Druckman, A., 260
Dubois, U., 149, 179, 182
Duca, J., 63, 152–153
Dumez, H., 45–46
Dunlap, R., 61
Dworkin, M., 76, 79–81, 84, 88–89, 262, 281
Dzialo, L., 176

E

Eadson, W., 135–136
Eames, M., 76, 79
Eaqub, S., 26, 146–147
Ecclestone, K., 103
Edwards, R., 107–108
Eirich, G.M., 182
Ellaban, O., 275
El-Laboudya, T., 210, 214–215
Elnakat, A., 174–175
Emmel, N., 100–103, 107–108
Erbach, G., 195
Evans, J., 303
Evans, T., 279
Eyre, N., 133–134

F

Fagan, C., 177
Farrell, N., 215–216
Faruqui, A., 275
Fernandez, R., 25, 67, 146–147
Finney, A., 106, 223
Fitzgerald, J., 68, 238
Fletcher, M., 126
Flues, F., 267
Fonseca, S., 54–55
Forman, A., 76, 79
Forrest, R., 26
Fortier, M.-O.P., 254, 259, 273–274
Foucault, M., 46, 55
Fouquet, R., 278–279

Freeman, M., 61
Freitas, M.A.V., 261
Frieling, J., 278
Frogneux, N., 76, 80, 226–227
Frondel, M., 191, 195–196
Fuerst, T., 136
Fulli, G., 275
Fylan, F., 126

G

Gabriel, M., 135
Galvin, R., 3–27, 31–48, 53–70, 75–93,
      145–168, 174, 221–234, 237–245, 262,
      269–271
Gardiner, L., 178
Gardner, B., 55
Garriga, A., 182–184
Gatersleben, B., 260
Gaventa, J., 58–59
Geels, F., 58–59
Gemoets, L., 275
Genovese, A., 252
Gentry, M., 129–130
George, S., 65
Giddens, A., 40–41, 44, 54, 57–58, 68
Gilberson, J., 135–136
Gilbertson, J., 100–103, 106–108
Gillard, L., 39
Gillard, R., 80, 100–104, 106–110, 222*np*
Gillingham, K., 260
Giordano, D., 304–305
Glew, D., 126
Gnoth, D., 53–55
Goebel, J., 209*np*
Goldthau, A., 79–80
Gomez, J.D., 174–175
Gonzales, L., 65
Gonzalez Pijuan, I., 176, 178–179
Goodley, S., 303
Gorse, C., 126
Gould, A., 24, 65
Gousy, H., 134–136
Graeber, D., 4–6, 21–22, 24–25, 32–33, 40
Graetz, M., 60–61
Gram-Hanssen, K., 54–55
Granovetter, M., 32, 54
Grösche, P., 214–215, 268
Großmann, K., 100–103, 106–107, 109–110
Grubler, A., 253–254
Guardian, T., 174, 177–178, 182–184
Gui, E.M., 275–276

Guiney, C., 101–102
Guivarch, C., 80
Guyatt, V., 150

H

Ha, S., 63–64
Ha, T., 177
Haar, L., 189–217
Haarstad, H., 80
Habermas, J., 86
Hake, J.-F., 259
Hall, S., 79
Halls, E., 305
Hamilton, I.G., 129, 132
Haneda, H., 262–263
Hargreaves, T., 54–55, 100–103, 107–108
Härkönen, J., 179, 182–184
Harnisch, M., 268
Harold, J., 242
Harré, R., 57
Harrington, B.E., 68, 106
Harris, D., 275
Harrod, R., 32
Hasell, J., 291
Hauertmann, M., 260
Healy, N., 80, 254, 259
Healy, S., 277
Heffron, R.J., 75–76, 78–81, 252–254, 259, 261–264, 269–271, 274
Heindl, P., 228–229, 254, 265–266, 268, 273
Hemmati, M., 174–175
Henggeler Antunes, C., 275
Hernández, D., 79–80
Herrero, S., 149
Herring, H., 260
Hey, C., 67
Heyman, A., 106
Heyman, B., 106
Hicks, J.R., 272*np*
Hill, K., 78–79
Hills, J., 104, 149, 228, 261, 264–265
Hirayama, Y., 26
Hitchings, R., 106
Hiteva, R.P., 271
Hledik, R., 275
Hofman, R., 25, 67, 146–147
Holzemer, L., 100–101, 109–110
Homer, C., 106
Hope, A.J., 136–137
Horne, R., 101–102
Horta, A., 54–55
Höwer, D., 255

Huang, X., 68, 238
Huber, M., 64
Hubert, M., 63, 152
Huebner, G., 129
Hughes, T., 191
Hulme, J., 129
Hymas, C., 305

I

Imbert, I., 228
Imran, M., 80
Ingham, D., 56, 61, 64–65
Ingham, G., 21, 23–25, 32–34, 36–37, 40–42, 44, 47, 56
Inman, P., 302–303
Isaac, N., 268

J

Jackson, T., 292
Jacobson, A., 252–254, 258, 263–264, 279
Jacques, P., 61
Jalovaara, M., 179, 182
James, W., 81
Jamieson, L., 107–108
Jenkins, K., 76, 78–81, 87, 260, 269–271
Jensen, J., 226
Jensen, P.A., 126, 133–134
Jeunemaitre, A., 45–46
Jewell, J., 253
Johansen, N., 19–20
Johansson, T.B., 253–254
Johnson, C., 176
Johnston, D., 126
Jones, B.R., 76, 78, 80–81, 84, 87, 271
Jones, R., 40
Jorgenson, A., 68, 238
Jouffe, Y., 80, 101–102

K

Kabeer, N., 10, 14, 175
Kahlheber, A., 100–103, 106–107, 109–110
Kaliampakos, D., 150, 228
Kammen, D.M., 252–254, 258, 263–264, 279
Kamp Dush, C.M., 177
Kanschik, P., 254
Kawachi, I., 12
Kearns, A., 103–104
Kelly, S., 129–130
Kennedy, B., 12
Kenning, P., 63, 152
Kenway, P., 149–150
Kerr, N., 100, 108–109

Keynes, J., 36
Khalid, R., 174
Khalilpour, R., 216–217
Kharb, R., 275
Kimberly, J.R., 252*np*
King, R., 65
Kiplinger, K., 287–288
Kirshner, J., 277
Klasen, S., 175, 177
Knight, K., 68, 238
Koh, S.C.L., 252
Kotikalapudi, C.K., 80–81, 253, 263
Kotz, D., 25
Krause, P., 209*np*
Kreycik, C., 194
Krishnan, R., 63–64
Kronsell, A., 174–175
Kumar, S., 275
Kurata, K., 63
Kurz, T., 55

L

Labandeira, X., 214–215
Labeaga, J., 214–215
LaBelle, M.C., 79–80, 262
Laclau, E., 58–59
Lacroix, E., 252*np*
Lafourcade, B., 263
Langdon, S., 149
Lauber, V., 31
Lavergne, S., 263
Lawson, R., 53–55
Laxroix, E., 252*np*
Lazar, N., 159, 162*b*
Lazzarato, M., 9–10, 25, 32–34, 36–37, 41–42,
    45–46, 48, 56, 62–64
Legendre, B., 228–229
Lehmann, J., 57
Lester, J., 78–79
Lewis, P., 261
Li, K., 261
Liang, X.-J., 261
Liddell, C., 101–102, 149, 228
Linares, P., 228–229
Lindley, S., 100–102, 226
Livingston, M., 122–123
Lloyd, B., 261
Lomas, K., 129–130
Longhurst, N., 39, 47–48
Loopstra, R., 292
López, X., 228–229
López-Oteroa, X., 214–215

Lovibond, S., 81–83
Lowe, R.J., 132
Lowrey, A., 234
Lucas, K., 101–102
Lucas, R., 153
Lusambili, A., 106
Luthra, S., 275
Luzecka, P., 110
Lyons, S., 215–216, 242

M

MacGill, I., 275–276
MacGregor, S., 174–175
MacInnes, T., 149–150
Madlener, R., 252–281
Magnusdottir, G.L., 174–175
Mahmood, A., 275
Mainali, B., 253
Majumdar, R., 175
Malm, A., 57, 64
Marchand, R., 252
Maréchal, K., 100–101, 109–110
Marsden, G., 101–102
Marx, J., 259
Marx, K., 40, 57
Maslesa, E., 126, 133–134
Massey, D.S., 182–184
Massot, M.-H., 101–102
Masuda, T., 262–263
Mayer, I., 149, 222–223
McAllister, P., 136
McArthur, D., 305
McCarthy, L., 101–102
McCauley, D., 75–76, 78–81, 87, 252–254, 259,
    261–264, 269–271, 274
McCormick, J., 22
McDaid, K.A., 106
McDonagh, E., 66
McDowell, L., 177
McGarvey, D., 25
McKenzie, S., 228
McKie, R., 90
McLanahan, S., 182–184
Md, F.A., 275
Mebratu, D., 75–76, 79–81, 252
Meckling, J., 67
Meier, H., 149, 179
Melzer, B., 67
Merleau-Ponty, N., 106
Meyer, S., 100–101, 109–110
Michelfelder, R.A., 216

Middlemiss, L., 31, 99–112, 149, 222*np*, 224–226, 228
Milanovic, B., 15
Miles-Shenton, D., 126
Milkie, M.A., 176
Mill, J.S., 273
Miller, C., 80
Millman, M., 135
Milman, A.D., 252–254, 258, 263–264, 279
Miraftab, M., 61
Mirasgedis, S., 149–150
Mizruchi, M., 40–41
Moezzi, M., 76, 80, 226–227
Monbiot, G., 292
Moore, R., 228–229
Moore, S., 149
Morelli, S., 291
Morgan, S., 65
Morris, C., 101–102, 228
Morrison, K., 40
Mortazavi, R., 214–215
Mouffe, C., 58–59, 81, 83–84, 86, 88
Mould, R., 100, 106, 109–110
Muellbauer, J., 63, 152–153
Mullen, C., 54–55, 79, 100–103, 107–108
Mundaca, L., 75–76, 79–81, 252–254
Mundell, R., 24, 34, 38*np*, 42
Munro, P., 277
Murata, K., 152–153
Murphy, A., 63, 152–153

N

Naeem, U., 275
Nahm, J., 67
Nakagami, H., 262–263
Nakhooda, S., 193–194
Nakicenovic, N., 253–254
Nanda, A., 136
Narda, L., 64
Nasra Haque, A., 174–175, 177
Neate, R., 303
Nelson, P., 106
Nersisyan, Y., 253
Neuhoff, K., 210, 214–215
Neutens, T., 101–102
Neves Alves, S., 277
Newbery, D.M., 211*np*
Newell, O., 31, 42–43, 47, 61
Nimal, E., 149, 222–223
Nobre, N., 54–55
Nogue, P., 149, 222–223
Nogues, P., 228

Norris Keiller, A., 222
Novokmet, F., 24–25
Nye, M., 54–55

O

O'Sullivan, A., 129
Oberst, C.A., 255, 260–261, 276
O'Leary, J., 80
O'Loughlin, D., 63
Oreszczyn, T., 129, 132
Orléan, A., 41, 44

P

Pachauri, S., 253–254
Palmer, G., 149–150
Palmer, J., 122–123
Pantzar, M., 54–55
Papada, L., 150, 228
Paravantis, J., 224
Parkhill, K.A., 110
Parsons, T., 40–41
Partington, R., 302–304
Pearce, J., 150
Pearson, P.J.G., 278–279
Pellicer, V., 100–101, 106–108, 110
Peltonen, M., 45–46
Percheski, C., 182–184
Pereira, M.G., 261
Perrons, D., 177
Pesch, U., 80
Petrova, S., 100–104, 106, 225–226
Pettit, J., 78–79
Philip, J.P., 132
Phillips, J., 31, 42–43, 47, 61
Phillips, Y., 137
Phimister, E., 228, 252*np*
Pichler, F., 228
Pickett, K., 146–147
Pieters, J., 175, 177
Pihlström, S., 81–84, 88, 89*np*
Piketty, T., 3–5, 10–11, 15–17, 21, 24–25, 31, 36–37, 44, 57, 60–61, 64–65, 68, 91, 145, 231*np*, 236–238
Pinder, J., 101–102
Pischner, R., 209*np*
Pizzigati, S., 288
Pollitt, M.G., 129–130, 193–194
Poruschi, L., 80
Powell-Hoyland, V., 106
Price, C., 149–150, 179
Price, H., 81
Prothrow-Stith, D., 12
Pye, S., 149

R

Raab, G., 63, 152
Rabindrakumar, S., 179, 181–182, 184
Rae, G., 228
Ramasar, V., 75–76, 79–81, 252
Ramsey, F., 217
Ramsey, J., 159
Rankin, J., 295
Rao, N., 253
Ratti, C., 59
Räty, R., 176
Rawls, J., 80–81, 85–86
Raworth, K., 39
Ray, K., 177
Razzaq, S., 275
Reames, T.G., 80, 173, 254, 259, 273–274
Redpath, T., 82
Reeves, A., 292, 305
Rehm, P., 66
Rehner, R., 76, 79–81, 269–271, 274
Reisch, L., 63, 152
Restrick, S., 100, 109–110
Riahi, K., 253–254
Ricci, O., 228–229
Richter, J., 80
Rieger, J., 65
Ritchie, N., 106
Robbins, R., 9–10, 25, 32–36, 44–45, 56, 62–64
Robert, M., 65
Robert, R., 24
Roberts, D., 228, 252np
Robinson, C., 100–102, 226
Robinson, J.P., 21–22, 176
Rogers-Vaughn, B., 65, 89np, 226
Rogner, H.-H., 253–254
Rohr, U., 174–175
Romero, J., 228–229
Røpke, I., 54–55
Roquet, C., 263
Rorty, R., 81, 86–88
Rosenow, J., 31, 133–134
Rosental, R., 275
Roser, M., 291
Ross, A., 63
Royston, S., 110
Rutter Pooley, C., 303
Ryan, T., 100–103, 107–108

S

Saez, E., 3–4, 15, 25, 31
Sagiroglu, S., 275

Sahr, A., 32, 34, 38, 38np, 47–48, 62–65
Sandilands, C., 176
Sanne, C., 174
Santamouris, M., 224
Sarasa, S., 182–184
Sareen, S., 80
Sarfi, R., 275
Saunders, J., 79
Savage, M., 25–26
Sayer, L.C., 176
Scanlon, K., 120
Schatzki, T., 54–55
Schirm, A., 159, 162b
Schlör, H., 259
Schlosberg, D., 78–79
Schmitz, H., 260–261, 276
Schoppe-Sullivan, S.J., 177
Schor, J., 68, 238
Schreiber, A., 259
Schröder, C., 214–215, 268
Schuessler, R., 228–229
Schuitema, G., 275
Schultz, I., 176
Schumpeter, J., 40
Schüssler, R., 254, 265–266
Schwanen, T., 101–102
Searle, J., 33, 83–84
Selby, J., 110
Sen, A., 259
Sevenet, M., 149, 222–223, 228
Sewell, W., 55–56
Seyfang, G., 39, 47–48
Shapiro, I., 60–61
Shaw, J., 179
Sherriff, G., 106
Shi, L., 291–292
Shimmi, S.L., 275
Shipworth, D., 129–130, 132
Shipworth, M., 129–130
Shove, E., 54–55, 110, 150, 174
Sibley, D., 217
Sidortsov, R.V., 76, 78–81, 84, 87, 271
Sieber, I., 209np
Silva, N.F., 261
Silveira, S., 253
Silvestre, J., 272–273
Simcock, H., 80
Simcock, N., 54–55, 76–77, 79–80, 87, 100–102,
    101np, 146, 226–227, 257, 259, 261–263,
    271, 280
Skea, J., 79
Skinner, E., 174–175

Smith, M., 80, 126
Smith, S., 277
Snell, C., 80, 99–103, 106–108, 149–150, 225–226
Soederberg, S., 9–10, 25, 32–33, 36, 41, 46–47, 56, 61, 63–64, 152
Sommer, S., 191
Sorrell, S., 260
Souza, M.N.M., 133–135
Sovacool, B.K., 75–81, 84, 87–89, 146, 174–175, 225–226, 245, 252–253, 262–263, 271, 281
Spitzer, M., 106, 223
Steemers, K., 59
Stephan, H., 76, 79–81, 269–271
Stephanides, P., 39
Stephenson, J., 53–55
Sterns, L., 40–41
Stevens, M., 106
Stewart, N., 152
Stiell, B., 106
Stiglitz, J., 4–5, 22, 91, 147
Stocks, A.J., 106
Stockton, H., 106, 224–225
Straver, K., 100–101, 106–108, 110
Strupeit, L., 253–254
Summerfield, A.J., 129, 132
Summers, C., 129
Sung, J., 65
Sunikka-Blank, M., 26, 36–37, 47, 54, 87, 91, 146–147, 151, 173–185
Swedberg, R., 32
Szmigin, I., 63

T

TA, D.P., 182
Taebi, B., 80
Tarasuk, V., 292
Tawney, R.H., 306
Taylor, G., 153
Teissier, O., 109–110
Teron, L., 254, 259, 273–274
Terry, N., 115–140
Therborn, G., 10
Thistlethwaite, S., 65
Thogersen, J., 275
Thomas, A., 267
Thomas, M., 23
Thomas, N., 215–216
Thomas, S., 194–195
Thompson, H., 225–226
Thomson, H., 99–102, 149–150
Thorogood, N., 106
Thorsnes, P., 53–55

Thuiller, W., 263
Tirado Herrero, S., 225–227
Tod, A.M., 100–103, 106–108
Todd, Z., 153
Toke, D., 31
Tomlinson, S., 297
Tourkolias, C., 149–150
Trishana Munardy, D., 254, 259, 273–274
Trotta, G., 133–135
Truninger, M., 54–55
Tuballa, M.L., 275
Tunstall, B., 178

U

Ureta, S., 79–80, 262
Ürge-Vorsatz, D., 227

V

Van de Graaf, T., 193–194
van den Bergh, J., 260
van der Horst, G., 277
Vance, C., 191, 195–196
Vandeschrick, C., 76, 80, 226–227
Vardiero, P., 275
Vassallo, A., 216–217
Vera-Toscano, E., 228, 252np
Verplanken, B., 55
Vinz, D., 174–175
Von Lampe, K., 24–25

W

Waddams Price, C., 227–228
Wagner, G.G., 209np
Walker, G., 53–55, 76–77, 79–80, 87, 100–104, 101np, 146, 149, 226–227, 259, 261
Walks, A., 63
Walsh, A., 135–136
Walzer, M., 86–88
Wand, W., 149–150, 179
Wang, S., 174, 176
Wang, W., 227–228
Waquant, L., 61
Ward, A., 215–216
Ward, K., 177
Wasserstein, R., 159, 162b
Waters, T., 222
Watson, C., 225
Watson, M., 54–55
Watson, P., 135
Waugh, E., 288
Weber, G., 255

Wei, C., 263–264
Wei, Y.-M., 261
Weller, S., 107–108
Werner, R., 32, 34
Western, B., 182–184
White, V., 106, 223
Whitehead, C., 120
Whitley, E., 103–104
Wicksell, K., 273
Wilhite, H., 262–263
Wilkinson, R., 146–147
Will, H., 277
Willand, N., 101–102
Williams, E., 194
Williams, K., 177
Williams, S., 261–263
Wilson, W., 134
Winters, J., 4–6, 11, 21, 25, 56, 59–61,
    64–65, 147
Wittgenstein, L., 54, 81–82, 89
Wlokas, H., 80–81, 253, 263
Wolff, E., 16–17
Wooldridge, J., 159
Wray, L., 253
Wray, R., 39
Wright, A., 129–130

Wu, S., 263–264
Wyatt, P., 136

Y

Yamaga, Y., 262–263
Yang, L., 24–25
Yavorsky, J.E., 177
Yeboah, G., 79–80
Yesilbudak, M., 275
Yuji, N., 193
Yurchenko, Y., 194–195

Z

Zafar, R., 275
Zamboni, L., 275
Zapp, P., 259
Zarazua de Rubens, G., 253–254, 259, 263–264
Zelezny, L.C., 176
Zheng, B., 264
Zheng, X., 263–264
Zhuang, H., 291–292
Zimmermann, G., 277
Zucman, G., 3–4, 15, 19–20, 24–25, 31, 60–61, 152,
    233

Subject index

Note: Page numbers followed by *f* indicate figures, *t* indicate tables, and *b* indicate boxes.

A

Actor-network theory, 53–54
Anti-money, 34–36
Atkinson index, 257, 264
At-risk-of-poverty indicator, 228
Australian Bureau of Statistics (ABS), 17–18

B

Bank loan money, 62–63
Bank of England, 38–39
Barter economy, 40
Broad-brush approach, 229

C

Cambridge energy poverty, 87
Cannibalistic capitalism, 63–64
Cash reserves, 38
Central bank money, 38–39, 43*b*
Clean Growth Strategy 2018, 138
Climate change, xix, xxiv–xxv, 174
    and energy, research on, 174
    gender-neutral, 174–175
Climate justice, 78–79
Coercive power, 60
Cold homes
    children, impact on, 120
    health impacts, 101–102, 120
    percentage of unheatable homes (*see* Percentage
        of unheatable homes (UH%))
Combined heat-and-power (CHP), 277
Commercial banking system, 61, 69
Commercial bank money, 38–39, 43*b*, 62
Communicative reason, 86
Community currency, 39, 47–48
Complementary currencies, 39
Constructive debt, 9–10
Contracts-for-differences (CfD), 194
Cox proportional hazard analysis, 63–64
Credit card debt, 63–64
Credit–debt relationship, 32–33, 35, 37, 39, 41–42,
    44–45, 47–48
Credit prediction models, 63–64
Credit Suisse, 20
Crisis community currency movements, 39

D

"Debt model" economy, 48
Decent Homes Standard, xxii–xxiii, 121, 122*b*
Delphi method, 255
Department of Work and Pensions (DWP), 222
DER. *See* Distributed energy resources (DER)
Disciplinary neoliberalism, 47
Discursive power, 58–59
Distributed energy resources (DER), 276
Distributional justice, 76–77
Distributive spatial inequalities, 262–263
DWP. *See* Department of Work and Pensions
    (DWP)

E

Economic inequality, xviii, xx, xxiii, xxv, 3–4, 146,
    173
    average income, 291
    Brexit, geography of
        donations, 296
        Leave voters, 297–299
        Remain campaign, 299–300
        UK Labour Party, 296
    Britain, 294–295, 297
    charity, 289
    in developed countries, 6
    down payment, 287–288
    economic crash and rise, in far-right politics,
        294–295
    egalitarian, 21–23, 26
    energy justice (*see* Energy justice)
    English educational system, 306
    equivalent statistics, 291–292
    estimation, 15–17
    extreme, 21–22, 25–26
    extremes, in Europe, 300–306
    fear, 292
    First World War, 5–6
    food insecurity, 292
    gender (*see* Gender)
    household surveys, 294
    human wellbeing, indicators of, 146
    improvements, in life expectancy, 292
    income *vs.* wealth inequality, 7–10
    investments, 289

Economic inequality *(Continued)*
  measurement, 10–14
    Gini coefficient, 10–14, 12–14*f*
    horizontal inequality, 10, 14
    Lorenz curve, 10–12, 11*f*, 14
    percentage shares, 11, 12–13*f*
    Robin Hood index, 10, 12–13, 14*f*
    Theil Index, 10, 14
    vertical inequality, 10, 14
  national income, 289
  negative equity, 287–288
  neoliberalism, 23–25
  oligarchs, 22
  percentage of unheatable homes (*see* Percentage of unheatable homes (UH%))
  policy and political changes, 291
  pollution, 300
  quality of life, 292
  structuration theory (*see* Structuration theory)
  taxation data, 294
  unemployment, 289
  wealth inequality, 291, 294
Economic poverty, xxiv
Economic reforms
  Regan government, 6
  Schröder government, 6
EFUS. *See* Energy Follow Up Survey (EFUS)
Egalitarian, 21–23, 26
EHS. *See* English Housing Survey (EHS)
Energy Company Obligation (ECO), 131–133
Energy costs money, xviii, 31
Energy cultures, 53–55
Energy Efficiency Commitment, 132
Energy Follow Up Survey (EFUS), 127–129
Energy inequality, justice and
  energy rebound and sufficiency, heterogeneity of, 260–261
  social life-cycle analysis of, 259–260
  spatial heterogeneity and primary energy use, 255
  structural economic change, spatial heterogeneity, 255–257
  temporal income, Germany and US, 257–259
Energy injustice, gender. *See* Gender
Energy justice, xix–xxii, 6, 53–55, 75–76, 173
  academic energy justice, 77–79
  affect-based notions, 87–88
  and economic inequality, 76, 91–93
  climate change, 87
  climate justice, 78–79
  communicative reason, 86

  deconstructionist, 81
  distributional justice, 76–77
  energy/fuel poverty, 76–77, 80–82, 87, 91
  environmental justice, 78–79
  gender perspective on (*see* Gender)
  global energy justice, implications for, 89–91
  Kant's "categorical imperative", 85
  moral claim, 78, 81
  moral commitments and beliefs, 82–84, 87, 92
  moral statements, Wittgenstein's reflections on, 82–83
  non-western moral frameworks, 84
  obligation, notion of, 88
  philosophical energy justice, 78
  pragmatist view, 81
  procedural justice, 76
  Rawls' approach, 80–81, 85–86
    moral commitment, 87
    rational metaphysical approach, 81, 84–85
    reasonable, definition of, 86
    western enlightenment assumptions, 86–87, 92
  recognition justice, 76
  renewable energy, transition to, 80
  sustainable energy production and supply, transition to, 80
  "thick" and "thin" morality, notion of, 88–89
Energy Performance Certificate (EPC), 124–126
Energy poverty, xix, xxii–xxiv, 101, 175, 179, 184–185
  10% indicator (*see* 10% indicator)
  CO_2 emissions, 226–227, 237–243
    poor households' incomes, 240–243
    progressive taxation, 238–240
  cold homes, health impacts, 101–102
  economic austerity, 223
  energy efficiency policy, 151
  energy inefficiency, 101–103, 146, 149
  energy inefficient buildings, 224–225
  energy justice, 76–77, 80–84, 87, 91–92
  energy services, 222–223
  high fuel prices, 146, 149
  and income poverty, 104
  income redistribution, 234–237
  lack of money, 31
  lived experience, 100
    John's experience, 105–107
    in UK, 99–100, 106–108
  low-income households, 222
  low incomes and high bills, 101–103, 149
  method and approach, 227–229

micro/targeting approach, 149
objective and subjective definitions, 149–150
percentage of unheatable homes (*see* Percentage
  of unheatable homes (UH%))
physical and mental health problems, 101–102
policy and practice, 110–112, 111*t*
politics of, 108–110
poverty, 146
single parent households
  black/minority ethnic background, 184
  children, 182–184
  gender payment gap, 181–182
  intergenerational immobility, 182–184
  low-income single parent families, 179
  poor housing conditions, 182, 183*t*
  prepaid meters, 184
  relative poverty, 179
  single parent *vs.* all households unable to heat
    homes, percentage of, 179–181, 180*t*, 181*f*
social and economic policy, 151
socially systemic approach, 102–104, 110
specific household energy needs, 149
targeting approach to, 225–226
Energy transition
  economic growth, 278–279
  energy inequality and justice (*see* Energy
    inequality, justice and)
  energy infrastructure, justice analysis, 252–253
  energy poverty
    energy/fuel poverty and energy justice, 261–263
    measurement issues, 263–266
    social/economic welfare considerations,
      272–274
    taxation, transfers, and subsidies, 267–271
  energy services, 253
  GHG emissions, 253
  global warming, 253
  heterogeneity, 253
  low-carbon systems, 252–253
  opportunity cost principle, 254
  productivity gains and structural changes,
    278–279
  smart system
    grids, enabling technologies, 275
    meters and RTP, 275
    MTPC, 277
    prosumer households and aggregate
      constructs, 276–277
    sustainable energy communities, 275–276
Energy Trilemma Index, 263–264
English Housing Survey (EHS), 120, 121*b*

Environmental justice, 78–79
EPC. *See* Energy Performance Certificate (EPC)
Exnovation, 252–253

F

Feed-in tariffs (FITs), 191, 194, 197–199
Financialization, 25
FITs. *See* Feed-in tariffs (FITs)
Fuel poverty. *See* Energy poverty

G

Gender, xxiii–xxiv
  climate change and energy policies, 174–175
  equality, improvement in, 175
  household practices and feminization of demand
    side response, 175–179
  inequality, 175–176
  residential energy consumption, 174–175
  single parent households, fuel poverty in
    black/minority ethnic background, 184
    children, 182–184
    gender payment gap, 181–182
    intergenerational immobility, 182–184
    low-income single parent families, 179
    poor housing conditions, 182, 183*t*
    prepaid meters, 184
    relative poverty, 179
    single parent *vs.* all households unable to heat
      homes, percentage of, 179–181, 180*t*, 181*f*
  wealth and income distribution, in Europe, 175
GHG. *See* Greenhouse gases (GHG)
Gini coefficient, percentage of unheatable homes,
    10–14, 12–14*f*, 20, 151, 157, 167, 234,
    238–239, 257–258, 258*f*, 264
  coefficient, 163, 165*t*, 166
  correlation between, 147–148, 148*f*, 166
  influence on, 147
  magnitude of, 154–156, 155*t*
  mean, maximum, minimum and standard
    deviation, 156, 156*t*
  post-redistribution Gini, ratio of, 166, 167*f*
  P-values, 159–162, 160–161*t*, 162*b*, 165–166,
    165*t*, 168
  regressed against log(UH%), 157, 158*t*
  t-statistics, 165–166, 165*t*
Gold coin, 42
Gold standard, 42
Government debt, 63–64
Great Recession of 2007–2008, 16–17
Green deal, 132*b*, 133–135, 137, 140
Greenhouse gases (GHG) emissions, 253

Gross Domestic Product per capita (GDP/capita), 153, 157, 167–168
  Ireland and Luxembourg, 152
  percentage of unheatable homes, 151–152, 157–158
    coefficient, 163–164, 165t
    magnitude of, 154–156, 155t
    mean, maximum, minimum and standard deviation, 156, 156t
    P-values, 159–160, 160–161t, 162, 165t
    regressed against log(UH%), 157, 158t
    t-statistics, 165t
    VIF scores, 158–159

H

HDDs. *See* Heating degree-days (HDDs)
Heating degree-days (HDDs), 153, 157, 164, 167–168
  coefficient, 164, 165t
  magnitude of, 154–156, 155t
  mean, maximum, minimum and standard deviation, 156, 156t
  P-values, 159–162, 160–161t, 165t
  regressed against log(UH%), 157, 158t
  t-statistics, 165t
Heritage Foundation, 60–61
HHSRS. *See* Housing Health and Safety Rating System (HHSRS)
HMO. *See* Houses of multiple occupation (HMO)
Homo debitus, 63
Household debt, 63–64
Household electrification, 261
Household energy services, 230–231
Houses of multiple occupation (HMO), 118
Housing Health and Safety Rating System (HHSRS), 137

I

'10% indicator', 228
  amounts, requirement of, 231–233
  analysis, logic of, 229–230
  high-income households, 233–234
  household energy services, 230–231
"10–50 index", 5–6, 5f
IFS. *See* Institute for Fiscal Studies (IFS)
Impoverishment (IMP), 265–266
Inadequate dwellings (ID), 153
  coefficient of, 164, 165t
  magnitude of, 154–156, 155t
  mean, maximum, minimum and standard deviation, 156, 156t

P-values, 159–160, 160–161t, 164, 165t
  regressed against log (UH%), 157, 158t
  t-statistics, 165t
Income inequality, 99–100, 104
Inside money. *See* Commercial bank money
Institute for Fiscal Studies (IFS), 302–303
Institutional power, 58–59
Interdisciplinary Cluster on Energy Systems, Equity and Vulnerability (InCluESEV), 79

K

Kantian approach, 85
Kenyan currency, 47
Keynesian economic model, xx
Kilowatt-peak power (kWp), 189–190

L

Landlords, 120, 135–137
Language game, 83–84, 88
Levelized Cost of Electricity Output (LCOE), 192–193
Loan money, 35–37, 62
Lorenz curve, 10–12, 11f, 14, 257, 263–264
Low income high cost (LIHC), 108–109, 228, 261, 265–266

M

Macro-level economic inequality, 47
Material power, 58–60
Mean-median ratio, 10
Methodenstreit, 40
Methodological individualism, 57
Microgrids, 276–277
Minimum Energy Efficiency Standard (MEES), 138–140
Minimum income standard (MIS), 228–229
Misrecognition, 262–263
Mobilizational power, 60–61
Money, xx–xxi, 64
  in ancient societies, 33
  and debt, social power of, 32–33
  anti-money, 34–36
  bank loan money, 35–37
  barter/neutral veil view of, 37, 40–41
  central bank *vs.* commercial bank money, 38–39
  chain store voucher, 39
  "chartalist" view, 44
  commodity, 42–43
  community currencies, 39
  complementary currencies, 39
  created out of nothing, 33–36

credit–debt relationship, 32–35, 37, 39, 44
crisis community currency movements, 39
effects within society, 40–41
energy costs, xviii, 31
goods and services, values of, 32, 35, 40–41
IOUs, gambling house chips, 39
material power, 60
Methodenstreit, 40
new economic sociology, 32
obligation and entitlement, relationship of, 32, 34
performative, 43*b*
power relationship, 34–35
promissory notes, 39
purchase point systems, 39
social relationship, 33–34
social science based energy research,
    implications for, 45–48
state's role, 44–45
temporary folk currencies, 39
Mortgage loans, 152–153
Motherhood, 177, 185
Multi-tenant prosumer concept (MTPC), 277

N

NeoKeynesianism, 65
Neoliberalism, 23–24, 36–37, 41, 47, 64–66
Neutral veil view, of money, 37, 40–41
New economic sociology, 32
Non-loan money, 62

O

"Offgrid solutions". *See* Microgrids
Office for National Statistics (ONS), 300–302
Office of Gas and Electricity Markets (OFGEM),
    197–198
ONS. *See* Office for National Statistics (ONS)
Ordinary least squares (OLS) multivariate
    regression analysis, 148, 151
Outside money. *See* Central bank money
Owner occupied homes, in UK, xxii–xxiii, 116
  vs. EU countries, 117
  MEES regulations, 138–140
  vs. private rented and social rent homes
    central heating and size, 123–124, 124*t*
    dwelling age by tenure, 122–123, 122*f*
    rising and penetrating damp issues, 123, 124*t*
    SAP cost rating, on EPC, 124–126, 125*f*, 125*b*,
      127*f*
    thermal comfort, 127–131
    wall insulation, 122–123, 123*f*
  trends in, 118–120

P

Percentage of unheatable homes (UH%), 145–146
  distribution of, 156, 157*f*
  material factors, 146–147
  panel data/year by year regressions, 154
  variables, 151–154, 167–168
    coefficients, 163–165
    distributions of, 156
    Gini index (*see* Gini coefficient, percentage of
      unheatable homes)
    magnitudes of, 154–156, 155*t*
    mean, maximum, minimum and standard
      deviation, 156, 156*t*
    P-values, 159–162, 160–161*t*, 162*b*
    Ramsey RESET tests, 159
    regressed against log(UH%), 157–158, 158*t*
    t-statistics, 164–165, 165*t*
    VIF test, 158–159
Performative, 43*b*
PIB. *See* position invariant burdening (PIB)
Political rights, 59
Position invariant burdening (PIB), 265–266
Poverty3–5, 9–12, 17–18, 25–26. *See also* Energy
      poverty
Power, 59
  of credit and debt, 61–64
  discursive power, 58–59
  institutional power, 58–59
  oligarchs
    coercive power, 60
    material power, 58–61
    mobilizational power, 60–61
    official positions, 59
    political rights, 59
    tax rate reduction, 60–61
Power purchase agreements (PPA), 193
Power relationship, money, 34
PPA. *See* Power purchase agreements (PPA)
Practice theory, 53–55
Precious metals, 42
Principle of Pareto optimality, 272–273
Private landlords, 120
Private rented homes, in UK, xxii–xxiii, 115–116
  energy efficiency measures
    ECO measures, 131–133
    landlords' view, 135–137
    split incentive issue, 133–134
    tenants' view, 134–135
  vs. EU countries, 117, 139
  heating efficiency
    central heating and size, 123–124, 124*t*

Private rented homes, in UK *(Continued)*
    condensation and mold, 123, 124*t*
    Decent Homes Standard, 121, 122*b*
    dwelling age by tenure, 122–123, 122*f*
    English Housing Survey, 121, 121*b*
    rising and penetrating damp issues, 123, 124*t*
    SAP rating, on EPC, 124–126, 125*f*, 125*b*,
        127*f*
    wall insulation, 122–123, 123*f*, 139
  landlords, 120, 139
  MEES regulations, 138–140
  thermal comfort, 127–131
  trends in, 118–120
Procedural justice, 76
Promissory note, 39
Prosumer rebound effect, 260–261
P-values, 159–162, 160–161*t*, 162*b*, 165*t*

R

Ramsey regression equation specification error test
        (RESET) test, 159, 166
Rawls' approach, energy justice, 80–81, 85–86
    moral commitment, 87
    rational metaphysical approach, 81, 84–85
    reasonable, definition of, 86
    western enlightenment assumptions, 86–87, 92
Recognition justice, 76
Redistributive social welfare systems, 66
Reduced rent homes, in UK. *See* Social rented
        homes, in UK
Reduced SAP (RdSAP) method, 125*b*
Regressive pricing, 191, 210–217
Reification, 57
Renewable electricity, European Union
    CfD, 194
    direct subsidies, 191
    dispatch priority, 193
    economic inequality, 191
    expenditure on, 195–197
    FiT support scheme, 194
    fossil fuel generation, 193
    growth in, 195
    investment and costing issues, 192–194
    marginal costs, 211, 211*f*
    photovoltaic electricity, 195
    pricing of, 211–214
      cost components, 197–200
    household electricity consumption, 209–211
    household size, 200–208
    regressive pricing, 191, 212
    socio-technical phenomenon, 191
Renewable Energies Act, 277
Robin Hood index, xx, 10, 12–13, 14*f*

S

SDG. *See* Sustainable Development Goal (SDG)
Silver coin, 42
Single parent households, fuel poverty
  black/minority ethnic background, 184
  children, 182–184
  gender payment gap, 181–182
  intergenerational immobility, 182–184
  low-income single parent families, 179
  poor housing conditions, 182, 183*t*
  prepaid meters, 184
  relative poverty, 179
  single parent *vs.* all households, unable to heat
        homes, percentage of, 179–181, 180*t*, 181*f*
Social and economic policy, fuel poverty, 151
Social determinism, 57
'Social life-cycle analysis' (S-LCA), 259
Socially systemic approach, 102–104, 110
Social mobility, 305
Social practice theory, 174
Social rented homes, in UK, xxii–xxiii, 115–116
    vs. EU countries, 117, 139
    MEES regulations, 138–140
    vs. private rented and owner occupied homes
      central heating and size, 123–124, 124*t*
      dwelling age by tenure, 122–123, 122*f*
      rising and penetrating damp issues, 123, 124*t*
      SAP rating, on EPC, 124–126, 125*f*, 125*b*
      thermal comfort, 127–131
      wall insulation, 122–123, 123*f*
    trends in, 118–120
Social science frameworks, 4
Sociotechnical systems theory, 53–54
Standard Assessment Procedure (SAP) ratings,
        xxii–xxiii, 124–126, 125*f*, 125*b*, 127*f*
Structuration theory, xxi, 56
    credit and debt, power of, 61–64
    mechanistic view, 54
    neoliberalism and welfare politics, 64–66
    oligarchs and power
      coercive power, 60
      material power, 58–61
      mobilizational power, 60–61
      official positions, 59
      political rights, 59
      tax rate reduction, 60–61
    practice theory, 54–55
    social science based energy research,
        implications for, 66–68
    social structure, 55
      duality of structure, 54, 58
      human action and discourse, 58
      income and wealth inequality, 57

Marxist perspective, 57
methodological individualism, 57
power, 58–59, 69
reification, 57
resources, 55–56, 58, 69
Sustainable Development Goal (SDG), 174–175
Sustainable energy transition. *See* Energy transition

T

Tariff equity law, 262
Tax havens, 27
Tax rate, 223, 235–237, 244
Tenant Electricity Act, 277
Ten percent rule (TPR), 261, 265
Theil's index, 10, 14, 257, 264
Total primary energy supply (TPES), 263–264
Toxic debt, 9–10, 152–153
TPES. *See* Total primary energy supply (TPES)

U

US central bank, 42
US dollar, 42
Utilitarian principle, 273

V

Value added tax (VAT), 237
Variable renewable energy sources (VRES), 275

Variance inflation factor (VIF) test, 158–159
VAT. *See* Value added tax (VAT)
Virtual power plants (VPPs), 276–277
Voluntary downshifting, 260
VPPs. *See* Virtual power plants (VPPs)
VRES. *See* Variable renewable energy sources
    (VRES)

W

Warm Front scheme, 225
Wealth and income inequality, 4–5
    10–50 index, 9
    bottom 10 percentile, 17–18
    constructive debt, 9–10
    effect of interest payments, 9–10
    percentage of German households "in danger of
        poverty", 9, 9f
    stating, 9
    tax havens, 19–21
    total national private wealth and income, 7–8, 8f
    toxic debt, 9–10
Wealth-defense industry, 25
WEC. *See* World Energy Council (WEC)
WID. *See* World Inequality Database (WID)
Wind turbines, 189–190, 192–193
World Energy Council (WEC), 263–264
World Inequality Database (WID), 5–6